How to Get Emergency Medical Help

In any case where poisoning is suspected, get the victim to emergency medical help immediately. If the person can be easily moved, take the patient to the nearest emergency medical center. If the person is seriously ill, call an ambulance for emergency medical help.

Take stricken animals immediately to the nearest veterinary hospital. If the animal is too large to move, call the veterinarian for assistance as soon as possible.

Medically trained persons can give a great deal of immediate relief for any poisoning, no matter what the poisonous substance happens to be. If poisoning from a plant is suspected, locate a botanist or a person knowledgeable about plants who can assist medical personnel in plant identification. Information on poisonous substances is available to medical personnel through the Poison Control Centers. If mushroom poisoning is suspected, the Poison Control Centers have names of persons trained in mushroom identification.

Persons who work with plants in agriculture can be of great assistance to veterinarians in determining the presence of poisonous plants and establishing as quickly as possible whether these plants have been eaten by the animals. New growth of plants rapidly masks any indication of having been grazed. Therefore, such investigations should be made immediately when there is suspicion that toxic plants are involved.

Poison Control Centers

FRESNO
209/445-1222
(collect calls
accepted)

Fresno Regional Poison Center
Fresno Community Hospital
P.O. Box 1232
Fresno and R Streets
Fresno, CA 93715

LOS
ANGELES
213/484-5151

Los Angeles County Medical
    Association
Poison Control Center
1925 Wilshire Boulevard
Los Angeles, CA 90057

OAKLAND 415/428-3000 415/547-2933 415/547-2928	Children's Hospital Medical Center 51st and Grove Streets Oakland, CA 94609

ORANGE
714/634-5988
	Orange County Poison Control
	   Center
	University of California Irvine
	   Medical Center
	101 City Drive South
	Building One, Route 78
	Orange, CA 92688

SACRAMENTO
916/453-3692
800/852-7221
	University of California Davis
	   Medical Center
	University of California Davis
	   Poison Control Center
	2315 Stockton Boulevard
	Sacramento, CA 95814

SAN DIEGO 619/294-6000	San Diego Regional Poison Center University of California San Diego Medical Center 225 West Dickinson Street San Diego, CA 92103

**SAN
FRANCISCO**
415/666-2845
800/792-0720
	San Francisco Bay Area Regional
	   Poison Control Center
	San Francisco General Hospital,
	   Room #1E86
	1001 Potrero Avenue
	San Francisco, CA 94110

SAN JOSE 408/299-5112 408/299-5113 800/662-9886 800/662-9887	Central Coast Counties Regional Poison Control Center Santa Clara Valley Medical Center 751 South Bascom Avenue San Jose, CA 95128

Poisonous Plants
of California

California Natural History Guides: 53

Poisonous Plants of California

**Thomas C. Fuller and
Elizabeth McClintock**

UNIVERSITY OF CALIFORNIA PRESS
Berkeley Los Angeles London

CALIFORNIA NATURAL HISTORY GUIDES
Arthur C. Smith, *General Editor*

Advisory Editorial Committee:
Raymond F. Dasmann
Mary Lee Jefferds
Don MacNeill
Robert Ornduff
Robert C. Stebbins

University of California Press
Berkeley, California
University of California Press, Ltd.
London, England

10 9 8 7 6 5 4 3 2 1

**Library of Congress
Cataloging-in-Publication Data**

Fuller, Thomas C.
Poisonous plants of California.

(California natural history guides ; 53)
Bibliography: p.
Includes index.
1. Poisonous plants—California.
2. Poisonous plants—Toxicology.
I. McClintock, Elizabeth May, 1912–
II. Series: California natural history guide ;
53. [DNLM: 1. Plant Poisoning. 2. Plants,
Toxic. WD 500 F968p]
QK100.C36P65   1986   581.6′9′09794
          85-29025
ISBN 0-520-05568-3 (alk. paper)
ISBN 0-520-05569-1 (pbk. : alk. paper)

Contents

Acknowledgments   vii
Introduction   1

Algae   11

Fungi   15

Ferns and Horsetails   56

Gymnosperms   61

Angiosperms

    Dicotyledons   64
    Monocotyledons   264

Plant Toxins and Derivative Drugs   305

Appendix A: Plants Causing Dermatitis   369

Appendix B: Plants Causing Hay Fever and Asthma   377

Appendix C: Plants Accumulating Nitrates   384

    General References   387
    Index of Common Names   391
    Index of Scientific Names   409
    General Index   425

Acknowledgments

We gratefully acknowledge the technical advice and assistance of Dr. Kenneth F. Lampe, Senior Scientist, American Medical Association, Chicago; Dr. Harry Thiers, Professor of Botany, San Francisco State University; Dr. Carlyn Halde, Associate Professor, Department of Microbiology, University of California, San Francisco; Rick Kerrigan, Santa Cruz, California; Greg Wright, Past President, Los Angeles Mycological Society; and Paul P. Vergeer, Richmond, California.

The professionalism of the veterinarians of the state of California is reflected in the material on animal poisoning presented here. We are especially indebted also to Mary S. Fuller for her medical knowledge and writing skills.

We wish to acknowledge, as well, the efforts of those who have supplied us with illustrations. With a few exceptions the color plates are photographs taken by the authors. Fred G. Andrews furnished Plate 1*a, b,* and *c.* Tokuwo Kono took the photograph for Plate 1*d.* Walter Knight supplied 8*b.* Ray J. Gill took the photographs of the seeds on Plate 9*a* and *b.* G. Douglas Barbe supplied Plate 12*a* and *b.*

We are indebted to Lucretia B. Hamilton for permission to use the black and white figures that were previously published in Schmutz and Hamilton (1979) and Parker (1958). These are Figures 1–4, 7, 10, 12, 14, 16, 18, 21, 22, 24–29, 31, 34, 36–42, 44–46, 48. Figures 6, 8, 9, 11, 13, 15, 17, 19 (upper left drawing), 20, 23, 30, 32, 43, and 47 are from Robbins, Bellue, and Ball (1951). Figure 5, all of Figure 19 except the upper left drawing, Figure 33, and Figure 35 are from Abrams and Ferris (1944, 1951).

Introduction

Many of the world's plants give us foods, drugs, fibers, and building materials; others add to the beauty of our surroundings. In addition to these useful plants there are some that we call poisonous. We define poisonous plants as those that contain substances capable of producing varying degrees of discomfort and adverse physical or chemical effects or even death to humans and animals when they are eaten or otherwise contacted. It should be emphasized that these adverse effects—unpleasant as they may be—are rarely fatal. Nevertheless, because a few plants can be fatally poisonous and others can cause discomfort or illness, everyone should become acquainted with these injurious plants.

The question is sometimes asked: How many plants are poisonous? About 20,000 species of seed plants are native or naturalized in the United States, and approximately the same number of cultivated ornamental and agricultural species occur here. J. M. Kingsbury (1964) in his *Poisonous Plants of the United States and Canada* discusses more than 700 species of plants known from case or experimental evidence to have toxic characteristics. Of these 700, however, only a small number are known to be so seriously poisonous as to cause fatalities to humans or animals. Seriously and fatally poisonous plants are listed in the sections titled "Human Poisoning" and "Livestock Poisoning."

Many poisonous plants have dual roles, and in appropriate doses or forms they can be useful. Digitalis and morphine are examples of drugs obtained from plants. Digitalis is a long-used and much prescribed cardiac medication, but when improperly used it can be fatally poisonous. The milky juice of the Opium Poppy is the source of important drugs including morphine and codeine. These drugs are used to relieve pain, but when indiscriminately used they are not only injurious but even fatally poisonous.

Some poisonous plants, given the proper circumstances or

preparation, may be edible. Occasionally a toxic substance is localized in a particular part of the plant. The leaf blade of Rhubarb is toxic but not the petiole, which is edible. Manioc, *Manihot esculenta,* has tuberous roots that yield a starch used as a staple food, but these roots contain a cyanogenic glycoside and are toxic in their natural state. When heat-processed, however, they are used as food in tropical regions. Tapioca is also prepared from Manioc roots after proper treatment.

In some plants the toxin is concentrated in the fruits or seeds. The seeds of Castor Bean, *Ricinus communis,* contain a highly toxic plant lectin, a toxalbumin, that is released when the seeds are chewed. Ingesting more than two or three Castor Bean seeds can be fatal.

Moreover, we use many plants to beautify our surroundings, but some ornamental plants contain toxic substances. Hence the public should become acquainted with the ornamentals that have poisonous properties. Usually poisonous plants are not hazardous to adults, but small children and household pets may ingest them and become ill.

Knowledge of the poisonous properties of plants has been acquired through the ages along with other information, such as their edibility and their uses in curing disease and promoting health. People around the world have learned which plants around them are useful and which are to be avoided. Initially, of course, this knowledge was learned by trial and error. Before recorded history it was found that certain plants caused death to animals and that a practical use could be made of this information by placing extracts of poisonous plants on arrows used for hunting animals for food, and also for eliminating enemies. In some parts of the world plants with toxic properties have been crushed and used to kill fish without rendering them inedible.

The Egyptians and the Greeks had knowledge of poisonous plants. The Egyptians knew of the toxic effects of Aconite, Henbane, and Poison Hemlock; the Greeks executed some of their political prisoners with poisonous plant extracts. In 399 B.C., Socrates was given a cup of Poison Hemlock, well known to be fatally poisonous. Nicander, a Greek of the second century B.C., wrote of poisons, many of plant origin, and also of the remedies for poisons. The Greek physician Dioscorides, who lived at the beginning of the Christian era, wrote an herbal that included about 600 medicinal plants, some of which were described as poisonous. This work became the authoritative source on medicinal plants throughout the Middle Ages and the Renaissance. No doubt some present-day folklore regarding medicinal and poisonous plants can be traced back to Dioscorides.

The Romans knew about poisonous plants, as well. Some Romans in high places who were victims of political intrigue met their death from poisoning, including that from plants. During the Middle Ages poisoning continued to be a means of murder and suicide. With the development of modern drugs and modern methods of causing death, plant poisoning has become less common and most of the precise knowledge needed for homicidal plant poisoning seems to have disappeared.

Concern for the poisoning of livestock by certain plants arose early in the history of the United States. Indeed, scientific agriculture and veterinary medicine began to develop under the Jefferson administration of the early 1800s. Progress in these fields was slow in the years following, however, because of the opening of new territories to the west and attendant problems. With the addition of new western states, agricultural colleges and experiment stations were founded. These institutions contributed new information on problems concerning livestock poisoning by plants.

Human Poisoning

Plant poisoning of humans is almost entirely accidental and can be avoided by taking a few precautions. Statistics show that small children, particularly those under 5 years (sometimes called young foragers), are most vulnerable to accidental poisoning.

How to Avoid Plant Poisoning of Humans

1. Become acquainted with the poisonous plants around homes, parks, streets, and recreational areas. Learn to recognize the twelve frequently occurring plants in the list that

follows and remember their botanical names. Common names are often confusing and inconsistent.

2. Do not eat plants (or their parts) growing around your home or in the wild, including wild mushrooms, unless you are *positive* of their identification. Birds, squirrels, and other wild animals or pets may eat a fruit or plant part, but that does not mean this plant material is safe for humans.

3. Keep plants away from infants; this includes plant parts easily picked up, such as seeds, small fruits, and bulbs.

4. Teach very young children not to put plants or plant parts into their mouths and impress upon them that certain plants are poisonous and potentially harmful.

5. Instruct young children not to suck nectar from a flower or make "tea" from any part of a plant.

6. Teach young children to recognize Poison Oak, stinging nettles, and other plants that may cause dermatitis.

7. Avoid handling plants with milky juice; some may be harmful, as certain members of the Spurge Family, Euphorbiaceae.

8. Make certain of their correct identification if you wish to eat wild plants. This is most important for wild mushrooms. The toxins of some plants, including mushrooms, are not always destroyed by cooking. Learn which plants may be eaten when cooked.

9. Collect and eat shellfish with care, for shellfish feed on certain algae containing toxins. Respect regulations against taking mussels and clams when prohibited.

Plants Seriously Poisonous to Humans

In California the following twelve frequently occurring seed plants may be seriously and sometimes fatally poisonous to humans when eaten. Everyone should learn to recognize them. Most of these plants are not native but are introductions from elsewhere.

1. *Abrus precatorius,* Rosary Pea; not grown in California, but its toxic seeds are often used in seed ornaments

2. *Brugmansia sanguinea,* Red Angel's Trumpet, and other species; introduced and growing in gardens and parks

3. *Cicuta douglasii*, Western Waterhemlock, and other species; California native
4. *Conium maculatum*, Poison Hemlock; introduced and naturalized
5. *Datura stramonium*, Jimsonweed, and other species; introduced and naturalized
6. *Digitalis purpurea*, Foxglove; introduced in gardens and naturalized
7. *Nerium oleander*, Oleander; introduced and widely planted
8. *Nicotiana glauca*, Tree Tobacco; introduced and naturalized
9. *Phytolacca americana*, Pokeweed; introduced and naturalized
10. *Ricinus communis*, Castor Bean; introduced in gardens and naturalized
11. *Taxus baccata*, English Yew; introduced and occasionally growing in gardens and parks
12. *Zigadenus venenosus*, Meadow Deathcamas, and other species; California native

Contact Dermatitis

A small group of seed plants in California cause allergic reactions on contact with the skin. Of these a widespread California native species, *Toxicodendron diversilobum*, Pacific Poison Oak, produces the most distress as well as economic loss through working hours lost.

Mushrooms

Of all poisonous plants in California, toxic mushrooms cause the most serious illnesses and nearly all the fatalities to humans. Those who gather and eat mushrooms should use every precaution to ensure that their mushrooms are correctly identified and known to be edible. The following California mushrooms are most seriously poisonous but only rarely fatal to humans when eaten:

1. *Amanita muscaria*
2. *Amanita ocreata*
3. *Amanita pantherina*
4. *Amanita phalloides*

5. *Clitocybe dealbata*
6. *Chlorophyllum molybdites*
7. *Entoloma rhodopolium*
8. *Gyromitra esculenta*
9. *Hebeloma crustuliniforme*
10. *Naematoloma fasciculare*
11. *Paxillus involutus*
12. *Scleroderma citrinum* [*S. aurantium*]
13. *Tricholoma pardinum*

Livestock Poisoning

The following seed plants in California are most seriously poisonous to livestock:

1. *Amsinckia douglasiana,* Fiddleneck, and other species; California native
2. *Asclepias fascicularis,* Whorled Milkweed, and other species; California native
3. *Astragalus lentiginosus,* Freckled Milkvetch or Spotted Loco, and other species; California native
4. *Centaurea solstitialis,* Yellow Starthistle, and other species; introduced and widely naturalized
5. *Cicuta douglasii,* Western Waterhemlock, and other species; California native
6. *Corydalis caseana,* Fitweed; California native
7. *Delphinium hesperium,* Western Larkspur, and other species; California native
8. *Halogeton glomeratus,* Halogeton; introduced and naturalized
9. *Hypericum perforatum,* Klamathweed; introduced and naturalized
10. *Lupinus latifolius,* Broadleaf Lupine, and other species; California native
11. *Nicotiana glauca,* Tree Tobacco, and other species; introduced and naturalized or native
12. *Prunus virginiana* var. *demissa,* Western Chokecherry; California native
13. *Pteridium aquilinum* var. *pubescens,* Bracken Fern; California native

14. *Sarcobatus vermiculatus,* Black Greasewood; California native
15. *Senecio jacobaea,* Tansy Ragwort, and other species; introduced and naturalized
16. *Tetradymia glabrata,* Littleleaf Horsebrush, and other species; California native
17. *Triglochin concinnum,* Arrowgrass, and other species; California native
18. *Veratrum californicum,* Corn Lily, and other species; California native
19. *Xanthium strumarium,* Cocklebur; California native
20. *Zigadenus venenosus,* Meadow Deathcamas, and other species; California native

Cattle losses from nitrate poisoning occur each year in California. Under special conditions nitrates are accumulated in lethal amounts in various species of rangeland plants and in those dried for hay. The greatest losses from nitrate poisoning occurred in the spring of 1952 when an estimated 2,000 to 3,000 head of cattle died as the result of the high amounts of free nitrates in forage plants. (See "Nitrates," pp. 305–307.)

In California the greatest single event causing loss of livestock from toxic plants occurred in the winter of 1963–1964 when more than 6,000 cows died from grass tetany; the animals did not have an adequate magnesium supply in rangeland plants, caused by unusual weather conditions. (See "Magnesium," pp. 312–314.)

Death of livestock from tannins in oak shoots and acorns is rare in California, yet two of the largest cattle losses in the state occurred in the northern Sacramento Valley in 1960 and 1985 after spring snowfalls covered the range plants forcing the cattle to forage solely on young shoots of oak trees. These cases are covered under *Quercus* in the Fagaceae.

Plant Toxins

The chemical compounds formed in plants, including nonpoisonous ones, are numerous and variable. Some are poorly known or even undescribed by standard chemical nomenclature. Thus not only are the myriad poisonous compounds formed in plants difficult to classify, but the specific organic compound involved in a case of poisoning may be unknown.

Poisons produced by plants are believed to protect those plants from insects or other plant-eating animals or to act against invasion by fungi or bacteria. When unripe or green, fleshy fruits may be distasteful or even poisonous; but when ripe the fruits become attractive and edible, and the mature seeds may then be dispersed. Some toxins are produced only when the plant tissues are invaded by a fungus, perhaps a protection against further infection. Bacteria may be repulsed similarly by the invaded plant tissues, but other compounds may bind beneficial bacteria in root nodules. Moreover, some plant species have both toxic and nontoxic forms that seem to survive equally well with no benefit in favor of one or the other. Some 85% of the species of flowering plants apparently flourish without compounds toxic to humans or animals.

Plant toxins harmful to animals are better known than those harmful to humans because animal toxicity can be investigated experimentally. Examples of toxicities caused by plant species are given under the several subdivisions of the plant kingdom where families, genera, and species are arranged alphabetically.

Identification and Naming

The poisonous plants discussed here are listed by their botanical names followed by one or more common names. Plants are named botanically according to an international set of rules and guiding principles formulated during the past 200 years. Through the application of these rules each species or kind of plant has only one correct botanical name, written in latinized form. Since botanical names are international, a plant occurring in different countries is known by the same name. As knowledge of plants increases, botanical names may sometimes change under the international rules of nomenclature. The most widely known synonyms (other, but not correct, botanical names) are listed after the correct botanical name and enclosed by brackets.

It is important to know the correct botanical name of a poisonous plant because it will lead to information about the plant in the literature. Moreover, knowing the correct botanical name means that new information relating to a poisonous plant can be published and added to the store of knowledge.

English common names, on the other hand, are often collo- quial. They are given mostly to well-known plants in a certain area, and these names often become integrated into that area's spoken language. The same plant in another area or country may have a different common name. The plant in the United States usually called Bermudagrass is known as Couchgrass or Stargrass in other parts of the world, for example, but all around the world it is known by its botanical name, *Cynodon dactylon*. Moreover, the same common name, such as laurel, nightshade, hemlock, or cedar, may be used for more than one plant species.

In times past, plants were the source of drugs and their study was part of the education of physicians and pharmacists. In more recent times drugs have come to be manufactured largely by pharmaceutical companies. Today health profes- sionals, although well informed in their special fields, have little knowledge of plants. Much of the confusion regarding poisonous plants lies in their identification. Improper identifi- cation may lead to inappropriate treatment.

Poison control centers, through their computerized micro- fiche file, the Poisindex, are provided with brief descriptions of poisonous plants and information about them. This descriptive material is helpful to the poison control personnel in identify- ing easily recognizable plants. Many others, including mush- rooms, are not readily identified. For these the poison control personnel should call on taxonomic botanists in their area. It is often possible to have mushrooms identified through profes- sional mycologists at educational institutions or knowledge- able members of mycological societies in various parts of California.

This book presents brief descriptions of the poisonous plants known to occur in California. The botanical terminol- ogy should be helpful to those wishing to identify poisonous plants. For plants native or naturalized in California, addi- tional botanical information can be found in *A California Flora* and *A Flora of Southern California,* both by Philip A.

Munz, and in various regional floras for areas within the state. A large number of the poisonous plants in California are not native but are ornamentals or naturalized weeds from other parts of the world with a similar climate. Most poisonous plants that occur in California as naturalized weeds are described in the two floras listed. Many poisonous plants are also described in *Weeds of California,* by Robbins, Bellue, and Ball. Most of the ornamental plants mentioned here are described in *Manual of Cultivated Plants* by L. H. Bailey and in *Hortus Third,* a publication of the Bailey Hortorium, Cornell University.

Frequently the same or similar toxic compounds occur in related species. Therefore poisonous members of the plant kingdom are presented under these headings:

Algae: Single-celled plants, multicellular plants of cells joined end to end with specialized cells, or multicellular plants with very simple tissues; spores as means of dispersal

Fungi: Single-celled or multicellular plants of cells joined end to end with specialized cells; spores as means of dispersal; food obtained from nonliving material or from living host cells

Vascular Plants: Plants with complex tissues conducting sugar or water and mineral salts

Ferns and Horsetails: Plants producing spores as means of dispersal

Seed Plants: Plants producing seeds as means of dispersal; spores contained within the plant

Gymnosperms: Plants with naked seeds, often formed on the surface of a bract within a cone

Angiosperms: Plants with covered seeds formed inside a fruit such as a pod or berry

Dicotyledons: Embryos in seeds with two seed leaves; leaves with netted veins

Monocotyledons: Embryos in seeds with one seed leaf; leaves usually with parallel veins

ALGAE

The abundant growth of blue-green algae on the surface of more or less stagnant water is called water bloom. Blue-green algae may occur as colonies of microscopic cells loosely held together. They may also grow as filaments of cells joined end to end, some cells with specialized functions. Masses of cells or filaments commonly are enclosed in gelatinous sheaths. Gelatinous clumps of these algae are blown by onshore winds and accumulate in large quantities at the edge of a pond or reservoir. Deaths of animals from drinking this contaminated water are known worldwide. Stagnant water is so distasteful that humans usually do not drink it. Human poisonings have occurred from accidentally swallowing or inhaling it, however. The principal genera of blue-green algae involved are *Anabaena, Aphanizomenon, Microcystis,* and *Oscillatoria.* Also implicated are the genera *Coelosphaerium, Gloeotrichia, Lyngbya, Nodularia, Nostoc,* and *Rivularia.* Most toxicities have been caused by two species: *Microcystis aeruginosa* [*Anacystis cyanea*] and *Anabaena flos-aquae.* Where poisoning from blue-green algae has occurred, a mixture of toxic and nontoxic strains has always been found.

Blue-green algae can convert the nitrogen gas in the air into nitrates and have been responsible for nitrate-nitrite poisonings of livestock as well as that from the algal toxins.

Toxic part: Entire masses of blue-green algae.

Toxin: Polypeptides or other toxic proteins.

Symptoms: In animals, weakness, muscular incoordination, foamy discharge from nose and mouth, thirst, nausea, vomiting, rapid weak pulse, blood-covered hard feces, convulsions, and death. With liver damage there may be jaundice and photo-sensitization. In humans, a hard cough and breathing difficulties may result from inhaling the spray of stagnant water. Although blue-green algae are normally present on the human body without causing any problems, they can occasionally cause skin disorders.

When blue-green algae become abundant in stagnant water, the growth of other algal species is suppressed by their excretions. Eventually, however, the blue-green algae are poisoned by their own excretions and the cells die. Where a fresh supply of water renews the area, self-poisoning by blue-green algae does not occur.

Fast-death symptoms, where death is the only result, appear in animals from within minutes to 1 or 2 hours after ingestion of blue-green algae. The toxic algae are found to be colonies of *Microcystis aeruginosa* and other species that have passed maturity and have been self-poisoned. Cells in the process of breaking down are the most toxic.

Slow-death symptoms may be seen in animals from 4 hours to as long as 2 days after ingestion of the blue-green algae. Lethargy and irregular breathing characterize this poisoning, perhaps with jaundice and photosensitization following. The toxic factor appears to be from the bacteria associated with the decay of *Microcystis aeruginosa* and not from the blue-green algal cells themselves.

The blue-green alga *Anabaena flos-aquae* can become abundant in stagnant waters in the summer months. Cattle, sheep, and waterfowl seem to be more readily poisoned by this alga than other animals. Death may occur from 1 to 20 minutes after ingesting it. Therefore the toxins have been named the very-fast-death factor to differentiate them from the fast-death toxins of *Microcystis*. Four different toxins have been identified from *Anabaena*.

In the early days of filling the San Luis Reservoir in western Merced County, sheep were pastured there to remove the grasses. In the middle of August 1967, a band of 1,000 sheep used the south side of the reservoir as their source of water. The water was vivid blue and green and of a thick consistency. Two hundred of the sheep developed incoordination and weakness in the hindquarters. The animals became prostrate and were unable to rise again. All recovered with a diet of hay and spring water instead of the water from the reservoir. The algae were identified as *Microcystis aeruginosa.*

In mid-August of 1979, the Sacramento District Veterinary Laboratory reported that a horse displayed a lack of muscular coordination for a period of 3 days. The veterinarian suspected the cause to be the water in the drinking trough. Microscopic examination found growths of the blue-green algae of *Anabaena* and *Gloeotrichia.*

Ducks show a differential response to species of blue-green algae. They are not affected by water bloom of *Microcystis* but die 1 or 2 hours after ingesting *Anabaena* bloom and display the same rubberneck symptom seen in death from botulism, a bacterial infection.

Dinoflagellates are one-celled organisms that have both plant and animal characteristics and therefore are sometimes placed in a phylum of their own rather than with the algae. Under warm conditions with an upwelling of deep ocean currents bringing a rich nutrient supply, dinoflagellates of the genus *Gonyaulax* become so abundant that seawater becomes red by day and phosphorescent at night. The red, or in some cases actually brown, color of the seawater is the result of the dinoflagellate pigments. Occurrences of a "red tide" or a "red sea" are known worldwide but mostly at latitudes higher than 30°. Although these conditions have long been regarded as warnings of shellfish toxicity, toxic levels of these one-celled organisms may occur without giving a red color to the water.

Three toxic species of dinoflagellates in the genus *Gonyaulax* are found along the coast of California: *G. acatenella, G. catenella,* and *G. polyedra. Gonyaulax polyedra* from the southern coast has a toxin, but this species has not caused any known cases of shellfish poisoning or fish poisoning.

From 1927 through 1984 there have been 505 cases of paralytic shellfish poisonings of humans in California; 32 of these cases resulted in death. The eating of toxic shellfish by humans results in paralytic poisoning of the central nervous system, followed in 1 to 12 hours by respiratory failure and death. The first symptom is a tingling, followed by a numbness of the lips, mouth, and fingers, and finally a numbness throughout the body. The toxin has a protein structure that has not been adequately described, although some characteristics of the purified toxin are known. Cooking may eliminate some of the toxin in the shellfish, but it is not completely destroyed.

Mussels and clams are not selective in their feeding. As they pump water through their bodies, they remove all minute organisms of the plankton including the toxic dinoflagellates. Dinoflagellate toxin is accumulated by mussels. Clams and scallops may also have toxic amounts, but they are only included in quarantines when adjacent to toxic mussels. Discarding the dark portion of the clam removes a great deal of the toxin, but the entire organism still remains poisonous.

Quarantines apply only to sport harvesting. Persons wanting to take mollusks should inquire locally at sport fishing shops to determine whether any quarantines are in effect. Mollusks sold commercially have no risk of poisoning. A quarantine prohibiting the sport harvesting of mussels along the entire California coast, including bays, harbors, and inlets, is imposed annually by the California Department of Health Services. In 1984, the annual quarantine was from May 1 through October 31, but additional quarantines were imposed beginning on March 2 for Sonoma, Marin, and San Francisco counties and on March 15 for San Mateo and Santa Cruz counties. These early starts were caused by unusual toxic blooms of dinoflagellates, the first time such nonquarantine blooms have been observed. Toxin at a level of 1,000 micrograms per 100 grams of meat can cause death of humans. Samples from Marin County tested as high as 4,000 micrograms. Significant amounts of toxin were found in mussels at Sausalito, Richmond, and Berkeley.

The Giant Kelp, *Macrocystis pyrifera,* a brown alga, is harvested commercially and has been used in feed supplements. This kelp has been shown to be mildly toxic to livestock, but with increased costs the commercial use of seaweeds as livestock feed is being abandoned.

FUNGI

Toxins in Foods

The organic compounds formed by the metabolism of living organisms and used in their growth and development are termed primary metabolites. Additional ones are called secondary metabolites. Secondary metabolites formed by bacteria and fungi are important to humans and animals. Normally bacteria are required in the digestive tract for the breakdown of food materials and to produce certain vitamins. Various antibiotics, such as penicillin, are secondary metabolites of fungi. Other metabolites of bacteria and fungi are very toxic, however, and their presence in foods is a separate study beyond the scope of this book.

Food poisonings from the toxins formed by bacteria of the genera *Clostridium, Staphylococcus,* and *Salmonella* are well documented. Because the toxins occur in minute quantities, their molecular structures are poorly known and appear to be diverse and complex.

Mycotoxins produced by toxic fungi have been known for years as being poisonous to humans and animals. They may cause damage to the liver, kidneys, central nervous system, and bone marrow. Smaller amounts of certain mycotoxins can reduce the normal immunity to infections.

The earliest records of human poisoning caused by fungi are those of fleshy fungi and ergot-infected grains. It was the outbreak of "turkey X disease" in England in 1961 that led to research in mycotoxins and eventually to the discovery and study of aflatoxins.

For many years moldy feedstuffs such as ground grains or

hay have caused death or distress to animals; the toxicity of these decaying foods varies greatly from toxic to completely nontoxic. Only in recent years has there been an equal emphasis on studying the effects of moldy foods on humans. Human diseases from mycotoxins are most likely to occur, however, only during food shortages. Experimental reproduction of the diseases caused by toxic fungi has been very difficult, and controversy still exists concerning the organisms that are responsible.

At a warm temperature, grain of 20% or more moisture content becomes moldy. Corn (maize) is the crop most likely to mold. Moldy corn fed to livestock has caused problems in the southeastern United States. Drying grains and hay properly to a moisture content of 13 to 15% and keeping them dry prevents toxic fungi from growing.

Cattle, sheep, horses, pigs, and fowl have all been affected by feeding on moldy stored grain or hay. (Plant material in fields decomposes rapidly and is rarely a cause of toxicity.) Fungi that have been considered responsible for poisonings in moldy feeds and forage belong to the genera *Aspergillus, Claviceps, Fusarium, Mycothecium, Penicillium, Phomopsis, Pithomyces, Rhizoctonia,* and *Stachybotrys*. The effects of such poisonings vary greatly with the particular fungus and the kind of plant material. Signs include drop in milk production, nervousness, blindness, convulsions, photosensitization, bleeding lesions in the skin or in the gut, liver or kidney damage, reduced fertility, and death. The distress and death of animals are caused by the poisons produced by the fungus and not by invasion of the animal tissue by an infectious fungus.

The structures of a number of the secondary metabolites in toxic moldy feed have been determined; they are all furofurans, either dihydro- or tetrahydro- forms:

furofuran

There are three subgroups of mold metabolites. Aflatoxins, the best known, consist of the furofuran heterocyclic rings joined to modified forms of the toxic lactone coumarin. The other subgroups have the furofuran rings attached to heterocyclic rings derived from xanthone or anthraquinone.

Aflatoxins

Aflatoxins are a group of related secondary metabolites produced by certain strains of *Aspergillus flavus* and *A. parasiticus*. Aflatoxin B_1, the most potent strain, causes both acute and chronic effects. It is carcinogenic and acts primarily on the liver of certain animals.

Aspergillus flavus grows readily when temperatures are warm and humidity is high, when the grains or ground meal have been improperly dried, or when moisture has been accidentally added. Milling the infected grains or seeds does not destroy the fungus, which can continue to grow in the moist material. Poisoning from aflatoxins is of great concern in the humid southeastern parts of the United States.

Aflatoxins affect the liver of animals and cause the same effects as pyrrolizidine alkaloids. The two groups of toxins are not related chemically, but damage to the liver appears the same.

Signs in cattle include walking in circles with frequent falls, blindness, elevated temperature, convulsions, and abortion. Of animals fed concentrates containing aflatoxins, calves 3 to 6 months old are the most susceptible to poisoning. The toxins can be passed through the milk, however, and affect even younger calves. Liver damage found in newborn calves is considered to be the result of aflatoxins passing through the placenta.

Death of cattle from aflatoxins may occur in 48 hours. Sheep have died in 18 hours under experimental conditions. Pigs are much more resistant and may not show any symptoms for at least 6 weeks. Poultry, especially turkeys and young ducks, are also susceptible to aflatoxins. Losses of turkeys, up to 40% of the flock, have been recorded. Selenium, added to the diet of turkey poults, has proved experimentally to protect against aflatoxin toxicity.

Ergot

The toxicity of ergot to humans and animals has been known for centuries. In California there are two widespread species of ergot fungi: *Claviceps purpurea* (fig. 1) infecting many native and cultivated grasses and *Claviceps paspali* on Dallisgrass, *Paspalum dilatatum,* and on Knotgrass, *P. paspalodes* [*P. distichum*]. *Claviceps cineraria* is found on James' Galleta, *Hilaria jamesii;* the grass occurs in California but the fungus has not been found here. Two species of *Claviceps* are found on Bermudagrass, *Cynodon dactylon,* in the southeastern United States but are not recorded from California.

FIG. 1. *Claviceps purpurea.* Ergot. Ergot bodies developed in place of normal grains in a seed head of barley.

The epidemic or spreading phase of the ergot fungus occurs at the time the grasses are in flower. Fungus spores infect the ovaries of the mature flowers of the host grass plants at the time of pollination, and there the fungus growth or mycelium develops. Following this initial infection a moist, sticky secretion, the "honeydew" stage, develops and causes the gradual decomposition of the tissue of the ovaries. Asexual spores (conidiospores) are formed at the tips of the mycelial threads and, carried by rain or insects, they infect other grass flowers. If there is warm, humid weather at this time, the ergot fungus may spread rapidly.

Parasitized flowers do not develop normal grains; the infected grains are mostly destroyed and replaced by purplish black ergot bodies called sclerotia. These hardened, compacted masses of fungal mycelium are often three to four times as long as the normal, uninfected grass grains. An infected young grain is replaced by the hyphae of the ergot fungus, but the epidermis of the ovary can still be found on part of the sclerotium, especially at the apex.

The ergot bodies fall to the ground where they may germinate at once or may overwinter and germinate the following spring. On germination the sclerotia produce long-stalked, globular fruiting heads within which are formed wind-borne spores that infect a new crop of grass flowers.

Toxic part: The sclerotia or ergot bodies.

Toxin: A number of alkaloids and amines; ergotamine is particularly active. Ergotamine is a polypeptide of lysergic acid and two amino acid groups.

Symptoms: In humans and animals, especially cattle, the common acute symptom is a convulsive stimulation of the central nervous system. Chronic symptoms include damage to the lining of the capillaries and contractions of the smaller arterial blood vessels resulting in gangrene of the extremities.

In humans, ergot poisoning occurs when contaminated cereal grains are eaten or made into flour. Two types of poisoning, gangrenous and convulsive, are known. The gangrenous type develops after ingestion of small amounts of ergot over a period of several weeks to several months. The symptoms of this form are tingling of the fingers followed by vomiting and diarrhea. Several days later gangrene appears in the fingers and toes. With chronic ergotism in humans, the limbs lose all sensation. In extreme cases the hands and feet rot away, but this form is quite rare. The convulsive type results from eating much larger quantities of ergot at one time. This form begins in the same way as the gangrenous, but it is followed by painful spasms of the muscles of the arms and legs and later generalized convulsions.

During the Middle Ages ergotism occurred in epidemics and was known as St. Anthony's Fire. In France during the years 944 and 1090 thousands of people died from this disease, which was caused mostly from eating bread made from rye grains contaminated with ergot. A comparatively minor outbreak was reported in France as recently as 1953. Baking does not destroy the poison. By the end of the sixteenth century such contaminated rye was recognized and discarded, thus preventing the disease. When harvests were poor, however, no grain would be wasted and epidemics again occurred. Human poisoning by ergot has been virtually eliminated through present-day regulations and inspection of grains.

The drug ergot made from ergot bodies has been used to control hemorrhaging, particularly during childbirth, and it has also been used in the treatment of migraine headaches. D-lysergic acid diethylamide (LSD), the most powerful psychotropic drug known, is a derivative of ergot alkaloids.

In the United States ergot poisoning in humans has been eliminated, but it still occurs in domestic animals. Cattle are most commonly affected, but effects have been seen in horses, mules, sheep, pigs, cats, dogs, and fowl. Individual animals show varying effects of ergot poisoning; moreover, some cows may become more or less addicted to ergot, seeking out infected seed heads of grasses throughout the pasture. Affected animals must be removed from infested fields immediately.

In mild cases of acute ergot poisoning cattle display intermittent blindness, deafness, and variations in sensitivity of the skin. In more severe poisonings convulsions are the main symptoms seen, usually followed by temporary paralysis and coma. Some animals may die with the first convulsions, others after several days. Muscle spasms in the legs result in goose-stepping, stiff-legged walking, and often jumping excitedly to the side when driven. Animals may fall in unusual positions— for example, with the legs extended straight out from the rear.

In cattle chronic ergot poisoning or gangrenous ergotism results from ingestion of smaller amounts of ergot over a long period of time. Cows may develop lameness in 10 days, but most cases of chronic poisoning develop over a period of several weeks or even months. Contraction of the smaller arteries reduces the blood supply to the extremities. The tail, ears, and limbs, particularly the hind legs, are affected. These areas manifest coldness, loss of sensation, and reddening followed by a bluish black color. The affected tissues develop gangrene and separate from the healthy parts. The hooves and the tips of the ears and tail may be lost.

Sheep have been experimentally fed ergot. They do not develop gangrene of the extremities, but they do suffer damage to the lining of the mouth, rumen, and small intestine. Pregnant sheep and pigs fed ergotized grains may lose their embryos or suffer a high mortality of the offspring.

Claviceps purpurea infects many species of grasses, including cereal grains: rye, wheat, and barley. Many native and introduced forage grasses on irrigated pasture and dry rangeland such as Wild Oats, ryegrasses, Orchardgrass, bromegrasses, fescues, and bluegrasses are infected.

Poisoning from another species of ergot, *Claviceps paspali,* on Dallisgrass, *Paspalum dilatatum,* can occur in irrigated pastures in the late summer when the seed heads are formed. This ergot is most toxic when maturing from the honeydew stage and entering the hardened sclerotial stage. Acute poisoning symptoms are the only ones seen from this species of ergot. Muscular tremors are common, noted only at first when the animals are driven. Later the tremors become continual and even prevent grazing. Affected animals readily recover when removed from an ergot-infected pasture.

In mid-November 1962, near Ballico, Merced County, 5 cows and 15 calves in a herd of 500 head displayed the nervous symptoms of ergot poisoning from Dallisgrass. The calves, which were very excitable, walked stiff-legged and often fell into ditches they were attempting to cross.

In Glenn County in 1980 a flock of sheep pastured on Dallisgrass infected with *Claviceps paspali* showed poisonings principally as abortions or premature delivery of weak lambs that did not survive. Two ewes showed signs of muscular incoordination but recovered in several days.

Secondary fungi often turn the honeydew stage of ergot infection into a sooty black mass. These secretions caused an unusual case near Redding, Shasta County, in September 1958. The black exudate from infected Dallisgrass adhered to the wool of sheep on the pasture. When the animals were bedded down in an adjacent newly plowed field, the accumulated exudate mixed with the soil caused the white animals to appear black in color.

Claviceps paspali also infects the seed heads of Knotgrass, *Paspalum paspalodes* [*P. distichum*], but these seed heads usually do not occur in sufficient numbers to cause any problem to livestock.

Some success in using ergot-infested fields has been obtained by mowing the infected seed heads with the mowing bar set high, raking up the seed heads, and leaving the forage for the livestock. Moreover, stocking the pasture with sufficiently large numbers of animals will prevent the formation of seed heads and therefore the growth of ergot bodies.

Dallisgrass presents a particularly difficult problem with ergot. Since it is less palatable than other forage plants in irrigated pastures, weedy Dallisgrass plants usually are left ungrazed and form seed heads. Also, because many seed heads of Dallisgrass are formed almost flat to the ground, below the height of a mowing bar, most infected seed heads cannot be removed.

The ergot fungus *Claviceps cineraria* on James' Galleta, *Hilaria jamesii,* is not known from California. The grass occurs in the state in the mountains of the eastern Mojave Desert and around Death Valley, however, and extends east to Texas where the ergot has been found. The ergot bodies, sclerotia, of this species are large ($\frac{1}{2}$ in. to $1\frac{1}{4}$ in. long) and dark gray at the base to light gray at the tip; the base is enclosed by the glumes or bracts of the grass spikelet.

Bermudagrass tremors of cattle in the southeastern United

States has been associated with a species of *Claviceps*, although in some outbreaks this fungus has not been found. These signs of acute ergot poisoning have not been reported for California. *Claviceps purpurea* and *C. cynodontis* have been recorded on *Cynodon dactylon* (both Common Bermudagrass and Coastal Bermudagrass). The black-colored secondary fungi have been present during the honeydew stage on the flower stalks in pastures where there have been outbreaks of bermudagrass tremors. This secondary fungal growth must not be confused with the smut fungus, also black-colored, commonly found in California on Bermudagrass, particularly on the first-formed flowering stalks in the summer.

Poisonous Mushrooms

Mushrooms are among the best known of the plant organisms called fungi. Their umbrella-shaped bodies, so often conspicuous and colorful, are largely ephemeral and live only for a limited time, mostly in forests and woods, in grassy areas and meadows, and on living or dead wood. Because they lack chlorophyll, fungi including mushrooms cannot live independently in the manner of green plants. They are either parasites, saprophytes, or mycorrhizal associates with green plants.

Fungal parasites derive their basic food from living hosts, mostly other plants, rarely animals. Many plant diseases are caused by parasitic fungi; wheat rust, Dutch elm disease, and ergot are examples. Fungal saprophytes obtain their food supply from dead organic matter such as leafy or grassy litter and accumulated duff in the soil or from decaying wood. The mycorrhizal association is a symbiotic or mutually beneficial relationship between a fungus and a green plant through the underground filamentous body or mycelium of the fungus growing in and on the feeder root system of the plant. The green plant supplies carbohydrates to the fungus, which in turn supplies water and mineral nutrients to the green plant.

The body of a mushroom consists of two portions: the visible reproductive part and the invisible vegetative part that grows belowground or within decaying or living wood. The visible reproductive portion is the fruiting body of the fungus and appears seasonally when the proper conditions of accumulated food, temperature, and moisture occur for its growth. The unseen vegetative part of the fungus consists of a vast, much-branched network of filaments called the mycelium that derives moisture and food from its surroundings and lives from year to year just as any aboveground, independently living, green plant.

The fleshy fungi, which include the poisonous as well as the edible mushrooms, belong to two major groups: the Ascomycetes or sac fungi and the Basidiomycetes or club fungi. The two groups are distinguished chiefly by their method of spore production.

The sac fungi, the Ascomycetes, produce their spores, usually eight, within a microscopic sac-like cell called the ascus. Cup fungi, morels, and truffles belong here. Although this is a large group of fungi it has only a few poisonous members, most of which are involved in only one kind of mushroom toxin (see "Gyromitrin Poisoning," pp. 31–32).

The club fungi, the Basidiomycetes, produce their spores, usually four, at the tip of a microscopic club-shaped cell, the basidium. Most of the edible and poisonous mushrooms belong to the club fungi, which are divided into several subgroups according to the shape of the fruiting bodies and the arrangement of the spore-bearing surfaces. The largest subgroup consists of the gilled mushrooms with the familiar cap borne at the top of the stalk. On the lower surface of the cap is a series of thin radiating partitions, the gills, and on these the spores are produced.

A small subgroup of Basidiomycetes is the pore fungi, also with a cap and stalk but lacking gills. The lower surface of the cap appears solid but is perforated by many minute pores in which the spores are produced. Pore fungi include the boletes and the polypores.

Earthballs and the closely related puffballs, which do not look like mushrooms, have round to oval fruiting bodies and when mature are filled with a powdery spore mass. Puffballs are edible but the mildly toxic earthballs are not.

In identifying mushrooms various features should be noted. The cap varies in size, shape, and color; its surface may feel viscid or smooth; its margin may be smooth, inrolled, or irregularly ragged. The gills on the lower surface of the cap may be free at the ends next to the stalk, attached to the stalk, or attached and decurrent, running down the stalk. The gills may be close together and crowded, or they may be widely separated and distant from each other. The height and shape of the stalk should be noted, as well, and whether it is hollow or appears to be stuffed.

In many mushrooms one or two layers of tissue called veils may be present on the undeveloped egg-shaped button stage of the young fruiting bodies. One of these, the universal veil, encloses the entire young button during its development. As the cap expands, the universal veil is broken and in some mushrooms it appears as a persistent often cup-shaped structure, called the volva, at the base of the stalk. When the volva is present, its shape varies in different species of mushrooms. It is cup-shaped and usually conspicuous in the deadly poisonous *Amanita phalloides*. There is also the partial veil that covers the young gills during development of the cap. As the cap matures, the partial veil breaks and may leave remnants appearing as a ring or annulus around the stalk; in a few mushrooms, it appears only as shredded patches on the margin of the cap.

Spores of fungi can be seen only with the aid of a microscope. When a mass of spores is deposited as a spore print, their color, which is an important characteristic, can be seen with the naked eye. To prepare a spore print, remove the mature mushroom cap from the stalk and place it on a piece of white paper with the spore side down. A wedge of dark paper may be put under a portion of the cap to show white spores. The mushroom cap should then be covered so that it will not become dry or subjected to air movements. Usually a few hours is sufficient for the spores to be deposited on the paper. After the specimen is removed, the paper should be allowed to air dry for about 10 minutes; then the spore print will be ready for examination.

Many mushrooms have distinctive odors and tastes. To determine a mushroom's taste, take a very small piece and chew for a minute or two, but do not swallow as the mushroom may be poisonous. Remove the piece from the mouth and clean the mouth with repeated washings of water. A strong burning or peppery taste, usually described as acrid, suggests the possibility of toxicity. Many toxic mushrooms have no distinctive taste, however.

Shortly after World War II a great interest developed in the United States in hunting, gathering, and eating wild mushrooms. Along with this interest, mycological societies have been formed in many parts of the country and professional mycologists have cooperated in disseminating information to their members. As a result many amateur mycologists have become well informed regarding characteristics of mushrooms and the intricacies of their identification. These efforts have been greatly helped by the publication of popular but informative, well-written, and well-illustrated books on mushrooms and their characteristics. Several are listed in the General References. In some mycological societies, special committees have been formed to study various aspects of mushroom poisoning. The Mycological Society of San Francisco has such a committee, the Toxicology Committee, which in 1977 published its Toxicology Monograph No. 1, *California Toxic Fungi.* But in spite of these efforts poisonings still occur.

In California the season for mushrooms begins with the first rains in the fall and continues through the winter. During the fall of 1981 central California had ideal weather conditions for the growth of mushrooms. Early spaced rains produced a continually moist but not overly wet soil and were accompanied by mild temperatures. From October 1 through the middle of December, the San Francisco Bay Area Regional Poison Control Center received ninety-six calls on mushroom-related poisonings—five times as many as had been received during the same period in the previous year. Ten of these cases were presumably from eating mushrooms of *Amanita phalloides;* three of the cases were fatal. This is a familiar story that is repeated every winter, especially in central and northern California. The number of poisonings and fatalities varies with each year's weather conditions. During winters with little rainfall or unusually cold temperatures there are fewer mushrooms and fewer poisonings.

Poisonings may be avoided by a positive identification of each individual mushroom gathered in the wild. Wild mushrooms, considered to be delicacies, present hazards especially to those who lack the expert knowledge needed to distinguish the edible, nontoxic ones from the few that are poisonous. Even persons with experience could be poisoned from eating a single immature poisonous mushroom that was growing in the same area with a large number of edible ones. When mushroom hunting, those who are inexperienced should accompany knowledgeable collectors to learn to distinguish edible mushrooms from poisonous ones. When on their own, inexperienced collectors should have their mushrooms carefully checked by someone thoroughly knowledgeable. There is no simple test for recognizing toxic mushrooms. Such bits of folklore as whether or not a cooked mushroom blackens a piece of silver should be dismissed for their lack of credibility. The only rule to remember is this: Identification of fleshy fungi is not easy. Careful examination is necessary, sometimes with a microscope. Whenever there is the least doubt about an identification, discard the mushroom. Remember the folk saying: "There are old mushroom hunters, and there are bold mushroom hunters—but there are no old, bold mushroom hunters."

The toxins found in poisonous mushrooms fall into four groups according to the physiological effects produced and the time required for the symptoms to appear after their ingestion.

I. Toxins causing cellular destruction and kidney and liver damage. These toxins may be fatal. Symptoms may have a delayed onset and usually do not appear for several hours (perhaps as much as 10 or 12 hours).
    1. Amanitin Poisoning. Caused by amatoxins and phallotoxins (cyclopeptides). Amatoxins, the more potent, are among the most lethal poisons known. Found mostly in mushrooms of the genus *Amanita,* these toxins are the cause of more than 90% of all fatal cases of mushroom poisoning.
    2. Gyromitrin Poisoning. Caused by monomethylhydrazine (MMH). This toxin is found in mushrooms of the genus *Gyromitra,* the false morels, and rarely causes fatalities.
II. Toxins mostly affecting the autonomic nervous system. Symptoms appear within 15 minutes to 6 hours after eating.

3. Coprine Poisoning. Caused by the toxin coprine. The mushroom *Coprinus atramentarius*, Inky Cap, contains this toxin. Symptoms are delayed up to 6 hours after eating but the reaction occurs only when alcoholic beverages are also consumed.

4. Muscarine Poisoning. Muscarine is the toxin found in mushrooms belonging to the genera *Clitocybe, Inocybe*, and *Omphalotus*. Symptoms appear rapidly—within 15 to 30 minutes after eating.

III. Toxins mostly affecting the central nervous system. Symptoms appear rapidly, within 20 minutes to 2 hours.

5. Ibotenic Acid and Muscimol Poisoning. *Amanita muscaria*, the Fly Agaric, one of the causes of this poisoning, has long been known for its hallucinogenic properties.

6. Psilocybin and Psilocin Poisoning. Mushrooms belonging to several genera, including *Psilocybe* and *Panaeolus*, cause this poisoning. Its most prominent symptoms are hallucinations.

IV. Toxins mostly causing gastrointestinal irritation. The largest number of toxic mushrooms are found in this group; the toxins of most of these mushrooms are not known.

Mushroom gatherers often find statements in the literature describing a mushroom's edibility. It should be remembered that even mushrooms said to be "edible" may not be edible for everyone. Just as a few people suffer allergic reactions to certain foods, mushrooms, including the commercial ones, may produce adverse reactions. Persons wishing to eat wild edible mushrooms for the first time should observe certain restraints and precautions: Eat only one kind at a time, eat only a small amount, and eat only fresh young mushrooms free from insect larvae.

Amanitin Poisoning

The toxins causing amanitin poisoning, the amatoxins, are complex polypeptide molecules. They are among the most lethal poisons known. Several species of *Amanita* cause more than 90% of all cases of fatal mushroom poisoning. Poisonous amanitas, large, conspicuous, fleshy mushrooms, are attractive to collectors who often mistake them for edible fungi. Amatoxins are also found in mushrooms belonging to other genera, *Galerina, Conocybe,* and *Lepiota.* Poisoning is rare, however, because their fruiting bodies are small in size, drab in appearance, and not as likely to be collected and eaten as those of the more conspicuous amanitas.

Symptoms following ingestion appear after a latent period of usually 10 or 12 hours, or sometimes less. They begin with nausea, severe and intermittent vomiting, abdominal pain, and profuse watery diarrhea. This period of gastroenteric symptoms lasts several hours and is followed by a remission phase, without symptoms, lasting from 1 to 3 days at which time the patient feels improved. The remission period is followed by severe, sometimes fatal, kidney and liver failure. Moreover, there may sometimes be complications involving other organs.

Amanita.

A large genus that includes the most deadly poisonous mushrooms, deliriant ones, and even a few edible ones. Because poisonous amanitas might be mistaken for edible mushrooms in other genera, mushroom collectors should learn the characteristics by which amanitas may be recognized. In collecting them, and other mushrooms as well, care should be taken to obtain the entire fruiting body, particularly the base of the stalk with the volva, which is very important in the identification of amanitas. In amanitas the cap separates easily from the stalk, the gills are white or pallid, free or nearly so, a volva is present, a veil or annulus may or may not be present below the cap, and the spores are white. The volva varies within the genus from a sac-like covering to a series of rings or scales or only as a dusting of powdery meal.

Amanita phalloides. Death Cap (pl. 1*c*).

Cap convex, to 6 in. or even 9 in. across, viscid, yellow green to greenish brown; stalk to 5 in. tall, bulbous at base; veil white, thin, hanging like a skirt beneath the cap; volva large, like a sac, enclosing the base of the stalk. As described by surviving victims the taste is good. The odor when fresh resembles raw potatoes but is not always detectable. Usually found under live oaks or in lawns near oaks; fall and early winter, may be abundant when there are early warm rains; central California south to Santa Barbara County.

Amanita ocreata.   Death Angel, Destroying Angel.

Differs from *Amanita phalloides* in having a white cap soon becoming buff, sometimes with pink, yellow, or orange discoloration; hanging membranous veil soon collapses and often disappears; voluminous volva is open and cup-like; taste is mild and slightly metallic. Mostly under coastal live oaks; usually from January into April; San Francisco Bay region south to San Diego.

In California mushroom fatalities had long been attributed to *Amanita phalloides,* but cases were reported for late winter and spring after *A. phalloides* had finished fruiting for the season. A study of amanitas in California showed that *Amanita ocreata,* which produces its mushrooms in late winter and spring and extends to the southern part of the state, had been overlooked as a possible cause of mushroom poisonings in the spring.

Conocybe filaris.

Cap bell-shaped, to $\frac{3}{4}$ in. across, brown to yellow, flesh thin, delicate; stalk to $2\frac{1}{2}$ in. tall; veil membranous, forming a delicate ring; spores brown. Usually found in well-watered lawns or other grassy places, often ephemeral; in fall; northern coastal California to Santa Cruz County and in the Los Angeles State and County Arboretum, Arcadia, Los Angeles County.

Galerina autumnalis.

Cap convex to flattish, to 2 in. across, viscid, yellow brown when wet, pale brown when dry; stalk slender, to $2\frac{1}{2}$ in. tall; veil forming a thin ring; spores rusty brown. In clusters on decayed logs and debris; fall and winter; northern California to the San Francisco Bay region and Fresno County in the Sierra Nevada and San Diego County.

Galerina marginata.

Similar in appearance to *Galerina autumnalis* but with a moist not viscid cap. In autumn on decaying wood in central California.

Lepiota josserandii [has been known in California as *L. helveola*]. Parasol Mushroom.

Cap sometimes umbonate, to 2 in. across, dry, brown scaly; stalk slender, to 3 in. tall, scaly; spores white. Cultivated ground or in oak woods; winter through spring; northern California coast and in Los Angeles County.

Lepiota castanea and *L. felina* are suspected of being poisonous.

Gyromitrin Poisoning

Gyromitrin poisoning is caused by *Gyromitra esculenta* and is suspected in other species of *Gyromitra* and in *Disciotis, Helvella,* and *Sarcosphaera.* Gyromitrin, first isolated from *Gyromitra esculenta,* is converted by hydrolysis to monomethylhydrazine (MMH). A volatile compound, gyromitrin is removed from the mushrooms by boiling, but the cooking water should be discarded because enough toxin remains in it to cause poisoning if it is consumed as a stew or soup. After ingestion the onset of symptoms is delayed, usually 6 to 8 hours, but may be less or more, up to 24 hours. Symptoms include fatigue, headache that may be severe, dizziness, abdominal pain or a feeling of fullness, and vomiting. Usually the patient recovers completely within 2 to 6 days. In severe cases, however, there may be liver and kidney failure and destruction of red blood cells. Fatalities have occurred.

Gyromitra esculenta. Brain Fungus, False Morel.

Fruiting body irregularly lobed with brain-like convolutions, interior hollow, 3 in. across, light to dark red brown, flesh thin; stalk somewhat compressed, hollow, to 2 in. tall, white to pink. In spring; northern California south to Santa Cruz County and in the middle Sierra Nevada.

Both *Gyromitra californica* [*Helvella californica*], California Elfin Saddle, and *Gyromitra infula,* Hooded Gyromitra, have more or less saddle-shaped fruiting bodies or caps in contrast to the convoluted cap of *G. esculenta.*

Mushrooms of the following three species, suspected of containing gyromitrin, are reported as toxic when eaten raw, although the toxins have not been isolated from them. All have cup-shaped fruiting bodies.

Disciotis venosa. Veiny Cup Fungus.

Cup irregularly shaped, brown, to 8 in. across, on a very short stalk, partly buried. Often under conifers; in spring; northern California.

Helvella acetabulum. Cabbage-leaf Helvella, False Morel.

Cup brown, shallow, to 4 in. across, outside with prominent ribs extending from base almost to upper margin. In woods; early spring; central California southward.

Sarcosphaera crassa. Pink Crown, Crown Fungus.

Fruiting body crown-shaped, to 4 in. across, inside pink to purple, outside whitish. Under conifers; spring; mountains of the north coast, Sierra Nevada southward to San Diego.

Coprine Poisoning

Coprine poisoning is caused by *Coprinus atramentarius,* the Inky Cap, a desirable edible mushroom that produces a toxic reaction in some people if it is eaten with or preceding alcoholic beverages. Ingestion of the mushroom sensitizes the individual to alcohol in a manner identical to disulfiram (Antabuse), used to produce aversion to alcohol in those with drinking problems. The toxin slows the body's metabolism of alcohol causing a toxic substance, acetaldehyde, to accumulate. Acetaldehyde is responsible for the severe nausea, vomiting, headache, and other responses. Symptoms recur if alcohol is again ingested after eating sufficient quantities of the mushroom. Usually it takes a few hours after eating the mushrooms for maximum sensitivity to occur. After that, sensitivity is diminished from 24 to 48 hours.

The onset of symptoms after ingestion is rapid, often within 30 minutes or up to 2 hours, and, as with Antabuse, appears to be related to the level of alcohol in the blood. Symptoms include flushing of the face, difficulty in breathing, a feeling of lightheadedness, throbbing of the large blood vessels, palpitations, and sometimes nausea and vomiting. Usually symptoms disappear in from about 30 minutes to 2 hours, and recovery is spontaneous. The toxin remains in the body for 2 days or longer, however, and if alcohol is again taken during that period symptoms will recur.

Coprinus atramentarius.   Inky Cap.

Cap oval, becoming bell-shaped, to 3 in. across and equally tall, gray, finely brown scaly; gills white, at maturity dissolving into an inky black fluid; stalk to 4 in. tall, hollow, white; spores black. The dissolving of the gills into an inky fluid is a distinguishing characteristic of all inky caps. In cultivated ground, lawns or gardens, or any disturbed ground with sufficient decaying matter; fall through spring; common throughout California. Because of its widespread occurrence in urban and other disturbed areas, Inky Cap has been referred to as a "mushroom weed."

Muscarine Poisoning

The toxin muscarine, an alkaloid, is the cause of muscarine poisoning. California mushrooms in the genera *Clitocybe, Inocybe,* and *Omphalotus* contain concentrations of the alkaloid high enough to cause poisoning in humans. Although muscarine was originally isolated from *Amanita muscaria* and also is present in *Amanita pantherina,* it occurs only in minute amounts in the two amanitas. Muscarine stimulates the parasympathetic nerve receptors of the autonomic nervous system causing increased secretions of perspiration, saliva, and tears. It has no effect on the central nervous system.

Symptoms begin soon, perhaps within 15 minutes after ingestion, with heavy perspiration and visual disturbances; there are also nausea, vomiting, abdominal cramps, and in some cases constriction of the pupils, blurred vision, diarrhea, slow pulse, and dizziness.

Clitocybe.

Mostly drab, small to large-sized mushrooms, caps gray to whitish, convex or depressed; gills adnate or decurrent; spores usually white; veil and volva absent; stalk usually fleshy. A large genus, about forty species in California, a few reported edible; edibility of others is not known, and a few are toxic.

Clitocybe dealbata.   Sweating Mushroom.

Cap dull gray white, convex to flattened, to 2 in. across, depressed in center; gills decurrent; stalk to 2 in. tall; spores white. Scattered to numerous; in pastures and other grassy places; fall and winter; rarely reported but may be more widespread in California than reports indicate. This small, rather nondescript mushroom sometimes occurs with the Fairy Ring Mushroom, *Marasmius oreades,* with which it might be confused. Because it often occurs in or near urban areas it is a potential hazard, particularly to small children. Fatalities have rarely been reported.

Inocybe.

Mostly small or medium-sized, drab, white to brown mushrooms, caps often conic to umbonate, with minute scales or fine silky fibers radiating from the center to the margins; spores in shades of brown. Most members of this large genus are poisonous. In addition to those described here, other species in California are toxic in varying degrees and should be avoided.

Inocybe geophylla.

Cap about 1 in. across, gray to white or lilac. Odor slightly unpleasant. Scattered or in groups; fall to early spring; northern and southern California.

Inocybe pudica.

Cap, gills, and stalk white or pallid, soon flushed with tan, salmon pink, or reddish brown or staining these colors when bruised. Cap 1 to 2 in. across; stalk somewhat stout, to $2\frac{1}{2}$ in. tall. Odor unpleasant. Scattered or grouped under conifers; late fall and winter; northern California.

Inocybe sororia.

Cap to 3 in. across, yellow, becoming darker; stalk to 3 in. tall. Odor of fresh green corn. Solitary or grouped; in woods; fall and winter; northern California south to Santa Diego County.

Omphalotus.

Mushrooms of this genus grow on wood and are brightly colored in shades of golden yellow to bright orange, gills decurrent.

Omphalotus olearius. Jack O'Lantern Mushroom.

Common name alludes to the phosphorescence given off by the gills. Cap convex to depressed, to 10 in. across; stalk solid, stout, 6 in. tall; spores white. Usually in clusters, on stumps of buried wood; fall and winter; fairly abundant and widespread in California.

The Jack O'Lantern Mushroom bears a resemblance to *Clitocybe dealbata* because of the decurrent gills. It also might be mistaken for the edible Chanterelle, *Cantharellus cibarius,* except that the Chanterelle grows on the ground and its thick-edged descending forked ridges differ from the unforked, sharp-edged gills of the Jack O'Lantern Mushroom.

Ibotenic Acid and Muscimol Poisoning

Amanita.

Mushrooms belonging to two species of the well-known genus *Amanita, A. muscaria* and *A. pantherina,* contain ibotenic acid and muscimol. Ibotenic acid, an amino acid, is unstable and breaks down when heated; with the loss of water and carbon dioxide, muscimol is produced. These toxins are deliriant substances that act on the central nervous system. Because these two known compounds do not account for the nausea that may follow eating of these mushrooms, other toxins may also be present.

Symptoms, usually appearing within 30 minutes to 2 hours after ingestion, include drowsiness, inconsistent nausea and vomiting, elation, confusion, visual distortions, mild euphoria, increased motor activity, sensation of floating, tremors, and visions. Symptoms alternate with drowsiness or sleep but generally disappear within 24 to 48 hours. Although muscimol is excreted in the urine, its effects may appear after most of the drug has been eliminated. There is no "hangover" effect as from alcohol. Deaths from these toxins are not known.

Amanita muscaria.   Fly Agaric (fig. 2; pl. 1*b*).

Cap to 15 in. across, almost globose at first, expanding in age, usually red, sometimes yellow to orange, rarely grayish white, with scattered small wart-like patches; stalk usually taller than the diameter of the cap, stout, thick, bulbous at base, white, covered with silky hairs; veil white, hanging skirt-like; volva a series of scaly, concentric rings above the bulbous base; spores white. Solitary or in clusters; in wooded areas; during fall and winter; widely distributed in California.

FIG. 2. *Amanita muscaria.* Fly Agaric. Cap red, orange, or yellow, rarely white; white warty patches on cap; white gills and stalk; volva a series of concentric rings on the base of the stalk.

A most attractive and conspicuous, brightly colored mushroom, the Fly Agaric contains one of the oldest intoxicants known and probably was used during ancient times in Europe and Asia. During the early eighteenth century its use as an intoxicant by people living on the Kamchatka Peninsula of eastern Asia was reported by European travelers. Even earlier records tell of orgies following use of this mushroom. Within recent years deliberate eating of *A. muscaria* and *A. pantherina* to induce hallucinations has increased greatly in this country.

Amanita pantherina.   Panther Mushroom, Panther Fungus, Panther Cap (pl. 1*a*).

Closely related to *Amanita muscaria, A. pantherina* is smaller, does not have the red cap, and is not so colorful. Cap to 6 in. across; surface pale to rich brown; stalk to 5 in. tall; volva sac-like with a collar-like ring at its apex that firmly covers the base of the stalk; spores white. Solitary or grouped; in woods; fall and through spring; throughout California. *Amanita pantherina* is reported as one of the common poisonous mushrooms in the Rocky Mountains and the Pacific Northwest and is frequently involved in poisonings.

Psilocybin—Psilocin Poisoning

Psilocybin and psilocin, indole alkaloids, are powerful hallucinogens that strongly affect the central nervous system. They are found in mushrooms of several genera occurring in California: *Gymnopilus, Panaeolina, Panaeolus, Psilocybe*. Mushrooms containing these hallucinogens have been used for centuries in Mexico and Central America in ritualistic ceremonies. Small mushroom icons found in Central America, estimated to be 3,500 years old or more, indicate their use that long ago. The Spaniards, in their attempts to Christianize the people of Mexico, were unable to suppress use of these mushrooms completely. Recent ethnobotanical and mycological studies in Mexico have shown that about a dozen different mushrooms are still being used today in divinitory and healing rites, sometimes in more or less secret ceremonies.

In recent years the use of hallucinogenic mushrooms has been taken up in the United States, and certain groups, particularly of young Americans, have begun to use these so-called recreational drugs to induce hallucinations. At first the mushrooms were sought in Mexico, but then it was found they could be obtained in this country, either wild or in cultivation. In 1970 the United States Congress passed the Comprehensive Drug Abuse and Control Act. It placed psilocybin and psilocin, psychoactive indole alkaloids, on the list of controlled substances, making it illegal to possess indole-containing mushrooms.

Symptoms of this poisoning have a rapid onset, usually within $\frac{1}{2}$ hour after ingestion, and generally last for not more than 4 hours. Individual reactions vary from feelings of pleasant relaxation to tenseness, anxiety, or dizziness. Nausea, abdominal pain, vomiting, and diarrhea are followed by various visual effects; outlines of objects are sharpened, and colors take on greater brilliance. Visual images increase in brilliance, often with bursts of color patterns, and dream-like shapes come and go. There is also elation or hilarity, uncoordinated movements, muscular weakness, drowsiness, and sleep. Various systemic effects occur, the result of stimulation of the central nervous system. Ingestion of large amounts of the mushroom can cause severe toxic reactions. Fatalities have been reported.

Gymnopilus spectabilis [*G. junonius, Pholiota spectabilis*].
Cap to 8 in. across, convex to flat, yellow orange to rusty orange; stalk to 10 in. tall, stout, colored as the cap; spores rust-colored. Solitary or usually clustered; on old stumps or logs; fall and winter; northern coastal California south to Monterey County and the Sierra Nevada. This mushroom has a very bitter taste that persists. Because those who ingest it sometimes break out into laughter and display foolish behavior, in Japan it is called the "Big Laughing Mushroom." The presence and concentration of psilocybin varies in different geographical areas resulting in variations in its effects. The Honey Mushroom, *Armillaria mellea,* which grows on living or dead wood

and is often densely clustered, might be mistaken for this mushroom.

Naematoloma popperianum.

Cap about 1 in. across, convex, later expanding, with an incurved margin, yellow tan, viscid; stalk 1 in. tall, whitish, turning bluish where bruised; spores rusty brown. Known only from San Francisco where it was found on a rubbish heap near woods. It is the only species in the genus *Naematoloma* known to contan psilocybin and psilocin. Another species, *Naematoloma fasciculare,* Clustered Woodlover, a densely clustered mushroom that grows on wood and has a bitter taste, does not contain psilocybin and psilocin but is otherwise seriously poisonous. (See "Gastrointestinal Irritants," pp. 41–55.)

Panaeolina foenisecii [*Panaeolus foenisecii, Psathyrella foenisecii*]. Mower's Mushroom, Haymaker's Mushroom, Lawn Mower's Mushroom.

Cap to 1½ in. across, reddish brown when young and moist, fading to pale buff, conic, in age becoming bell-shaped, often umbonate; gills purple to chocolate brown; stalk 4 in. tall, brittle, whitish, darkening in age, with soft short hairs; spores purple brown. Scattered or in groups; in grassy areas; appears during most of the year whenever there is sufficient moisture; throughout California. This is one of the most common lawn mushrooms in North America. Psilocybin is present in mushrooms of this species from the eastern United States, but reports of its presence in the West are conflicting. It has been reported by some authorities as edible but by others as containing psilocybin and to be a mild hallucinogen.

Panaeolus sphinctrinus. Bell-cap Panaeolus.

Cap to 2 in. across, bell-shaped, red brown or dark brown to gray brown, surface smooth, moist to dry, margin with hanging toothed fragments, remnants of the veil; stalk to 3 in. tall, slender, smooth, brown, covered with white powder; spores black. Solitary to several; on dung in pastures and manured lawns or in gardens; central California; summer. Mildly hallucinogenic. Closely related to *Panaeolus campanulatus,* and by some the two are considered a single species.

Panaeolus subbalteatus.   Belted Panaeolus, Girdled Panaeolus.

Cap to 2 in. across, convex, expanding to become almost flat, usually umbonate, brown with a dark belt extending ring-like around the margin; stalk to 4 in. tall, red brown, covered with white hairs, occasionally bruising blue at base; spores black. Scattered or clustered; on dung or manured soil, often on compost heaps or in gardens; February through May; most of California. Strongly hallucinogenic.

Psilocybe baeocystis.

Cap to 2 in. across, conic, margin incurved, expanding to convex and wavy, viscid, olive brown to gray, bruising blue; stalk to 3 in. tall with whitish fibers; spores dark purple. Gardens, lawns, or under trees; fall and winter; rare; northern California. A strong hallucinogen.

Psilocybe coprophila.   Dung Psilocybe.

Cap to 1 in. across, convex to bell-shaped, brown or reddish brown; gills purple brown; stalk to 2 in. tall, slender, whitish at first, darkening to brown, not bruising blue; spores purple brown. Solitary or grouped; on dung, often on "cow pies"; almost anytime during the year following rain or heavy fog; northern California. Mildly hallucinogenic.

Psilocybe cyanescens.

Cap to 2 in. across, convex to nearly flat, margin incurved, becoming flat and somewhat wavy, chestnut brown when young, fading to yellowish, viscid; stalk 3 in. tall, whitish, with silky fibrous hairs; spores purple brown. Cap and stalk bruise blue. Solitary or grouped; under conifers in humus among leaves and twigs or on wood chips; fall and winter; northern California but not common.

Psilocybe pelliculosa.

Cap to 1 in. across, bell-shaped to conic, yellow brown to gray brown; a thin gelatinous covering can be removed when the cap is wet; stalk to 4 in. tall, hairy, slender; spores purple brown. Cap and stalk bruise bluish. Scattered or grouped; in or at the margins of cool, shady woods; fall and winter; northern California. Mildly hallucinogenic.

Gastrointestinal Irritants

This group consists of a miscellaneous assortment of fleshy fungi that produce varying degrees of gastrointestinal irritation. The chemistry of their toxins is little known, and the toxicity of the mushrooms in this group varies considerably. Moreover, there are individual reactions to mushrooms of the same species growing in different areas. Some mushrooms cause discomfort if eaten raw or in large amounts; others may be eaten after they have been cooked; still others are toxic regardless of their preparation. Mushrooms in this group rarely cause fatalities. Symptoms usually appear within 15 minutes to a few hours following ingestion.

Agaricus.

Mushrooms of this genus are usually edible. The commercial mushroom, *Agaricus bisporus,* belongs here. The Meadow Mushroom, *Agaricus campestris,* its close relative, is widespread in fields and grassy areas throughout North America. Mushrooms belonging to *Agaricus* may be recognized by the fleshy cap, white to gray brown, with free, closely arranged, pale or pinkish gills, later turning chocolate brown; stalk with a partial veil but usually no volva; spores brown. Mushrooms known to be toxic have an odor of phenol when fresh and an astringent or metallic taste; they may bruise yellowish, particularly toward the base of the stalk. A reliable test to show this color change is the spot application of a 3 to 10% solution of potassium hydroxide, KOH, to the surface of the cap. A positive KOH test gives a yellow color and will aid in recognition of toxic members of *Agaricus*. Some people eat these mushrooms without any problem while in others they cause nausea, diarrhea, and sometimes headache. Probably because they so closely resemble their edible relatives, the phenol-smelling mushrooms are a frequent cause of poisoning.

The four species of *Agaricus* described below are fairly widespread in California. In addition, however, there are others less common and even undescribed that have a phenolic odor and stain yellowish when bruised or treated with KOH. All mushrooms of this genus should be carefully examined before considering them to be edible and must be sampled with restraint. There are occasional reactions to non-phenol-smelling mushrooms by a few people who apparently have problems with mushrooms that almost everyone else can eat.

Agaricus californicus.   California Agaricus.

Cap to 4 in. across, shining, dry, more or less scaly; stalk to 4 in. tall, stout; veil remaining as a persistent hanging ring. Scattered or in groups; in urban areas, especially lawns, gardens, and parks; most abundant in fall but year-round in wet places; known only from California.

Agaricus hondensis.

Cap to 7 in. across, with pale pinkish brown fibrillose scales, darkening in age; stalk to 7 in. tall, abruptly bulbous at base; veil forming a felt-like ring. An attractive mushroom found in groups or rings in woods in accumulated litter; late fall and winter; Mendocino County south to Monterey County, Yuba County, and southern California.

Agaricus praeclaresquamosus [sometimes called *A. meleagris* and *A. placomyces*].

Cap to 5 in. or more across, white, ivory, or gray to smoky black, with fine gray to smoky black fibrillose scales; stalk to $4\frac{1}{2}$ in. tall, colored as the cap; veil more or less floccose, cream to dull brown. Solitary or grouped; in woods or in the open; fall to early spring; northern coastal California south to Santa Cruz County and in the middle Sierra Nevada.

Agaricus xanthodermis (pl. 1*d*).

Cap almost round, becoming convex and flat, to 10 in. across, white, later turning gray or buff to brown; stalk to 7 in. tall, sometimes enlarged at base, smooth, white, with flat patches forming a thick ring immediately below the cap. Scattered or in groups; often in cultivated areas or in disturbed areas along roads and paths or in woods or pastures; fall through spring; San Francisco Bay region and southward to southern California.

Amanita.

This genus, which contains the world's most poisonous mushrooms as well as some edibles and hallucinogens, has been discussed under "Amanitin Poisoning" (see pp. 28–31) and under "Ibotenic Acid and Muscimol Poisoning" (pp. 35–37).

Amanita chlorinosma.

Reports for California of chlorine-smelling mushrooms such as this one may be based on misidentified specimens. According to a recent study of amanitas there are none with such an odor in the state.

Amanita cokeri [has been called *Amanita solitaria*, a name that belongs to a European species not known in the United States].

Cap to 6 in. across, ivory white, usually covered with small brownish warts, convex to flat, viscid; stalk to 6 in. tall, white, stout, tapering below ground level; veil hanging skirt-like; volva of irregular scaly patches at the base of the stalk; no odor. Solitary or grouped; fall to spring; north coast south to San Mateo County.

Amanita rubescens. The Blusher.

Cap to 6 in. across, red brown, when young covered with warty patches; stalk to 6 in. tall, stout, white, staining pink to red, enlarged at base; veil a fragile, irregular skirt-like ring; volva of indistinct red scaly fragments; bitter aftertaste; mild odor. Solitary or grouped; usually under live oaks; late winter to spring; widely distributed in California. The Blusher is variable with respect to its coloring and the shape of the warts on the cap. It might be confused at certain stages with such poisonous amanitas as *Amanita pantherina*. Although it is said to be edible when cooked, it is best not to eat this mushroom.

Amanita vaginata.   Grisette.

Cap to 6 in. across, gray to brown, viscid, sometimes with a white patch, margin striate; stalk to 6 in. tall, somewhat slender; veil absent; volva sac-like; odor mild. Solitary to scattered or in groups; in woods; fall and winter, sometimes to early spring; in the northern part of the state only. Reported as edible when cooked but care should be taken to identify it correctly.

Armillaria mellea.   Honey Mushroom, Shoestring Mushroom, Bootlace Mushroom, Oak Root Fungus.

Cap to 4 in. across, yellow brown (honey-like) to rusty brown, convex, becoming flat and depressed in center, sticky, with dark hair-like scales mostly in the depressed center; stalk to 6 in. tall; veil usually a persistent ring; spores white; taste bitter. In small to large, often dense, clusters at the base of living trees as a parasite or as a saprophyte on dead and decaying stumps of hardwoods and conifers; usually in fall and winter; widespread in California. It is a serious and destructive parasite of trees both in the wild and in cultivated gardens and orchards.

The Honey Mushroom produces black string-like runners or cords called rhizomorphs under the bark of trees or on roots beneath the soil. These runners can travel considerable distances to parasitize adjacent trees. When cooked, this mushroom has been regarded as edible. Even when cooked, however, it causes stomach upsets in a few people.

Boletus.   **Bolete.**

Several boletes that are mildly poisonous are found in California. They have red pores and may or may not turn blue when bruised. In some people they cause gastrointestinal symptoms.

Boletus erythropus.

Cap to 6 in. across, convex when young, in shades of brown, staining blue black when bruised; pores red, turning blue; stalk to 5 in. tall, stout, bulbous, yellow, covered with red granules. Solitary under oaks or in oak-madrone woods but uncommon; fall and early winter; scattered from northern and central California south to Santa Barbara County. Reported as poisonous to some people.

Boletus piperatus.

This small bolete, probably the smallest in California, has a cap only 1 to 2 in. broad, yellowish to reddish brown, does not bruise blue, has reddish brown pores and a peppery taste. In rare cases mildly poisonous.

Boletus satanas.

Cap to 8 in. across, whitish or pallid, later pink, convex, somewhat viscid; pores red turning blue; stalk short, bulbous. Solitary or grouped; in humus under live oaks; one of the first boletes to appear in October and November; northern California and the northern part of southern California. This mushroom is poisonous, at least when raw, causing vomiting, diarrhea, and severe cramps. When thoroughly cooked it is said to be edible, at least to some people.

Chlorophyllum molybdites [Lepiota morganii, L. molybdites]. Green-spored Parasol Mushroom, Green-gilled Parasol Mushroom (fig. 3).

Cap large, often as much as 12 in. across, oval to convex, becoming somewhat flat, dry, white, with numerous brownish scales; gills white but turning color slowly to green, bruising yellow to brown; veil white, thick, fringed along margin; stalk to 10 in. tall, smooth, white with brownish stains at base, separates easily from cap; spores green. Usually in groups, sometimes in rings; in meadows, lawns, or other grassy places, occasionally in urban gardens; spring and summer during moist

FIG. 3. *Chlorophyllum molybdites.* Green-spored Parasol Mushroom. White-colored mushroom with brownish scales on the cap.

warm weather; Contra Costa and San Mateo counties, the southern Sierra Nevada, south to San Diego County. This mushroom is often involved in poisonings. Vomiting is a prominent symptom, along with diarrhea and occasionally mental confusion. Persistent, severe symptoms may require hospitalization. The toxin, which is unknown, is destroyed by heating to 212°F. Before the spores mature and color the gills green this mushroom is mistaken for the edible *Lepiota rhacodes.*

Cortinarius.

Mushrooms of mostly medium size, brightly colored, and often conspicuous. When young they may be recognized by a cobwebby partial veil, the cortina, that splits on expansion of the cap, usually leaving a webby fringe of hairs on the inner margin of the cap or a hairy ring on the stalk (or both). *Cortinarius orellanus,* not in California but in the Pacific Northwest, is one of the deadly poisonous mushrooms. It contains orellanine, but because the onset of its symptoms may be delayed for days, it was long overlooked as being dangerously toxic.

The following four species from the northwest coast of California are reportedly mildly poisonous:

> *Cortinarius camphoratus;* cap to 3 in. across, violet; stalk to 4 in. tall; autumn; Humboldt, Mendocino, and Sonoma counties
>
> *Cortinarius purpurascens;* mushrooms lilac; flesh and gills turn purple when bruised; in redwood, fir, or aspen-fir forests; autumn; Amador, El Dorado, and Mendocino counties
>
> *Cortinarius traganus;* cap to 6 in. across, lilac, convex at first; gills cinnamon brown; stalk to 4 in. tall, lilac, thick, bulbous at the base; pungent odor; in coniferous woods; fall and winter; Humboldt and Mendocino counties
>
> *Cortinarius vibratilis;* cap to 2 in. across, pale brown; stalk to 3 in. tall, tapering upward; bitter taste and slight odor; in woods; fall and winter; Humboldt, Mendocino, and Sonoma counties

Entoloma rhodopolium.

Cap to 6 in. across, drab tan to gray; stalk to 5 in. tall. In woods; mid-autumn through winter; north coast to Santa Cruz County. It may cause severe gastrointestinal poisoning. Other entolomas may be poisonous and should be avoided.

Gomphus floccosus [*Cantharellus floccosus*].

Fruiting body vase-shaped, hollow, to 6 in. broad and 8 in. tall, orange yellow, with coarse scales on the inner surface, ridged on the outer. Autumn; at scattered localities in the state. It bears a superficial resemblance to the related but edible Chanterelle, *Cantharellus cibarius*. For some people it is a gastrointestinal irritant. Mushrooms of other species of *Gomphus* also should be avoided.

Hebeloma crustuliniforme.   Poison Pie.

Cap to 4 in. across, creamy white, darker later; stalk to 3 in. tall, enlarged at base. Fall through spring; at scattered localities in the state. *Hebeloma mesophaeum* and *H. sinapizans,* both in northern California, are reported to be similarly poisonous.

Hygrophorus. Waxy Caps.

A large genus with many species in California of which some are brightly colored and attractive. Their mushrooms usually have caps with a waxy surface, and the soft, fleshy-textured gills are waxy in feel and appearance. The edibility of many species is not known; some are considered edible, but a few may cause illness.

Hygrophorus conicus. Witch's Hat.

Cap to 3½ in. across, sharply conical, eventually expanding, scarlet to orange or yellow, blackening in age or when bruised; stalk to 6 in. tall; spores white. Usually in woods; fall to early spring; northern and central California on the coast and in the Sierra Nevada, uncommon in southern California. Reports are conflicting regarding the edible or poisonous qualities of this mushroom.

Hygrophorus puniceus.

Mushrooms of this species have a bright red, viscid cap that is not so sharply conical as that of *H. conicus* and not blackening. North coastal California south to Santa Cruz County. Usually reported as edible but known to cause illness in some people.

Lactarius. Milk Caps.

These mushrooms exude a white or colored fluid when the flesh is cut or broken. In some the fluid is white at first but then turns yellow; their taste is acrid. In some people they may cause a mild gastroenteritis of short duration.

Lactarius chrysorheus.

Cap to 5 in. across, pink to brownish; stalk to 3 in. tall with a white fluid that quickly changes to yellow. Found during the rainy season; throughout the state. Several other milk caps with similar characteristics are *Lactarius insulsus, L. resimus,*

L. rufus, L. uvidus, and *L. zonarius* in northern California and *L. trivialis* throughout the state.

Laetiporus sulphureus [*Polyporus sulphureus, Grifola sulphurea*].   Sulfur Shelf, Chicken-of-the-woods.

Fruiting body or cap irregular in shape, shelf-like, to 14 in. across, soft, becoming tough and brittle when dry, deep yellow fading to dull white, pores minute; spores white. One of the polypores, a nongilled mushroom, usually with several caps occurring in a series of overlapping, fan-shaped shelves growing on tree trunks, stumps, or logs, often on Eucalyptus; northern California south to San Francisco and scattered in the Sierra Nevada. This fungus should not be eaten raw. It is usually considered edible when cooked but should be eaten with caution. Reported to cause gastrointestinal upset; symptoms appear shortly after ingestion and are of short duration.

Lepiota.   Parasol Mushrooms.

These mushrooms resemble *Agaricus,* with free gills, a veil, no volva, and the stalk separating cleanly from the cap, but the spores are white. Mushrooms of this genus are reported to cause allergic reactions, and at least the two following, found in California, cause gastrointestinal symptoms:

Lepiota clypeolaria, Ragged Lepiota; cap to 3 in. across, white, margin ragged, with numerous brown scales except for the smooth brown center; stalk to 7 in. tall; in woods; fall and winter; northern California south to Santa Cruz and in the Sierra Nevada

Lepiota cristata; mushrooms are smaller than those of *L. clypeolaria*

Leucoagaricus naucinus [*Lepiota naucina*].

Cap to 6 in. across, ovate or conical, becoming flat, dry, white or whitish buff to grayish; stalk to 6 in. tall, slender, enlarged at base; veil collar-like; spores white or faintly pink. Common in lawns or other grassy areas, among ice plants along freeways, occasionally in woods; fall or winter; scattered localities in the state. Usually reported as edible but has caused poisonings in the Pacific Northwest.

Leucocoprinus birnbaumii [*Leucocoprinus luteus,*
Lepiota lutea].

Cap to 2 in. across, ovate or conical, later collapsing like a half-open umbrella, bright yellow, golden or greenish yellow, striate; veil collar-like or disappearing; stalk to 4 in. tall, enlarged toward base. Found singly or in groups in planter boxes and flowerpots and in greenhouses, where it is conspicuous; indoors appears throughout the year, outdoors in summer and early fall; Contra Costa, Marin, and San Mateo counties south to San Diego. Mildly poisonous, as are the closely related lepiotas, and should not be eaten.

Morchella. Morel.

Morchella is related to *Gyromitra, Helvella,* and *Disciotis* discussed under "Gyromitrin Poisoning" (pp. 31–32). Easily recognized, their honeycombed caps have raised ridges and deep pits or chambers. The interior of the cap is hollow, and the base of the cap is seated at the apex of the stalk. Until recently the edibility of morels was never questioned, particularly if well cooked.

Morchella elata [has been called *M. conica* and
M. angusticeps]. Black Morel.

Cap to 4 in. long, more or less conical, dark gray to black; stalk to 4 in. tall or more, white, becoming yellow or pale brown. In urban areas in gardens, at the edge of woods or along streams, sometimes after the snow melts in the mountains; mostly during spring; common in coastal California. Although edible after being cooked, this morel causes gastrointestinal upsets in some people, particularly if large amounts are eaten, if undercooked, or if eaten along with an alcoholic beverage.

Naematoloma fasciculare. Clustered Woodlover, Sulfur Tuft.

Cap to 4 in. across, yellowish, orange yellow to olive yellow, moist; stalk to 5 in. tall. Found on decaying wood; fall to spring; throughout California. These mushrooms are severely poisonous. A study of fatalities caused by them showed that gastrointestinal symptoms are delayed as much as 9 hours following ingestion, and there may be extensive kidney and liver damage.

Panaeolus retirugis.

Small cap to 1 in. across or slightly more, globose or bell-shaped, margin incurved when young, brown to gray brown, wrinkled to reticulate with raised ribs; stalk slender, to 3½ in. tall, whitish to purple; spores dark brown. Grassy areas and pastures, sometimes on dung; winter; reported as occurring in northern and southern California. This little brown mushroom, related to *Panaeolina foenisecii* discussed under "Psilocybin–Psilocin Poisoning" (pp. 37–41), is suspected of being poisonous.

Panellus stipticus [Panus stipticus].

Cap semicircular, to 1⅓ in. across, convex or flattened, orange buff to brownish; spores white. Caps grow in a shelf-like arrangement on fallen logs and stumps; autumn; scattered localities in northern California and the Sierra Nevada. Fresh specimens show luminescent gills. They have an unpleasant odor and astringent taste and are reported to be poisonous.

Paxillus involutus.   Poison Paxillus.

Cap convex, becoming depressed, to 6 in. across, sticky, drab reddish brown to olive brown; gills decurrent; stalk to 4 in. tall; spores yellow brown. Autumn; coastal northern California to Santa Cruz County, rare in Los Angeles County. This mushroom has been listed as edible but is now known to be potentially dangerous. It may induce allergic sensitivity so that eating it more than once can result in a serious, sometimes fatal, hemolytic anemia, characterized by excessive destruction of red blood cells.

Pholiota aurivella.

Cap to 6 in. across, convex, orange, with scattered spot-like scales, viscid; gills becoming brown with age; stalk to 3 in. tall, yellowish brown, scaly below; veil hairy, leaving a thin ring; spores brown. On dead logs or stumps, usually in groups; autumn; Amador, Mendocino, Nevada, and Sierra counties.

Some pholiotas are edible but a number are not, and this one is suspected of being poisonous.

Pholiota squarrosa.   Scaly Pholiota.

Cap 1 to 5 in. across, convex to bell-shaped, brown, covered with yellow-brown scales; stalk to 5 in. tall; spores brown. Odor faintly onion-like; taste is rancid. In dense clumps at the base of trees; fall and winter; Sierra Nevada. Reported as edible, but it causes gastric upsets in some people. Drinking an alcoholic beverage while eating this mushroom or soon after may contribute to the poisoning.

Ramaria formosa [*Clavaria formosa*].   Salmon Coral Mushroom.

Fruiting body not a cap and stalk but a cluster of erect finger-like or club-shaped branches, fleshy, pliant, pinkish or salmon-colored above and white below; spores yellow. Usually beneath hardwoods, often tan oaks; fall and winter; Fresno, Mendocino, San Mateo, Tehama, and Yuba counties. Some coral fungi are edible, but this one has a bitter taste and is reported to cause illness in some people.

Russula.

Mushrooms of this genus are related to *Lactarius,* the milk caps, but their fruiting bodies do not exude a colored juice. The caps of both genera are usually depressed, dry and brittle, and of a rigid texture, and the spores are white. In *Russula* the caps of different mushrooms are colored, mostly red, yellow, and green. The white gills become brown, black, gray, or red. The stalk is stout, rigid, brittle and breaks like chalk. Mushrooms have an acrid taste and are not recommended as being edible.

Russula emetica.   Emetic Russula, The Sickener.

A colorful mushroom with a red sticky cap to 4 in. across; stalk to 3 in. tall, white. Found in fall and winter; north coastal areas of the state. Reported as poisonous because, as its name suggests, it induces vomiting. Other russulas listed as causing illness are *Russula albonigra, R. foetentula, R. fragilis, R.*

laurocerasi, R. nigricans, R. rosacea, R. subfoetens from north coastal California, and *R. densifolia* and *R. veternosa* from northern to southern California.

Scleroderma citrinum [*S. aurantium*].   Earthball, Hard-skinned Puffball, Pigskin Poison Puffball.

Fruiting body round, hard, with yellow brown, wart-covered, thick skin, inside filled with white to lilac flesh, the spore mass, very soon becoming dark purple to black and powdery. Usually in groups on the ground or wood debris; autumn; scattered localities in California. Earthballs are related to the edible puffballs, *Lycoperdon,* which in the edible stage have white flesh. Earthballs contain toxins that, in more than very minute quantities, can be harmful, causing stomach pains, weakness, nausea, and sometimes a tingling sensation.

Stropharia ambigua.

An attractive shaggy mushroom with a light yellow brown cap, to 6 in. across; white cottony veil remnants hang from the margin of the cap; stalk to 6 in. tall, white, with cottony scales. In rich, damp humus; fall to spring; northern California south to Santa Cruz County and east to the Sierra Nevada. It is suspected of being poisonous.

Stropharia coronilla.

Cap to 2 in. across, golden brown to tan or nearly white, viscid; gills becoming purple brown or gray black with age; stalk to 2 in. tall, stout; veil leaving a persistent membranous ring; spores dark purple brown. In grassy areas, lawns, pastures, or under open stands of trees; year-round but mostly in the fall, often occurring with *Agaricus campestris* and species of *Panaeolus;* northern California. Reported to be poisonous.

Tricholoma.

A large and widely distributed genus with medium-sized fleshy caps; stout fleshy stalks and white spore prints. Several species are known to be poisonous, others are reported as edible, while for others edibility and toxicity are unknown or questionable. Species are difficult to determine, and because the edibility of some is questionable, none should be eaten unless their identity is known with certainty.

Tricholoma pardinum.

Cap convex to flat, to 4 in. across, dirty white or gray, covered with small scales; gills white; stalk to 4 in. tall, chalky white. On the ground; often under conifers; fall and winter; on the north coast as far south as Santa Cruz County and in the Sierra Nevada. Causes severe gastrointestinal upset. Other tricholomas having white or gray caps with white gills should be avoided.

Tricholoma saponaceum.   Soapy Tricholoma.

Cap to 6 in. across, slightly viscid, grayish olive to olive brown; gills white; stalk to 7 in. tall, white becoming brownish stained. Soap-like odor and taste. Found in woods, especially oaks and madrones; fall and winter; throughout the state. Unpalatable and suspected of being poisonous.

Tricholoma virgatum.

Cap conical, becoming convex and umbonate, to 3 in. across, grayish or purplish gray, streaked with fine dark fibers; stalk white, smooth, to 4 in. tall; acrid taste. Found in woods, often conifers; fall and winter; northern California south to Santa Cruz and east to the Sierra Nevada. Somewhat resembles the more poisonous *Tricholoma pardinum,* from which it may be distinguished by its slightly smaller, smooth cap.

Verpa.   Thimble Cap.

This genus of nongilled fungi belongs to the Ascomycetes and is related to *Morchella.* The two genera may be distinguished by the attachment of the inside of their caps to the stalks. In *Verpa* the thimble-shaped cap is attached to the tip of the stalk only, while in *Morchella* the cap is attached to the stalk at various places along its inner surface.

Verpa bohemica.   Wrinkled Thimble Cap.

Cap 1 to 2 in. long, thimble-shaped, surface irregularly longitudinally ridged and wrinkled, not pitted, pale to dark yellowish brown; stalk to 5 in. tall, whitish. In moist woods, rich soils, sometimes along streams; spring; central California from coast to Sierra Nevada. Edible but with caution; some people become ill from it, especially if large amounts are ingested. It has been suggested that a gyromitrin-like toxin may be produced by this fungus. Therefore, those who wish to eat it should boil it first and pour off the water before eating. See "Gyromitrin Poisoning" (pp. 31–32).

FERNS AND HORSETAILS

Polypodiaceae, Polypody Family

Pteridium aquilinum.   Bracken Fern, Bracken, Brake
(fig. 4; pl. 2*a*).

This fern has a worldwide distribution, and a number of geographical races have been described. The plants native throughout California are a part of var. *pubescens*. Bracken Fern is widespread in California, particularly in the mountains; it becomes weedy under grazing but has been used as an ornamental. This genus is sometimes placed in a segregate family, Pteridaceae, or an even smaller family, Hypolepidaceae. These plants have long creeping underground rhizomes; coarse leaves appear singly aboveground to 3 ft tall; leaf blades are repeatedly pinnatifid from the erect petioles. Bracken Fern reproduces by spores that are produced by sporangia covered by an inner membrane under the recurved margin of the leaf.

Toxic part: Leaves, fresh or dried, and the rhizomes.

Toxin: Thiaminase (an enzyme that breaks down thiamine, a vitamin of the B complex), which is toxic to horses and pigs, and also unknown substances toxic to cattle and sheep. From the dried young leaves and rhizomes more than thirty compounds, either sesquiterpenes or combined with sugars as glycosides, have been isolated and are toxic to cells or are carcinogens causing tumors in rats.

Symptoms: In horses, cardiac irregularities, incoordination, swaying from side to side with front legs crossed and hind legs spread wide apart, muscular tremors, and finally prostration. Injection of thiamine dramatically reverses the symp-

FIG. 4. *Pteridium aquilinum*. Bracken Fern. Inset: Lower side of leaflet with marginal sporangia.

toms. Pigs have been poisoned experimentally by feeding them Bracken Fern rhizomes.

Cattle and sheep are resistant to thiaminase toxicity because the bacteria present in the rumen usually produce sufficient amounts of thiamine. Bracken Fern poisoning in cattle, from unknown toxins, causes decreased activity of the bone marrow and hence lowers the number of blood cells. Blood escapes from the vessels and bacterial infections commonly follow. Symptoms may include elevated temperatures, edematous swellings, difficult breathing, nasal bleeding, and also blood in the feces. Since injections of thiamine, vitamin B_{12}, and folic acid do not prevent this disease in cattle, the toxin is not related to thiaminase. When used experimentally the fern rhizomes were found to contain about five times more of the unknown toxin than the leaves. The toxin has been found in significant quantities in cow's milk.

Poisoning from Bracken Fern is highly fatal to cattle. Death is usually from internal hemorrhages and secondary bacterial infections. This condition has been reproduced in sheep only by feeding large amounts of fern material over long periods of time.

From the few known cases of livestock poisonings and the abundant worldwide distribution of the species, Bracken Fern poisoning appears to be rare. Hungry animals should not have access to young shoots of this fern, especially following a fire when there is no other forage available. Moreover, cattle must not be allowed to graze in a newly plowed field where the exposed underground rhizomes could be eaten. Young fern leaves included in hay are the most common source of difficulty because animals cannot avoid them.

In the Sierra Nevada, experimental feedings of Bracken Fern to cattle gave quite variable results. After a number of years only a single death was produced experimentally. In October 1953, however, 30 head of a band of 1,550 sheep died perhaps from Bracken Fern poisoning. The sheep had been trailed for 4 days in the Sonora Pass area, and at Niagara Creek they were loaded into trucks on a very hot day. At the time of loading there were no symptoms of poisoning, but later, on the valley floor, the animals showed incoordination, particularly a tendency to move sideways after being held. The black-faced sheep, the most aggressive foragers, were the ones affected. The animals stood alone and depressed; breathing was rapid and difficult with considerable mucus from nose and mouth; eyes were stained yellow; blood appeared in the feces. Following death, autopsies showed a general yellow coloring with extensive subcutaneous and visceral hemorrhages.

In many areas of the world where the fern grows as a native plant, the young fronds or fiddleheads of Bracken Fern are used as food for humans. Two recently published popular books on edible wild plants in California suggest ways for preparing the young fronds. Heat destroys the enzyme thiaminase. Bracken Fern is not recommended as edible, however, since there are carcinogens present. Tumors in rats have taken as long as 802 days to be formed. Until these cancer-producing compounds are better understood, it would be wise not to eat the fronds of Bracken Fern.

Equisetaceae, Horsetail Family

Only one genus, *Equisetum,* in the family; the plants reproduce by spores and are considered to be distantly related to the ferns.

Equisetum. Horsetail, Scouring Rush, Jointed Rush.

Herbaceous perennials with creeping rhizomes; stems erect, rush-like, grooved, jointed, hollow except at the nodes, sometimes branched at the nodes; branches whorled and as numerous as the leaves; leaves are minute, scale-like, often black at the tips, and fused at the sides to form a connate sheath at the node; spores borne in a short terminal spike resembling a cone.

The jointed stems can be separated at each node. In this respect they superficially resemble the slender stems of trees in the genus *Casuarina,* a flowering plant. Stems of *Equisetum* contain many crystals of silicates. Horsetails have no economic value, although in the past the rough stems were used for polishing wood and scouring utensils.

Of the six species native in California, three are reported to have caused poisoning in livestock. These three species are widespread throughout the state except the deserts:

Equisetum arvense, Field Horsetail, Common Horsetail; has two kinds of stems; one is not green, unbranched, and fertile with spore-bearing spikes at the tips, and the other is sterile, green, with many whorls of slender branches

Equisetum hyemale, Scouring Rush, Common Scouring Rush; evergreen stems to 4 ft tall, unbranched or rarely with a few slender branches, with the spore-bearing cones at the tips of the large stems

Equisetum laevigatum, Smooth Scouring Rush; similar to *E. hyemale* but somewhat shorter, to 3 ft tall, and the sheathing whorls of leaves longer than broad

Toxic part: Entire plant.

Toxin: Thiaminase, an enzyme that destroys thiamine and causes deficiency of vitamin B_1 (also in Bracken Fern, *Pteridium aquilinum*, Polypodiaceae). There are probably other unknown toxins.

Symptoms: In horses, incoordination and slow pulse, the result of cardiac irregularity, are the principal signs of thiamine deficiency. First there is swaying from side to side and then incoordination with the forelegs crossed and the hind legs spread apart. The animal's back becomes arched and it appears to be crouching. Following muscular tremors and prostration the horse is unable to rise. The terminal signs of this deficiency include backward spasms of the head and legs accompanied by convulsions of the large intestine. An affected animal eats well until the later stages of the disease when it stops eating because of sleepiness. Additional signs may develop in the digestive and circulatory systems, but their relationship to thiamine deficiency is not clear. Cattle and sheep do not develop these symptoms. *Equisetum* plants are not palatable to animals; poisoning can occur only from inclusion of plants in hay where they cannot be avoided.

These signs disappear with the injection of thiamine. Thiamine acts as a coenzyme with carboxylase in the metabolism of carbohydrates, proteins, and fats. The brain is highly dependent upon the energy released from carbohydrate metabolism. Continued deficiency of thiamine results in the accumulation of pyruvates. See also "Polioencephalomalacia" (pp. 366–367).

Similar complexes of signs including such paralysis may develop from ingestion of Tansy Ragwort, *Senecio jacobaea*, or Perennial Ryegrass, *Lolium perenne*. In these cases, however, there is additional liver necrosis and elevation of temperature; affected horses do not respond to thiamine therapy. In thiamine deficiency there is elevation of temperature only in the very terminal stages of the disease.

GYMNOSPERMS

Cupressaceae, Cypress Family

Cupressus macrocarpa. Monterey Cypress.

Native to the Monterey Peninsula, widely cultivated along the coast of California. In cows abortions and deaths have been reported in New Zealand since 1943; suspected of poisoning cattle in England as early as 1905. This toxicity is unknown in California, but care should be taken in the use of these trees as windbreaks along the coast where cattle might have access to the foliage.

Cycadaceae, Cycad Family

Cycas revoluta. Sago Palm, Japanese Sago Palm.

Entire plant has toxic glycosides and a toxic amino acid. Opportunities for cycad poisoning are rare. Seeds might be eaten by children, but most plants in California are male. Livestock should not be permitted to eat these plants, particularly the soft young leaves.

Pinaceae, Pine Family

Pinus ponderosa. Western Yellow Pine, Ponderosa Pine.

Native in the mountains of southern California north through the Coast Ranges and the Sierra Nevada to British Columbia and extending to the Rocky Mountains, Texas, and Mexico in different forms. Sometimes cultivated. Tree to 200 ft tall but commonly shorter, in dense stands; bark reddish orange in color, breaking off in irregular plate-like segments; leaves needle-like in bundles of three subtended by papery bracts; female cones to 8 in. long with spreading scales; seeds winged.

Toxic part: Leaves and young shoots.

Toxin: Unknown but considered to be of two different components, probably an anti-estrogen and a toxic terpene. Pine needles have the greatest toxicity during the winter months.

Symptoms: Abortion in cattle particularly in the seventh or eighth month of pregnancy; excessive bleeding from the uterus and retention of fetal membranes are common. There are some indications that sheep and deer may be similarly affected.

Records of abortions from cows browsing fallen trees and pine needles seem to be rare in California. Apparently abortions in northeastern Shasta County in early winter of 1958–1959 were the result of the cows eating pine needles. Even with adequate forage available, cows will eat the foliage of fallen trees. In British Columbia this cause of abortions has been well documented; as much as 5 pounds of pine needles per animal per day was consumed. Although drying does not destroy the toxin, it is destroyed by heat.

After eating pine needles, cows may abort within 2 days or up to 2 weeks later, even after having been removed from the pine material. Excessive uterine bleeding, weak delivery contractions, and incomplete dilation of the cervix are found. A nauseating foul odor is characteristic of these abortions. The placenta is retained and may be the only sign of toxicity. Cows having retained placentas require careful medical treatment; there is a high mortality rate of untreated cows. Those that died from Ponderosa Pine needle toxicity have massive discolored hemorrhages outside the larger cardiac blood vessels. Depending on the age of the fetus at the time of abortion, the calf may be weak or die shortly after birth. If the calf is near full term, however, it may be born in good condition.

Taxaceae, Yew Family

Taxus baccata. English Yew.

Native of Europe, North Africa, Western Asia. Widely cultivated in California and elsewhere in gardens and parks. *Taxus cuspidata,* the Japanese Yew from Japan, Korea, and Manchuria, is less frequently cultivated. Evergreen shrub or tree to 60 ft tall; leaves alternate, stiff, linear to $1\frac{1}{2}$ in. long; fruit a single seed embedded in a fleshy red aril.

Toxic part: Entire plant, but especially the seeds. The juicy sweet red aril is said not to be poisonous. It is attractive to children, however, and in chewing it they might also ingest the toxic seed.

Toxin: Taxine alkaloids, highly toxic.

Symptoms: Taxine alkaloids cause irritation of the intestinal mucous membranes. Within an hour after ingestion, dizziness, dry throat, dilation of the pupils, abdominal pain, nausea, and vomiting appear. Muscular weakness occurs; a rash appears on the skin; the patient may become comatose, and the heart rhythm becomes irregular. Taxine alkaloids are absorbed rapidly from the intestine, and if large amounts of the plant are ingested, sudden death from circulatory or respiratory failure can occur.

In Sonoma County in March 1969, in a group of 20 ewes and 28 lambs, 5 ewes and 1 lamb died soon after ingestion of clippings of the Irish Yew, *Taxus baccata* 'Stricta' (pl. 2*b*), that had been thrown into the pasture. The animals ran around excitedly and dropped; one prostrate animal did recover. Twigs and leaves of *Taxus* were identified in the stomach contents. In 1980, in the same general area, a single cow died suddenly; seeds of *Taxus* were found in the rumen.

Taxus is a small genus of about eight species. Not all are toxic, but *Taxus baccata* is considered one of the most poisonous of the native trees or shrubs in England. *Taxus brevifolia,* Western Yew, native to northern California and the Pacific Northwest, lacks the toxic alkaloids.

ANGIOSPERMS

Dicotyledons

Aceraceae, Maple Family

Acer rubrum. Red Maple, Scarlet Maple, Swamp Maple.

Occasionally cultivated. Leaves and bark of the twigs have caused the death of horses in eastern North America from the New England states to Georgia; effects include anemia and free hemoglobin in the blood plasma, jaundice, enlargement of the spleen, and blackened kidneys. Horses became ill within 2 days after eating the leaves and death followed 5 to 7 days later; deaths of cattle are also recorded. The toxic principle is unknown, but the effects suggest that it is an oxidant. Such poisonings are unknown in California, but care should be taken not to discard trimmings from Red Maple or to plant these trees within reach of cattle and horses.

Amaranthaceae, Amaranth Family

Amaranthus retroflexus. Redroot Pigweed, Rough Pigweed.

Introduced from tropical America; in California a widespread weed of cultivated ground and waste places. Annual rough hairy plants to 4 ft tall with a characteristic red or pink taproot; flowers are borne in dense terminal and axillary spikes. The plants contain an unknown toxic agent that causes accumulation of edema fluid around the kidneys. Sheep, pigs, and calves are affected. Adult cattle and horses are resistant. This poisoning is not reported for California. These plants may accumulate oxalates to 30% of their dry weight and may also accumulate toxic amounts of nitrates.

Anacardiaceae, Sumac Family

This family is well known for its members causing contact dermatitis. Pacific Poison Oak is widespread in California, but Poison Ivy and Poison Sumac of the eastern United States do not occur in the West.

Mangifera indica. Mango.

The juice of the stem at the base of the fruit contains a toxin similar to that of *Toxicodendron* discussed below. The skin of this fruit, common in California markets, may have toxic sap on it and should be carefully removed and not eaten.

Schinus molle. Pepper Tree, Molle, California Pepper Tree, Peruvian Pepper Tree, Peruvian Mastic Tree.

Native of the Andes of Peru; cultivated as an ornamental, occasionally naturalized. Evergreen tree, spreading and with graceful weeping habit, leaves pinnately compound, fifteen or more leaflets; tiny yellowish white flowers in much-branched panicles; fruit a small rose-red berry.

Schinus terebinthifolius. Brazilian Pepper Tree, Christmas Berry Tree.

Similar to the Peruvian Pepper Tree but with more erect habit and fewer (usually about seven) but larger leaflets. Not planted as frequently in California as the Peruvian Pepper Tree and rarely reported as naturalized. Both pepper trees cause contact dermatitis, but the symptoms from the Brazilian Pepper Tree are more severe. Ingestion of large amounts of the berries causes gastroenteritis, vomiting, and diarrhea. The red berry-like fruits of the Brazilian Pepper Tree sometimes are sold and used as an imported spice called "pink berries" or "pink peppercorns." No information is available regarding the toxicity of these berries when used in small amounts as a spice.

Toxicodendron diversilobum [*Rhus diversiloba*]. Pacific Poison Oak, Western Poison Oak (fig. 5; pl. 2*c, d*).

Native to California and north to southernmost British Columbia, south to northern Baja California. Common in shrubby thickets and wooded areas below 5,000 ft elevation. A variable species in habit and in leaf shape, it is usually an erect deciduous shrub 3 to 6 ft tall, occasionally taller; can be a vine, climbing the trunks of trees, or may be a small tree. Leaves deciduous, crowded on branches, with three glossy leaflets, rarely five, turning a vivid red or reddish violet in late summer and autumn; inconspicuous small greenish flowers in panicles in spring; fruit a creamy white drupe with black stripes, $\frac{1}{8}$ to $\frac{3}{16}$ in. across.

Pacific Poison Oak may be confused with another member of the Anacardiaceae, the Squaw Bush, *Rhus trilobata* (fig. 5), in which the current year's stem and red fruits are finely pubescent or felty hairy. Shrubs or vines of the Poison Oak have stems that are glabrous, lacking hairs, and white fruits with black stripes that often fade with age.

Toxic part: Entire plant, except pollen, throughout the year even during the winter months when the branches are leafless.

Toxin: Urushiol, a nonvolatile phenolic allergen, contained within the plant's resin canals and released when the tissues are injured.

Symptoms: A person becomes sensitized after the first direct skin contact with any part of the plant. Probably seven out of every ten people in the United States can develop this allergy if exposed to the toxin a second time. Symptoms from the first exposure to the allergen in the various species of *Toxicoden-*

FIG. 5. Left: *Toxicodendron diversilobum.* Pacific Poison Oak. Right: *Rhus trilobata.* Squaw Bush.

dron are extremely rare. With subsequent exposures, however, symptoms may develop 24 to 72 hours later or in less time in highly sensitive individuals. Early manifestation of contact dermatitis caused by this toxin is redness and swelling of the skin, accompanied by itching that continues to be a constant feature of the disorder. This redness is followed by the eruption of small blisters where the skin has been brushed by the plant. In severe cases the blisters enlarge and exude a nonallergenic fluid and may become secondarily infected by scratching. After several days the blisters become crusty and scaly. The dermatitis rarely lasts longer than 10 days. Only humans and closely related primates get this dermatitis. Persons who are particularly sensitive to urushiol may experience swelling of the eyes and throat, stomach cramps, nausea, vomiting, and diarrhea.

This dermatitis is caused by brushing against bruised or broken stems or leaves of the plant or by touching objects that have come in contact with the plant such as clothing or tools or the fur of a pet. It is easily spread to other portions of the body by fingers or clothing. Droplets of toxin are carried in smoke on particles of incompletely burned leaves and stems of Poison Oak plants.

Urushiol, the allergen, is a long-lived substance that penetrates the skin within 10 minutes after contact with the plant. Washing with soap and water is recommended as soon as possible to remove excess toxin. The affected areas should be kept clean and exposed to the air. Any furry pets, clothing, and tools which have been contaminated should also be thoroughly cleansed with soap and water as measures against further spread.

In treating the dermatitis, calamine and similar lotions reduce the itching and dry the blisters. Cold, wet compresses reduce the inflammation. Corticosteroids used under a doctor's direction give considerable relief; antihistamines, ineffective against the rash, are useful for their sedative action. Oily ointments or alcoholic solutions should not be used, as urushiol is soluble in fats and alcohol. Other products to be avoided are those containing local anesthetics (benzocaine), irritants (phenol, camphor), or metals (iron, zirconium).

During a single year in California, the working hours lost as a result of this dermatitis, and the discomfort experienced by sensitive individuals, make this plant one of the most hazardous of all the poisonous plants in the state. The California State Compensation Commission Fund gives disability payments to workers so affected.

Apiaceae, Parsley Family

Also called Umbelliferae, the large family Apiaceae (fig. 6) contains several poisonous plants including the waterhem-

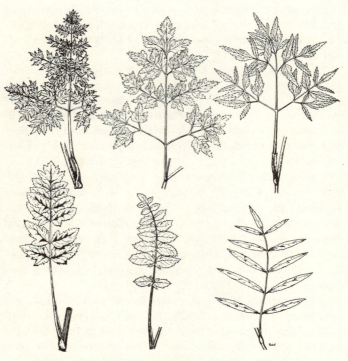

FIG. 6. Leaves of different species of Apiaceae, Parsley Family. Upper left: *Conium maculatum*. Poison Hemlock. Upper center: *Oenanthe sarmentosa*. Water Parsley. Upper right: *Cicuta douglasii*. Western Waterhemlock. Lower left: *Pastinaca sativa*. Parsnip. Lower center: *Berula erecta*. Cutleaf Water Parsnip. Lower right: *Sium suave*. Water Parsnip.

locks, *Cicuta,* which are among the most violently poisonous plants native of North America, and Poison Hemlock or Spotted Hemlock, *Conium maculatum,* native of Europe, reputed to have poisoned Socrates in 399 B.C. In ancient times when Athens was a city-state, Poison Hemlock was administered to those condemned to death for certain offenses against the state.

Several members of the family found in California contain furocoumarins and produce hypersensitivity to sunlight:

Ammi majus, Bishop's Weed, Greater Ammi
Apium graveolens, Celery
Daucus carota, Carrot
Heracleum mantegazzianum, Giant Hogweed
Pastinaca sativa, Parsnip (fig. 6)

Toxic part: Juice of the plant.
Toxin: Furocoumarins, a group of fluorescent substances.

Symptoms: Furocoumarins take in wavelengths of light usually not absorbed and transmit the energy into the skin which, about 48 hours after exposure, results in blistering and reddening similar to a sunburn. A deep purple pigmentation may follow that persists for a long time or may become a permanent disfiguration.

Hypersensitivity to sunlight is produced in humans and in livestock and poultry. Exposure to long-wave ultraviolet light is necessary for this reaction. Persons with dark pigmented skin do not develop the dermatitis as readily as those with light skin. Animals develop the severe reddening, blistering, and peeling of the skin only on unpigmented areas.

Celery plants produce furocoumarins when infected with the fungus *Sclerotinia sclerotiorum,* which causes pink rot of celery. Healthy celery plants do not produce these furocoumarins, and the fungus in culture does not produce the toxic substances. This disease of celery is controlled rather carefully in California celery fields, which may account for the few cases of dermatitis seen. Moreover, many workers simply do not pick the diseased plants, thereby lessening the incidence of the disease.

Ammi visnaga.   Toothpick Ammi (pl. 4*a*).

Native of Europe, becoming more widespread in California because this species is resistant to some of the soil sterilants used in roadside weed control. Winter annual plants to 3 ft tall; very similar to the Bishop's Weed, *Ammi majus,* but the leaves are dissected into linear or thread-like divisions; numerous white flowers in compound umbels borne on a stem swollen at the top, rays of the umbel compacted into a dense cluster at maturity. The plants, especially the seeds, contain two fuo-chromones, khellin and visnagin, which are medicinally active in relaxing smooth muscle, useful as vasodilators or broncho-dilators. Livestock will not eat these bitter plants, but a single case of photosensitization has been reported in poultry.

Berula erecta.   Cutleaf Water Parsnip (fig. 6).

Widespread, but not abundant, throughout California, even rarely on the deserts, extending to Baja California, British Co-lumbia, and east to Ontario, Michigan, and Oklahoma. Peren-nial plants growing in water, often stagnant, 1 to 3 ft tall; leaves pinnate with five to nine pairs of leaflets, margins serrate or incised; small white flowers in umbels to 2 in. across; fruits flattened, about $\frac{1}{16}$ in. long. The toxicity of this species needs further investigation. Even though widely distributed in cen-tral and western North America, plants of Cutleaf Water Pars-nip have been implicated in livestock poisonings only in sev-eral early reports in California and a single case in British Columbia. In South Africa *Berula thunbergii* has been re-corded as not always toxic, but it is known to be poisonous to cattle especially in the spring and only on certain soils.

Cicuta.   **Waterhemlock.**

Widespread genus native to North America. California has two similar species; both are large perennials to 8 ft tall with thickened underground stems or tubers. When cut lengthwise the tubers show hollow horizontal chambers and a bright yel-lowish orange juice that, on exposure to the air, changes to red-dish and finally reddish brown. Other members of this family may show the chambered pith of the tuber, but when cut open they do not reveal the yellowish orange fluid. Although rooted in soil under water or in mud, the stems are erect, branching, and hollow except for plates at the nodes; leaves large to 1 ft long and across, usually two or three times pinnately compound with numerous leaflets; individual leaflets lanceolate, 1 to 3 in. long, edges toothed with numerous veins ending at the edge of the leaflets in the notches between the teeth; numerous small white flowers in rounded compound umbels; fruits smooth, almost round, ¼ in. long, with prominent longitudinal corky ribs alternating with oil tubes indenting the fruit when seen in cross section.

Cicuta douglasii. Western Waterhemlock (fig. 6; pl. 3*c,d*).

Lateral corky ribs of the fruit much broader than the oil tubes, which scarcely indent the fruit; in the leaflet blades a network of veins encloses elongated vein islets. In California, *Cicuta douglasii* is native to the central and northern parts of the state below 8,000 ft elevation in freshwater streams, marshes, and ditches; it extends along the western coast of North America to Alaska and inland from central California to Nevada and western Montana.

Cicuta maculata [including *C. bolanderi* and part of what has been called *C. douglasii* in California]. Spotted Waterhemlock.

Fruit markedly indented by the oil tubes, which are as broad as or much broader than the corky ribs; a network of veins encloses rounded vein islets in the leaflet blades. *Cicuta maculata* is native to central and southern California with a single location in Modoc County in northern California. It is widespread in North America from southern Mexico, across the continental United States and Canada, and into Alaska. Three varieties of this species—var. *angustifolia,* var. *bolanderi,* and var. *maculata*—are present in California.

Toxic part: Entire plant.

Toxin: Cicutoxin, an unsaturated aliphatic alcohol.

Symptoms: The toxin acts directly on the central nervous system. From ¼ to 1 hour after eating there is nausea, profuse salivation, tremors, abdominal pain, and convulsions, followed by paralysis and death from respiratory failure.

In the United States many human fatalities have been recorded from eating tubers and leaves of waterhemlock. Livestock poisonings have been fairly common in California. In fact, the native waterhemlocks are considered to be of primary importance among poisonous plants in the state.

In Modoc County on a ranch with a history of poisonings caused by waterhemlocks, cattle losses have occurred roughly every 15 to 17 years, especially in dry years when forage was scarce. In January 1962 near Ft. Dick, Del Norte County, in a herd of first-calf heifers, 50% of the calf crop was aborted. There had been definite grazing of waterhemlock plants. In March 1965 near Arcata, Humboldt County, the deaths of twenty-three mature cows were caused by waterhemlock. These cows were new to the pasture area.

Conium maculatum. Poison Hemlock, Spotted Hemlock, Deadly Hemlock, Snakeweed, Poison Parsley, Poison Stinkweed (fig. 6; pl. 3*a, b*).

Native of Europe but widely naturalized throughout California; especially abundant along the drying edges of waterways and along roadsides. Biennial forming a rosette of fern-like leaves in the first year, to 3 ft tall and across, the leaves having a mouse-like odor when crushed; in the second year the flowering stem reaches 9 to 10 ft tall, branching at the top, the stout, smooth stem having irregular purple spots often with a clear center; numerous small white flowers in compound umbels at the ends of the branches; fruits oval, about ⅛ in. long, prominently ribbed at maturity. Poison Hemlock can be readily distinguished by the purple spots on the stems and the petioles of the leaves.

Toxic part: Entire plant, especially root and seeds.

Toxin: Coniine and other alkaloids. A second alkaloid, lambda coniceine, is seven to eight times more toxic than coniine and is present in larger quantities during the growing season when the fruits are green.

Symptoms: The primary action is on the central nervous system and is similar to nicotine poisoning. Onset of symptoms is usually rapid with irritation of the mucous membranes of the mouth and throat, increased salivation, nausea and vomiting, scant abdominal pain or diarrhea, thirst, headache, dilation of the pupils, sweating, and dizziness. Convulsions occur in severe cases and may be followed by coma or death, the result of respiratory failure.

Coniine was one of the first alkaloids to be discovered (it was found in 1827) and was the first to be prepared synthetically. *Conium* was formerly used in medicine as an antispasmodic, a sedative, and a pain-reliever.

Because of its common occurrence in California, Poison Hemlock is a potential hazard to children or adults who might mistake it for harmless plants of other species of the family such as parsley or dill or mistake its seeds for those of anise. Ingestions of this kind occasionally occur. In March 1983, two cases were reported to the San Francisco Regional Poison Control Center. Both persons were given emergency medical treatment and recovered.

Along the Truckee River near Reno, Nevada, seedlings of Poison Hemlock appeared spontaneously in a row where parsley seeds were planted. Poison Hemlock plants are abundant along this river course. Fortunately these poisonous plants were identified before any harm resulted from their use.

The toxicity of *Conium maculatum* plants is not as great as that of the waterhemlocks, *Cicuta*. *Conium* toxicity varies throughout the year. The plants are most poisonous during the active growing season and flowering period. Much of its toxicity is lost when the plant is dry. This variation in toxicity was known to the ancient Greeks who later added other substances such as opium to their "poison cups."

Poison Hemlock is distasteful to livestock and usually is not eaten. A drop in milk production was seen in Marin County where Poison Hemlock plants were eaten by dairy cows being held in corrals before and after milking. Poison Hemlock plants were untouched on the pasture side of the fences but were eaten down to the ground on the corral side where there was no other forage.

An occurrence of deformed calves in northern Utah in 1971 implicated *Conium maculatum* as the probable cause. These cases were similar to the deformities produced by lupine feeding of pregnant cows who gave birth to deformed or "crooked" calves. Similar deformities in newborn swine were attributed to Poison Hemlock in Missouri.

Cymopterus. Cymopterus, Wild Carrot.

Eight species are native to eastern California on the deserts and at higher elevations of the Sierra Nevada and the White Mountains; none have been recorded as toxic in this state. *Cymopterus watsonii*, Spring Parsley, native of northern Nevada and parts of Oregon, Utah, and Colorado, contains toxic furocoumarins and has caused photosensitivity in sheep.

Sanicula bipinnata. Poison Sanicle.

There are no current records of the toxicity of this species, but the common name Poison Sanicle is found in all references. In 1917, when horses were more commonly utilized, P. B. Kennedy stated that some if not all species of *Sanicula* are poisonous to stock, particularly horses. Plants of the various species of *Sanicula* are biennials or perennials with rosettes of basal leaves; leaves deeply palmately lobed, the lobes incised or pinnately divided; yellow or purple flowers are formed in compound umbels on naked stems, rarely with a stem leaf; fruits covered with hooked bristles or tubercles.

Sphenosciadium capitellatum. White Heads, Swamp White Heads, Ranger's Button.

Native of California, widespread in wet soils at elevations from 3,000 to over 10,000 ft from the San Jacinto Mountains, Riverside County, north on both sides of the Sierra Nevada into the North Coast Ranges, also in the White Mountains and eastward. Stout thick-rooted perennial to 5 ft tall; leaf blades to 18 in. long, once to twice divided into linear or lanceolate leaflets, petioles the same length and clasping the stem; small flowers in tight heads, the heads in an umbellate arrangement.

Toxic part: Aboveground parts of the plant.

Toxin: Unknown.

Symptoms: In cattle, respiratory distress with neck distended, weakness, slight salivation, and apparent abdominal pain. After death the animals appear bloated with protrusion of the anus and a bloody discharge from the nose and mouth. In horses, photosensitization with avoidance of light and cloudiness of the eyes was produced experimentally.

Plants of *Sphenosciadium* are not usually eaten by livestock. In Tuolumne County, however, Wolfin Meadow, also called Poison Meadow, situated on the western side of the Sierra Nevada at an elevation of 5,000 ft, has a long history of cattle poisoning; the meadow has been fenced from grazing for many years. *Sphenosciadium* is abundant in this meadow but not in any of the other meadows in the general area.

For 30 days in June 1966, cattle were permitted access to Wolfin Meadow during the day and removed at night with no apparent harm. But when the animals were allowed free access to the meadow, three animals died within a few days. Six of the remaining animals died the next day even though they had been removed from the meadow. In September of that same year cattle broke through the fence around the meadow and five animals died at this time. A heifer, which died later, manifested respiratory distress, some salivation, cervical extension, weakness, and abdominal tenderness. A mature cow, which died after several days, had normal respiration but some incoordination of the hindquarters, extension of the neck, and bulging eyes. The nursing calf of this cow appeared normal and survived.

Experimentally the dosing of calves with a stomach tube using ground-up green plants of *Sphenosciadium,* 10 to 16 grams per kilogram of body weight, produced respiratory distress within 2 to 4 hours. The animals died in $4\frac{1}{2}$ to 60 hours after receiving the plant material. Lesser amounts of plant material caused mild symptoms in calves but were not enough to cause death.

In horses, dosing with *Sphenosciadium* plant material produced photosensitization. The animals developed photophobia, discharge of tears from the eyes, blindness with cloudiness of the cornea, and swollen red lesions on the white areas of the nose. The skin lesions healed within 2 weeks, but the eyes did not clear for 2 months.

Apocynaceae, Dogbane Family

Many members of this large family contain cardiac glycosides. Others yield useful drugs such as reserpine from *Rauvolfia.*

Acokanthera oblongifolia.   Wintersweet.
Acokanthera oppositifolia.   Bushman's Poison.

Both native of South Africa; cultivated in gardens. Evergreen shrubs with milky juice; flowers with long tubular corollas; fruit a black drupe, larger than an olive.

Toxic part: Entire plant except for the flesh of ripe fruits.

Toxin: Cardiovascular glycosides (ouabain, G-strophanthin, and acokantherin). These are the principal ingredients of the South African Bushman's arrow poisons.

Symptoms: Nausea and vomiting. Can be fatal as a result of changes in cardiac rhythm. In South Africa, a person struck by a poisoned arrow died within an hour.

Adenium obesum.   Desert Rose, Mock Azalea.

Rarely cultivated in California; contains cardiac glycosides; used as fish and arrow poisons in eastern Africa.

Allamanda cathartica.   Yellow Allamanda, Golden Trumpet.

Cultivated rarely in California in the warmer sections; milky juice is a mild to strong cathartic and may cause stomach upset, but its toxicity is not serious.

Apocynum androsaemifolium.   Spreading Dogbane.
Apocynum cannabinum.   Indian Hemp, Dogbane Hemp, American Hemp.

Both native to California, north to Canada and east to the Atlantic Coast. Perennial herbs with somewhat woody stems, milky juice; leaves opposite, flowers white or pinkish, corolla small, urn-shaped, five-lobed.

Toxic part: Entire plant.

Toxin: A cardiac glycoside, apocynamarin, and other glycosides and resins; as little as 15 to 30 grams of fresh leaves may be lethal to a cow or horse.

Symptoms: In animals, increased body temperature but coldness of extremities, rapid pulse, sore mouth with discoloration of mouth and nostrils, dilated eye pupils, loss of appetite, and death within a few hours or 1 day.

As native plants of rural areas, both species can cause poisoning and sometimes death to cattle, sheep, and horses. But because the milky juice is bitter and sticky or rubbery, the plants are usually avoided.

Two horses died in southwestern Modoc County in April 1981 from eating alfalfa hay that included a large quantity of Indian Hemp. These plants grew along a creek bank at the margin of the field and were mowed in the second cutting of the alfalfa. Tops of plants to 3 ft tall, including flowers, were found in the hay. No cases of human poisoning are known.

Catharanthus roseus.   Periwinkle (pl. 4*b*).

Cultivated; the source of several drugs used for leukemia and other diseases, discussed under "Alkaloids" (see pp. 340–341); reported as toxic to cattle.

Nerium oleander [*N. indicum*].   Oleander (pl. 4*c*).

Native of the Mediterranean region; widely cultivated in California, especially in warm sections, used as a highway divider and light screen for freeways. No other ornamental will keep the leaves on the shrubs down to the ground level and live on such minimal watering. Evergreen shrub 5 to 20 ft tall; leaves in threes or opposite; flowers in showy clusters, pink, rose, white, or yellow, corolla funnel-shaped, with five lobes, occasionally double.

Toxic part: Entire plant. Smoke from burning wood can be toxic.

Toxin: Oleandrin, a glycoside that has a digitalis-like action on the heart muscle.

Symptoms: Toxic if ingested, causing nausea, vomiting, dizziness, and irregular cardiac action. It may be fatal.

Oleander is known to be toxic to livestock. Numerous records document deaths of cattle and horses from its ingestion, particularly when dried leaves are mixed with green forage or dried hay. Dried leaves are not as bitterly repugnant as fresh ones. In January 1982, in San Bernardino, Oleander leaves apparently caused the death of two pit bull puppies.

Oleander has been widely used in ornamental plantings in California both in home gardens and along hundreds of miles of highways. A special study of Oleander covering a period of 20 years was made by the California Department of Food and Agriculture while collecting reports of possible injury and damage from agricultural chemicals and products. Every report was carefully checked, and no valid records of human deaths from Oleander in California were found during that time. In 1982, however, in Contra Costa County the death of a 96-year-old woman was reported. Knowledgeable of the toxic effects of Oleander, she apparently ingested leaves of the plant in an attempt to take her own life. Although she was given emergency medical treatment, the amount taken was sufficient to cause death.

The sap of Oleander is reported to be irritant, but its allergenic properties have not been adequately studied. Considering the widespread planting of this shrub and the numerous possible exposures, few cases have been reported.

Children have become ill from sucking nectar from the flowers and chewing the leaves. Smoke from incompletely burned stems and leaves has caused poisonings. Reports of the death of soldiers using Oleander sticks to barbecue meat are varied; cases are attributed to soldiers of Napoleon or Alexander the Great. Honey produced from the nectar is reported as toxic, but bees do not normally visit these flowers.

Thevetia peruviana [*T. neriifolia*].   Yellow Oleander, Lucky Nut, Be-still Tree, Flor Del Peru (pl. 4d).
Thevetia thevetioides.   Giant Thevetia.

Both species native of tropical America; occasionally cultivated in the warm sections of California.

Both species are shrubs or small trees with milky juice; leaves alternate; flowers yellow or orange, tinged with pink, large and showy, corolla funnel-shaped, five-lobed; fruit a drupe, broader than long with two to four flat seeds.

Toxic part: Entire plant, especially the seeds.

Toxin: Cardiac glycosides (thevetin, cerebrin, and neriifolin), which produce symptoms similar to digitalis.

Symptoms: Nausea, vomiting, and irregular pulse. Can be fatal. One fruit can cause the death of an adult. The milky juice is irritating and may cause dermatitis in those susceptible.

Aquifoliaceae, Holly Family

Ilex. **Holly.**

Evergreen shrubs or small trees; leaves alternate, leathery; very small greenish or whitish flowers; berries usually bright red, rarely yellow. The following species, as well as several others, are cultivated in California:

> *Ilex aquifolium,* Holly, English Holly, European Holly; native of Europe; many named cultivars
> *Ilex cornuta,* Chinese Holly (fig. 7); native of China; several named cultivars, including 'Burfordii', Burford Holly
> *Ilex crenata,* Japanese Holly, Box-leaved Holly; native of Japan
> *Ilex vomitoria,* Yaupon, Cassine; native of southeastern United States

Toxic part: Berries.

Toxin: The toxic substance has not been identified. Extract is called ilicin.

Symptoms: Ingestion of the berries of some species causes nausea and vomiting, occasional diarrhea, and stupor when eaten in quantity. Most dangerous to small children.

The leaves are not toxic in some species and are brewed for their caffeine content. The leaves of *Ilex vomitoria* contain a considerable amount of caffeine and were formerly used in the southeastern United States for yaupon tea and earlier by the Indians for their "black drinks."

Araliaceae, Ginseng Family

The family is of minor importance as a source of plant toxicities.

Aralia spinosa. Devil's Walking Stick, Hercules' Club, Angelica Tree, Prickly Ash.

Sometimes cultivated as an ornamental; suspected of poisoning livestock in Maryland; experimental feeding of the seeds to guinea pigs caused their deaths.

Brassaia actinophylla [*Schefflera actinophylla*]. Australian or Queensland Umbrella Tree.

From Australia; grown as houseplants or rarely outdoors; reported in Minnesota as having poisoned a dog.

Hedera helix. English Ivy.

Native of Europe, western Asia, and North Africa; frequently cultivated in California and occasionally naturalized. One of the commonest vines, not to be confused with other plants called ivy, it is variable, and many leaf shapes and

FIG. 7. *Ilex cornuta*. Chinese Holly. Upper left: Spiny leaf of Chinese Holly. Center left: Male flower. Lower left: Flowering stem of *Ilex cornuta* 'Burfordii'. Burford Holly. Upper right: Fruiting stem of Burford Holly. Lower right: Seed, berry, and female flower.

growth forms have given rise to numerous cultivars, some of which are unstable and difficult to identify. Woody climber in trees and on walls, climbing by means of small adhesive rootlets on the stems, and when without a support makes a ground cover; leathery juvenile leaves usually three to five-lobed, 2 to 3 in. long; adult stem leaves ovate, entire, and appearing on short, erect, nonclimbing flowering and fruiting branches; flowers small, greenish, in umbels; fruit a black berry.

Toxic part: Leaves and berries.

Toxin: Hederin (a saponin glycoside) and hederagenin.

Symptoms: Usually simple gastroenteritis with vomiting and diarrhea; allergic contact dermatitis, sometimes severe, with linear and vesicular lesions, similar to those caused by Poison Oak.

English Ivy is considered harmful to livestock when eaten in large quantities.

In susceptible humans, handling (as when pruning) can cause contact dermatitis. The following species of *Hedera,* also cultivated in California, may produce similar reactions:

Hedera canariensis [*H. helix* subsp. *canariensis*], Algerian Ivy, Canary Islands Ivy, Madeira Ivy
Hedera colchica, Persian Ivy
Hedera nepalensis, Nepal Ivy

Asclepiadaceae, Milkweed Family

Araujia sericofera [*Physianthus albens*]. White Bladder Flower, White Moth Plant, Cruel Plant.

Cultivated as an ornamental, rarely naturalized. Poisoning of animals from this species is not known in California; seeds of a related species have caused deaths of poultry; normally livestock do not eat it.

Asclepias. **Milkweed, Silkweed, Butterfly Flower.**

About twelve species native in California, several additional ones cultivated as ornamentals. Perennial herbs or shrubs with a milky juice; leaves simple, alternate, opposite or whorled; flowers modified for specialized pollination, sepals mostly separate, petals united into a corolla, a five-lobed crown usually with a column between the petals and the stamens with a horn-like projection in each concave hood surrounding the column; five stamens attached by their anthers, the anthers winged at the base; pollen united into a mass, the pollinium; fruit consists of a follicle usually splitting on one side when mature; seeds with a copious tuft of hairs at one end. The following native species are known to be toxic to livestock.

Asclepias eriocarpa [*A. fremontii*].   Indian Milkweed, Kotolo.

Native from Shasta and Mendocino counties south to Baja California at middle and lower elevations; leaves usually in whorls of three, the horn enclosed in the hood in the flower.

Asclepias fascicularis [*A. mexicana*].   Whorled Milkweed, Narrowleaf Milkweed (pl. 5*a*).

Native throughout California, common west of the mountains at low and middle elevations, rare on the deserts, widespread in western North America; glabrous narrow leaves in whorls of three, sometimes upper leaves opposite, horns exerted from hoods, fruit erect at maturity.

Asclepias speciosa.   Showy Milkweed (pl. 5*b*).

Native from Solano, Fresno, and Inyo counties northward, widespread beyond the state; woolly leaves opposite, hoods conspicuously projecting from flower, fruit markedly reflexed.

Any species of *Asclepias* must be regarded as potentially toxic. *Asclepias cordifolia,* Purple Milkweed, in the North Coast Ranges and western slopes of the Sierra Nevada, the plant a purplish green color, leaves mostly opposite, has often been suspected in cases of livestock poisoning.

Several species are cultivated as ornamentals in California:

> *Asclepias curassavica,* Blood Flower; attractive orange yellow flowers
>
> *Asclepias fruticosa* [*Gomphocarpus fruticosus*] and *A. physocarpa* [*Gomphocarpus physocarpus*], Swan Plant; both native of Africa, with pale creamy white flowers, cultivated for their inflated greenish fruits

Toxic part: Entire plant.

Toxin: A resinoid (galitoxin), several glycosides, and an alkaloid.

Symptoms: Depression and weakness followed within a day or two by convulsions, coma, and death.

In California a number of cases of livestock poisoning have been recorded from different species of milkweed.

In the summer of 1957, range forage was very scarce south of Eagleville, Modoc County. The cattle ate the seed pods of *Asclepias speciosa,* Showy Milkweed, which was growing in dense patches along the creek beds; dead animals were found in the creek beds in or near milkweed plants. Imprints beside one of the cows indicated that before death the animal, while lying on its side, had pawed the ground.

Asclepias fascicularis, Whorled Milkweed, has been implicated in a number of livestock deaths from both green and dried plants. Fresh plants are distasteful and have caused death or distress to livestock only when on short rations. Dried milkweed plants in hay are not so unpalatable to livestock, however, and the animals cannot be as selective of plants included in hay. No human poisonings have been reported for these species of milkweeds.

Asteraceae, Aster Family

Also called Compositae, this large, widely distributed family has many toxic members.

Ambrosia. Ragweed.

The ragweeds that cause such hay fever distress to humans in the central and eastern parts of the United States do not occur in sufficient numbers to be of concern in California. People sensitive to the ragweeds generally find relief in this state. *Ambrosia trifida,* Giant Ragweed, has not become established in California, and the few waifs that have occurred have been eradicated. *Ambrosia artemisiifolia* [*A. elatior*], Short Ragweed, has also occurred in several places but has not persisted. *Ambrosia psilostachya,* Western Ragweed (fig. 8), is native to California but produces smaller plants, 1 to 2 ft tall, and never has the great numbers of plants in one location as the ragweeds in the central and eastern parts of the United States.

FIG. 8. *Ambrosia psilostachya*. Western Ragweed. A. Head of staminate flowers. B. Achene.

Baileya. Desert Marigold.

Three species of *Baileya* are native to the deserts of California; white-woolly herbs readily recognized by the persistent, reflexed, yellow ray flowers. Plants 1 to 2 ft tall; leaves alternate, once or twice pinnate on the lower stems, simple above. *Baileya multiradiata*, Desert Baileya, native in eastern San Bernardino County, extending to Utah, and western Texas; biennial or perennial, basal leaves persisting, twenty-five to thirty ray flowers in each head. *Baileya pauciradiata*, Colorado Desert Marigold, native in the Colorado Desert, eastern Mojave Desert, adjacent Arizona, and Mexico; annual or persisting a second year, basal leaves not persisting, five to seven ray flowers in each head. *Baileya pleniradiata*, Woolly Marigold, native in Inyo County south to Riverside County, east to Utah, western Texas, and adjacent Mexico; annual or perennial, basal leaves not persisting, twenty to forty ray flowers in each head.

Toxic part: Whole plant.

Toxin: Seven sesquiterpene lactones occur in the three species.

Symptoms: In animals, depression, weakness, frothing at the mouth, rapid pulse, and death may result.

Sheep and goats are the only animals affected by these plants and only if large amounts of the plants are eaten over a long period of time. Cattle and horses eat these plants without any apparent difficulties. The cause of the "desert disease" of bees remains unknown, but it occurs almost every year. Pollen grains of *Baileya* and *Oenothera* are found on these dead bees in both the Coachella and Imperial valleys.

Balsamorhiza sagittata. Balsamroot, Arrow-leaved Balsamroot.

Native in northeastern California, extending in the United States to the north and east; stout perennial with large arrow-shaped leaves in rosettes at ground level; flower heads showy, resembling sunflowers; suspected of being poisonous to livestock but probably harmless unless large amounts are eaten; plants usually increase in numbers under grazing.

Centaurea solstitialis. Yellow Starthistle, Barnaby's Thistle (pl. 5c).

Southern Europe; widespread weed in central and northern California in waste ground and disturbed areas. Erect, branched, summer annual, cottony pubescent; upper leaves alternate, linear, entire, decurrent bases giving the stem a winged appearance; lower leaves in a rosette, irregularly pinnatifid; flower heads with stiff yellow spines to 1 in. long; disk flowers only, yellow.

Toxic part: Entire plant with leaves and flowers, usually in May through December.

Toxin: Unknown; toxic only to horses.

Symptoms: Called "chewing disease" because the horse is unable to swallow and will continue incomplete chewing of the same mouthful of food for hours. The animal becomes depressed and may die of starvation or thirst. The toxin causes lesions on the brain that result in this behavior. Large amounts of Yellow Starthistle, roughly the weight of the animal, must be eaten over a period of time, but the effects of the toxin are cumulative.

In September 1963, near Centerville, Fresno County, a rancher who had five mares with foals on a river-bottom pasture noticed their difficulty in grazing, chewing, and swallowing. The affected mares had a tense, taut upper lip to the extent that the upper teeth left an imprint on the inside of the lip. When fed some good-quality hay, the animals would grasp a small amount in their mouths and then stand with head depressed unable to masticate and swallow. These animals had been on this pasture for several months, and the poisoning by the Yellow Starthistle had been cumulative. Most of the pasture consisted of various weeds including Yellow Starthistle; the Bermudagrass in the pasture had been closely grazed.

Centaurea repens, Russian Knapweed, is a widely distributed weed in California, but there are no known cases of chewing disease from this weed in the state. Cases from Washington and Colorado are known. Russian Knapweed has produced the disease experimentally. The plants are very bitter and are not usually eaten by horses.

Chamaemelum nobile [*Anthemis nobilis*].   Chamomile, Garden Chamomile, Russian Chamomile.

Native of the western Mediterranean region and the Azores; cultivated in California, commonly grown for the dried flower heads used in herbal teas, sometimes used in place of a grass turf. Creeping perennial to 12 in. tall; leaves to 2 in. long, pinnately divided two or three times, light yellow green; flower heads to 1 in. across, ray flowers white, disk flowers yellow, formed on a conical receptacle.

Tea made from this species has caused inflammation of the mucous membranes of the nose and even anaphylactic shock in persons known to be allergic to ragweed pollen. See also *Matricaria recutita* in this plant family.

Chrysothamnus nauseosus.   Rabbitbrush, Rubber Rabbitbrush.

Native in California east of the Sierra Nevada, widespread in the cold desert areas of western North America; shown experimentally to be poisonous to livestock but so unpalatable it is not eaten in sufficient quantities to be a problem; under grazing it increases to become quite weedy.

Eupatorium adenophorum [*Ageratina adenophora*]. Maui
Pamakani, Crofton Weed, White Thoroughwort, Mexican
Devil.

Native of Mexico; naturalized in areas of permanent water
in the San Gabriel Mountains north of Los Angeles and at scattered localities along the coast north to Marin County where
this weed is established on dry ground in the fog belt. An unknown toxin causes a dense congestion of the lungs only in
horses that have eaten it for several months. Poisoning unknown in California but has occurred elsewhere.

Grindelia squarrosa. Gumweed, Resinweed.

Native of the Great Plains east of the Rocky Mountains;
sparingly introduced into California in the Antelope Valley and
the western Mojave Desert. This and several other species of
Grindelia are secondary selenium absorbers.

Gutierrezia microcephala. Threadleaf Snakeweed.
Gutierrezia sarothrae. Broom Snakeweed.

Both species also have the common names of Matchweed,
Broomweed, Resinweed, Turpentineweed. The two species occur in the desert areas of California and eastward. *Gutierrezia
microcephala* differs from *G. sarothrae* in having fewer flowers in the heads that are noticeably narrower. Perennial sub-
shrubs, resinous, much branched from the base; leaves alternate, linear, entire; flower heads clustered, each with a few
yellow ray and disk flowers.

Toxic part: Whole plant.

Toxin: Saponin.

Symptoms: Abortion in cattle, less common in sheep and
goats. Listlessness and loss of appetite resulting in weight loss,
bloody urine, nasal and vaginal discharges, and mucus in the
feces.

Plants are most toxic during early stages of growth and
when grown on sandy soils. Snakeweeds are relatively unpalatable, and livestock eat only small amounts unless there is a
shortage of good forage. The dried plants remain toxic in hay.
Annual losses in Texas from Threadleaf Snakeweed poisoning
have been estimated between $2 million and $3 million.

Helenium amarum [*H. tenuifolium*].   Bitter Sneezeweed,
Fine-leaved Sneezeweed.

Native to the southeastern United States; at one time sparingly established in California, not known to be present at this
time. Entire plant, fresh or dried, contains tenulin, a sesquiterpene lactone. Horses, mules, and sheep may die from eating
this plant. When it is ingested, cows produce bitter milk that is
unpalatable but not toxic to humans. There was one occurrence
of this weed in irrigated pastures in a dairy area of Stanislaus
County, but the plants were not eaten by the cows.

Helenium autumnale.   Western Sneezeweed, Common
Sneezeweed, False Sunflower, Swamp Sunflower, Oxeye,
Yellow Star.

Native in California only in the three northernmost counties, northward to British Columbia and eastward to the Rocky
Mountains; cultivated as an ornamental. Perennial to 4 or 5 ft
tall, densely covered with minute hairs; leaves decurrent as
wings on the stems; flower heads showy, daisy-like, ray flowers
yellow; disk globose, pompon-shaped, with yellow disk flowers. Cultivated forms may have orange or reddish ray flowers.

Toxic part: Entire plant.

Toxin: Sesquiterpene lactones, variable in occurrence
throughout the range of the species, helenalin or tenulin, and
minor amounts of other lactones.

Symptoms: Highly irritating to the nose, eyes, and stomach
of livestock; may cause acute poisoning and death in horses,
mules, and cattle; also toxic to dogs.

Western Sneezeweed occurs as unpalatable plants in the
meadow pastures and has not been a source of livestock losses
in recent times in California. It is of concern to the dairy industry because ingestion by cows gives the milk a bitter taste.

Helenium hoopesii [*Dugaldia hoopesii*].   Orange
Sneezeweed, Bitterweed (pl. 5*d*).

Native in eastern California from Tulare and Mono counties
northward from 7,500 to 10,500 ft elevation in the Sierra
Nevada and at lower elevations in the Warner Mountains of
Modoc County, extending into Oregon and east to the Rocky
Mountains. Stout perennial, stems and leaves hairy, stems not

winged; leaves clasping at the base; flower heads showy, daisy-like, yellow ray flowers, brown disk flowers in a dome-shaped head.

Toxic part: Entire plant, fresh or dried.

Toxin: Hymenovin, a sesquiterpene lactone. Formerly the toxin was thought to be dugaldin, but the early studies actually were investigating hymenovin.

Symptoms: "Spewing sickness" of sheep because of the vomiting that occurs. Depression with tremors following any disturbance is also typical.

Usually Orange Sneezeweed plants are not eaten; they increase in mountain meadows under grazing conditions.

Hymenoxys odorata. Bitter Rubberweed, Western Bitterweed, Fragrant Bitterweed.

Native to southwestern United States but found in California only in the lowlands along the Colorado River from Parker Dam to Yuma; contains hymenovin, a mixture of two sesquiterpene lactones. The cumulative toxicity from this plant is not known for California, but it is a serious poisonous range plant in many parts of the Southwest.

Iva acerosa [*Oxytenia acerosa*]. Copperweed.

Death Valley region of California and east; unknown toxin in fresh or dried plants causes weakness, coma, and death in cattle and sheep; such poisoning not recorded for California, but in other parts of the Southwest it has caused animal deaths particularly in the fall.

Lactuca virosa. Bitter Lettuce, Wild Lettuce, Lettuce-opium.

Naturalized in California at scattered places around San Francisco Bay, this biennial lettuce is cultivated in Germany, France, and England for the dried milky juice called lactucarium; reported to be used as a mild sedative and diuretic, a home remedy, replacing opium as a cough suppressant.

Lepidospartum squamatum. Scale Broom.

Contains an unknown toxin; reported to have been the cause of a severe allergic reaction to a resident of Lancaster, Los Angeles County, in February 1957, a cold time of the year. Cattle refused to eat barley hay that contained dried Scale Broom plants. California native.

Matricaria recutita [*Chamomilla recutita*].   Sweet False Chamomile. Sometimes called *Matricaria chamomilla,* but that name is a synonym of *Tripleurospermum maritimum* subsp. *inodorum.*

Native of Europe and western Asia; cultivated in California and naturalized on the Pacific Coast, but more widely so in eastern North America. Annual plants to 30 in. tall; leaves 2 to 3 in. long, twice pinnately cut; flower heads 1 in. broad, ray flowers white and often reflexed, disk flowers yellow and formed on a conical receptacle.

As with *Chamaemelum nobile,* chamomile tea from this species can cause inflammation of the nasal membranes and even produce anaphylactic shock in persons known to be allergic to ragweed pollen.

Osteospermum ecklonis.   African Daisy.

Native to South Africa; cultivated in central and southern California gardens, rarely spontaneous. The genera *Arctotis* and *Dimorphotheca* have species also known as African Daisies. Somewhat shrubby plants to 4 ft tall and across; leaves to 4 in. long, toothed on the margins or at least at the tips of the leaf blades, covered with sticky glandular hairs; daisy-like flower heads 3 in. across, ray flowers white above and blue to lavender below, disk flowers dark blue.

Toxic part: Entire plant (stems, leaves, and flowers), most toxin in the leaves.

Toxin: Linamarin (phaseolunatin, a cyanogenic glycoside) and a saponin.

Symptoms: Labored breathing, tremors and muscular contractions, sometimes followed by convulsions, paralysis, and loss of consciousness. May be fatal.

When the plant cells are damaged, linamarase, the enzyme already present, causes linamarin to be split into glucose, hydrocyanic acid, and acetone. Poisoning of livestock from *Osteospermum ecklonis* has been reported from South Africa and

Australia. Livestock should not have access to this ornamental shrub or its prunings.

Osteospermum fruticosum.   Trailing African Daisy, Freeway Daisy.

Native of South Africa; much more commonly grown in California than *O. ecklonis.* Selected cultivars have ray flowers white above and purple beneath, solid purple, or variations of these colors. This species is listed as a possible cause of stock poisoning in South Africa. The wilted leaves contain 1.4% hydrocyanic acid derived from linamarin. Similar care should be taken to prevent livestock from getting to plants of this species.

Psathyrotes annua.   Desert Velvet, Velvet Rosettes.

Native of southeastern California in the Mojave Desert, but extending to Idaho, Arizona, and adjacent Mexico. Low-growing, many-branched winter annual with a strong odor, gray scurfy pubescent; leaves alternate, small deltoid blades, long petioles; flower heads small, inconspicuous, lacking ray flowers. This species has been shown experimentally to be toxic to livestock; but since animals do not eat these unpalatable plants, poisonings are unknown.

Psilostrophe cooperi.   Paper Flower.

Common desert shrub found in California in the eastern Mojave Desert and the northern part of the Colorado Desert and extending to the east and south. Has been shown experimentally to contain 0.2% psilotropin, a sesquiterpene lactone, in the aboveground parts of the much-branched, white woolly plant. There are no records of toxicity from plants of this species, which are not usually eaten by livestock. Three other species of *Psilostrophe* cause economic losses in the Southwest, particularly in sheep on winter ranges.

Senecio.   **Groundsel.**

One of the large genera of flowering plants with more than 2,000 species widely distributed in temperate regions around the world. Thirty-eight species are listed as occurring in California, including four introduced weeds; several other species are cultivated as ornamentals.

Toxic part: Entire plant.

Toxin: Pyrrolizidine alkaloids.

Symptoms: Liver damage in sheep, cattle, horses, and humans. The liver tissue becomes extremely hard and fibrous. Debilitation, weakness, and diarrhea are the beginning signs of this toxicity. Liver failure causes accumulations of ammonia in toxic amounts since its normal conversion to urea does not occur. Damage to the central nervous system and brain results in manifestations of aggressiveness and incoordination, disorientation, sleepiness, and coma. Ammonia also causes a direct increase in the breathing rate that accelerates the accumulation of ammonia in the brain. If there is chronic liver damage, the animal usually will not survive. The most toxic North American species of *Senecio* do not occur in California.

Senecio cineraria [*Senecio bicolor* subsp. *cineraria*].   Dusty Miller, Cineraria.

Cultivated as an ornamental; much-branched white woolly perennial. Leaves pinnately lobed; flower heads showy, in clusters; ray flowers yellow. A cause of contact dermatitis; not known to be toxic to animals in California but reported as lethal to sheep in Iraq.

Senecio confusus.   Mexican Flame Vine, Orange Glow Vine.

Rarely cultivated in warm areas of California. Vigorous vine with alternate, thickish, sparsely toothed leaves, showy flower heads, each about ½ in. across with orange to orange red ray flowers; causes contact dermatitis.

Senecio douglasii.   Shrubby Butterweed.

Native of California, widespread and common in the chaparral below 5,000 ft. Branched shrub; leaves to 2 in. long, pinnately divided into narrow revolute lobes, whitish beneath; flower heads with showy yellow ray flowers.

Plants of Shrubby Butterweed usually are not eaten by livestock. In October 1963, however, four horses on a ranch near

Santa Susana, Ventura County, displayed definite nervous symptoms. One horse stood on a hillside in a depressed state and the other animals were very excited when forced to move. There was no forage available except native shrubs. The top halves of the Butterweed shrubs had been extensively browsed.

Senecio jacobaea.   Tansy Ragwort, Ragwort, Stinking Willie (pl. 6*a*).

From Europe; introduced into coastal northern California in northern Mendocino County to Del Norte County where it grows as a weed in pastures and waste ground; also an abundant weed in Oregon and Washington. Coarse, erect biennial or short-lived perennial to 6 ft tall; stems in part purplish red; leaves in basal rosettes and alternate up the stems, deeply and irregularly two to three times pinnately divided; flower heads showy, in somewhat flat-topped clusters, ray flowers yellow.

Tansy Ragwort plants contain pyrrolizidine alkaloids that can be fatal to livestock, causing damage to the liver and lungs. Nonlethal amounts of these alkaloids may cause slight liver injury that may not be detected in the animals for several months or even years.

Cows and horses usually avoid eating Tansy Ragwort plants, but they are known to be fatally poisoned by them. These plants are readily eaten by sheep, but they may do poorly on them. Sheep have been poisoned experimentally. Although dairy goats have shown more resistance than cows to poisoning from Tansy Ragwort, under stress they indicate a decline in general health, become abortive, and die. Therefore goats must be considered susceptible and not as resistant as sheep to poisoning from this plant.

The alkaloids are not lost on drying, and when present in hay they cause poisonings because animals cannot reject the toxic plants.

The pyrrolizidine alkaloids of Tansy Ragwort have been found in the milk of dairy cows and goats. Neither the calves nor the kids raised on such milk have shown any toxicity from these alkaloids. The conclusion was reached that alkaloids in the milk were not carcinogenic, and the amount of the alkaloids was not sufficient to cause serious damage or lesions in the liver of the offspring.

Pyrrolizidine alkaloids from Tansy Ragwort have been discovered in honey produced in areas of western Oregon and western Washington. Beekeepers easily recognize the toxic honey as the frames in the hives and the wax combs are excessively stained yellow from Tansy Ragwort pollen. The honey is off-color and very bitter in taste and is therefore not usually sold commercially. Instead it is used during the winter as food for the bees. The amount of alkaloids in the honey is so low that humans cannot eat a sufficient quantity to develop acute symptoms of poisoning. The effects of pyrrolizidine alkaloids are cumulative, however, and therefore such foodstuffs must not be considered safe over long periods of time.

Senecio mikanioides.   German Ivy, Parlor Ivy, Water Ivy, Cape Ivy.

Introduced into California as an ornamental; now an invasive weed naturalized along the coast in canyons and moist areas. Glabrous, half-succulent perennial climber with long, flexuous stems; palmately lobed, ivy-like leaves; small flower heads in terminal clusters, ray flowers lacking. No cases of livestock poisoning are known in California, but this species is reported as having caused one case in New Zealand and another in New South Wales, Australia. Native of South Africa.

Senecio vulgaris.   Common Groundsel (fig. 9).

Also called Old-man-of-the-spring because this weed is one of the first plants to grow in the spring; when the flower heads mature, the white parachute-like pappus of the seeds gives the plants a superficial resemblance to an old man's gray beard.

A European native species; now a widespread and common weed in temperate regions around the world. Annual, sparingly branched; leaves pinnatifid with irregular lobes; clusters of pale yellow flower heads with only disk flowers, involucral bracts black-tipped.

Poisoning of cattle has been reported in California from *Senecio vulgaris* when animals have been forced to eat this weed. Dairy cows were poisoned when staked in Sutter County

FIG. 9. *Senecio vulgaris*. Common Groundsel. A. Receptacle. B. Achene.

orchards where these were the only plants available in the early spring.

Near Hilmar, Merced County, during 1962 and 1963, deaths in a herd of dairy cows occurred only in calves 3 to 8 months old. Hay containing Common Groundsel was part of their diet. The animals developed nervousness, followed by incoordination, pushing on fences, and prostration, and died within 2 to 5 days. The remaining hay was fed to mature cows that manifested no symptoms. Calves from cows that were fed the contaminated hay during pregnancy died the following autumn although they themselves had had no exposure to the Common Groundsel hay. The same effect of these alkaloids on offspring has been shown to occur in rats that were fed the toxins during pregnancy.

In 1970, calves were poisoned from eating over a period of a number of months first-cutting alfalfa hay that contained 5% to 10% Common Groundsel.

Near Los Banos, Merced County, in March 1975, from a herd of twenty-four cows and calves, thirteen died and three additional animals became ill from eating hay containing large amounts of Common Groundsel. Signs included very disturbed behavior, walking in circles, and blindness with glazed eyes.

Four horses died in January 1983, near Tranquility, Fresno County, because they had been fed hay containing a large amount of *Senecio vulgaris*.

Solidago spectabilis.   Basin Goldenrod.

Native in California in Inyo and Mono counties and extending into the Great Basin; abundant in wet alkaline soils, sea level to 7,500 ft elevation. Perennial plants 1 to 4 ft tall; terminal, dense, flower heads in oblong panicles 8 to 10 in. tall. Livestock poisonings from Basin Goldenrod are not known in California, but losses in sheep and cattle occurred in Nevada where investigations found a hemolytic toxin in plants of this species.

Tagetes.   Marigold, Marygold.

Different species of *Tagetes* have been demonstrated to contain substances in the roots that are toxic to some but not all species of root-knot nematodes, *Meloidogyne,* and root-lesion nematodes, *Pratylenchus.* Marigold plants do not actually secrete substances into the soil to kill nematodes; they are killed only in the roots of the plants. The nematode-killing compounds are derivatives of thiophene.

Tagetes minuta, Wild Marigold, is highly susceptible to the root-knot nematode *Meloidogyne arenaria,* less susceptible to *M. hapla,* and resistant to other species of root-knot nematodes. Plants of *Tagetes patula,* French Marigold, are as effective as plants of *Tagetes minuta* in reducing nematode populations in the soil but unlike *Tagetes minuta* do not become serious weed problems.

Tanacetum vulgare.   Tansy, Golden Buttons (fig. 10).

Native of Eurasia; cultivated and sparingly naturalized in California. Tall erect perennial, aromatic; leaves deeply pinnately lobed, fern-like, lobes irregularly toothed; flower heads lacking ray flowers, grouped in loose flat-topped clusters. The entire plant is strongly scented and has a bitter taste from a series of toxic essential oils (monoterpenes, mostly thujone). Organic acids are also present. Thujone acts on the central

FIG. 10. *Tanacetum vulgare*. Tansy. Lower right: Disk floret.

nervous system; continued use may result in convulsions and death.

Aqueous decoctions or "teas" made from Tansy have been used medicinally, sometimes to induce abortions. The effects of using thujone are cumulative. For centuries oil of tansy has been used as a vermifuge. In this species there are a number of races that are genetically fixed and characterized by the presence of a certain essential oil.

Tetradymia canescens.   Spineless Horsebrush, Gray Horsebrush (pl. 6*c*).
Tetradymia glabrata.   Littleleaf Horsebrush, Smooth Horsebrush, Spring Rabbitbrush, Coal-oil-brush. Rat Brush, Dog Brush (pl. 6*d*).

Both species native to the Mojave Desert, the Tehachapi Mountains, the east side of the Sierra Nevada, and north to British Columbia, east to Utah and Idaho.

Plants of these species are low shrubs; leaves alternate, linear; flower heads in terminal clusters, each head narrow with yellow disk flowers only. The slender leaves of *Tetradymia canescens* are borne singly while those of *T. glabrata* are borne together, several in a cluster.

Toxic part: Any part of the plant, but apparently the young buds are the most toxic to sheep. A great deal remains to be learned; eating *Artemisia* plants before eating the *Tetradymia* may be part of the complex toxicity.

Toxin: Tetradymol and even more toxic compounds derived in the animal body from this compound.

Symptoms: Liver damage followed by photosensitization or death and "bighead," a swelling of the sheep's head.

In the western states these plants have a long history of poisoning and resultant death of sheep. As little as $\frac{1}{2}$ pound of the branch tips of Littleleaf Horsebrush can cause photosensitization in a 100-pound sheep.

In May 1960, about 200 head of sheep developed bighead while being grazed on a high arid tableland in the southern Surprise Valley in northeastern Lassen County. The shrubs had started to flower at this time, and the sheep had eaten the tips of the branches including the flower buds.

Tetradymia stenolepis.　Mojave Horsebrush.

During the first week of July 1965, a band of 2,000 sheep was driven down the draws of the western Mojave Desert near Boron, Kern County. Sixty-five sheep died and another forty-five displayed signs of toxicity. Mojave Horsebrush occurred only along the draws, and the tips of the branches had been eaten by the sheep. These shrubs have slender inch-long spines on the stems, but the trailing sheep ate the flower heads off the ends of the branches. The owner stated that if the sheep had been driven on the ridges instead of in the draws much of the poisoning might have been avoided. This species had not been previously known to produce toxicity and requires further study.

Verbesina encelioides.　Crownbeard, Golden Crownbeard, Butter Daisy.

Native of central and western United States; var. *exauriculata,* lacking auricles at the base of the petiole, is naturalized in California in Monterey, Fresno, Kern, Los Angeles, and Ventura counties and in Washoe County, Nevada, adjacent to California. Annual plants to 4 ft tall, branched above, grayish white with fine hairs; leaves broadly oval, toothed, white hairy on the lower surface; flower heads daisy-like, 2 in. across, ray flowers yellow, deeply notched or lobed, yellow disk flowers raised on a dome-shaped receptacle; achenes flattened with a corky rib on each edge.

Toxic levels of nitrates are known to accumulate in plants of Crownbeard in the United States, but the plants are eaten by animals only when other forage is lacking. In Australia a toxin called galegine, 3-methyl-2-butenylguanidine, has been found in naturalized plants, the same toxin found in *Galega officinalis*. Sudden death is produced in sheep, cattle, and pigs most frequently during drought years and a few days after a summer rain.

Xanthium spinosum.   Spiny Clotbur, Spanish Thistle (fig. 11).
Xanthium strumarium.   Cocklebur (pl. 6*b*).

Cosmopolitan weeds of waste places, old pastures, and roadsides; native and widespread in California. Tall, much-branched, coarse summer annuals; leaves alternate, lobed; flower heads with either male or female flowers, the male flower heads above the female, the female flower heads with only two flowers; involucre becomes a hard bur with two beaks and covered with hooked prickles; contains two seeds.

The two species may be distinguished readily by the conspicuous inch-long three-forked spines in the leaf axils of the Spiny Clotbur, which are not found on the stems of the Cocklebur.

Toxic part: Seeds and seedlings.

Toxin: Carboxylatractyloside, a highly toxic glycoside. Hydroquinone, which has been listed as the toxic glycoside in plants of these species, apparently is not present.

Symptoms: In animals, weakness, rapid pulse and respiration, vomiting, low temperature, muscular incoordination, prostration, spasms, and death, which may occur before other symptoms develop. In humans, *Xanthium* plants sometimes cause allergic contact dermatitis.

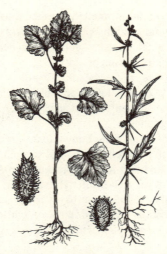

FIG. 11. Left: *Xanthium strumarium.* Cocklebur. Plant and spiny fruit. Right: *Xanthium spinosum.* Spiny Clotbur. Plant and spiny fruit.

Seeds are the most toxic part of the plants, but they are only eaten by livestock if included in grain in which there has been enough mechanical damage to the seed pods to remove the hooked spines. Only the very young seedlings of the Cocklebur, about the first 2 weeks of growth, cause toxicity to livestock. This toxicity is rapidly lost as the plants grow older. The mature plants are very tough and coarse and are not eaten even by hungry livestock.

In 1938, in Yolo County, twenty cows from a herd of eighty died from eating the young sprouts of Cocklebur that were the only plants growing in an overgrazed pasture. In an adjacent field thousands of Cocklebur plants gave an indication of what the animals had eaten.

In the Sacramento area during March 1938, about thirty hogs died in an orchard where the Cocklebur seedlings formed a mat on the ground. In this case the animals were observed eating the seedlings. Several animals became paralyzed; others, not so affected, were vomiting a light green substance. The hogs were then turned into an alfalfa pasture free of Cocklebur, and after 2 days no more deaths occurred.

Berberidaceae, Barberry Family

Berberis. **Barberry.**

About 450 species, widely distributed. Many species are grown as ornamentals in California. Shrubs, deciduous or evergreen, usually spiny; leaves simple, margins often spiny; small yellow flowers, often numerous, in showy clusters; berries red or black.

Mahonia. **Holly Grape, Oregon Grape** [sometimes referring only to *M. aquifolium* of the Pacific Northwest].

About 100 species widely distributed, several native in California; these and others cultivated as ornamentals. Spineless shrubs closely related to *Berberis,* from which it differs chiefly in its pinnately compound leaves. Sometimes included as part of the genus *Berberis*.

Toxic part: For both genera, the entire plant, especially the roots, except for the berries that are supposedly not toxic.

Toxin: Berberine and related isoquinoline alkaloids. Berberine is a very strong base that decomposes to a resinoid.

Symptoms: In humans, use of the drug berberis containing the bitter alkaloids causes elevated temperature, inflammation of the mucous membranes from the mouth to the intestine, marked inflammation of the kidney with blood in the urine, and contractions of the involuntary and voluntary muscles. Uterine contractions result from this use, but they are not as long-lasting as those from ergot. Animals are not affected by either the dried root preparations or the purified berberine alkaloid.

The drug berberis is obtained from the dried roots of *Mahonia aquifolium*. This species and the European *Berberis vulgaris* contain berberine, oxyacanthine, berbamine, and other isoquinoline alkaloids. The fresh berries of *Berberis* have been reported as emetic and cathartic, yet those of *Mahonia aquifolium* are described as being used to make a delicious dark blue jelly. Children have been poisoned from eating the roots of barberry plants.

Barberry spines are sharp and can cause annoying mechanical injury; they may also introduce fungal organisms into the skin, a condition sometimes referred to as "barberry poisoning" by nursery employees.

Boraginaceae, Borage Family

The pyrrolizidine alkaloids found in *Heliotropium, Echium, Symphytum,* and *Amsinckia* cause liver and lung damage. *Amsinckia* and *Plagiobothrys* accumulate nitrates.

Amsinckia douglasiana. Fiddleneck, Douglas' Fiddleneck, Fireweed, Tarweed, Buckthorn.

Native in California in Monterey and San Benito counties to Kern and Santa Barbara counties. Erect annual; leaves and stems bristly hairy; flowers orange, formed along one side of the stem, which is sparingly forked or branched and coiled at the tips.

FIG. 12. *Amsinckia intermedia.* Fiddleneck. A. Nutlet.

Amsinckia intermedia.   Fiddleneck, Rancher's Fireweed, Tarweed, Buckthorn (fig. 12; pl. 7*a*).

Found throughout most of California; north to Washington, east to Idaho and Arizona, south to northern Baja California. Bristly hairy erect annual; flowers orange yellow, borne in the one-sided raceme that is characteristic of the family.

The two species may be separated by technical characteristics of the flowers and mature nutlets.

Toxic part: Seeds (actually mature nutlets).

Toxin: Pyrrolizidine alkaloids that cause liver damage to horses, cattle, and swine.

Symptoms: Weakness, a widespread stance, incoordination and a goose-stepping gait when made to move, apparent blindness with pushing against fences, chewing on objects or themselves, and jaundice with a yellowish color of the mucous membranes. Death from liver damage and accumulation of fluid in the lungs. Less than lethal amounts of the pyrrolizidine alkaloids may produce liver damage that is not detected for several months or even years after ingestion of the plant material.

Livestock are poisoned after several days or weeks of ingesting fiddleneck seeds in harvested grain or in mature plants present in hay. Plants of these species frequently occur as abundant weeds of first-year alfalfa plantings, native hay fields, and grain fields. Sheep are relatively resistant to this poisoning.

During the winter of 1962–1963, in the central San Joaquin Valley of California, in a lot of thirty dairy calves $2\frac{1}{2}$ to 3 months old, ten died after eating weedy hay for $2\frac{1}{2}$ to 5 months. The hay was a first-cutting alfalfa and oat mixture; about 5% to 10% of the plants were *Amsinckia intermedia*. Adult cattle fed this hay at the same time were not affected. That fall, however, calves that had been carried by these cows died shortly after birth; the alkaloids had been transferred through the placenta.

In California investigations have demonstrated toxic levels of nitrates in plants of both these species of *Amsinckia,* particularly in periods of cold cloudy weather or drought. The loss of 2,000 to 3,000 cattle from this toxicity is described under "Nitrates" (see pp. 305–307). One of the popcorn flowers, an unidentified species of *Plagiobothrys* in this plant family, is re-

ported in both California and Arizona as accumulating high
levels of nitrates.

Cynoglossum officinale.   Hound's Tongue.

Native of Europe and Asia; naturalized in California south
of McCloud, southern Siskiyou County, and into adjacent
Shasta County; abundant in fencerows and in cut-over timber-
land. Biennial forming a rosette of basal leaves to 12 in. long
the first year and flowering stalks to 3 ft tall the second year;
plants covered with silky hairs and having a mouse-like odor;
flowers tubular with spreading lobes to $\frac{1}{4}$ in. across, reddish
purple in color; nutlets covered with short, barbed prickles.
The entire plant contains the alkaloids cynoglossine and consolidine, which can be lethally toxic to cattle. Signs include
bloat, rapid or labored respiration, diarrhea with dark green
feces, and increased thirst. Poisoning from Hound's Tongue has
not been reported for this small infested area of California
north of Shasta Lake even though cattle are pastured there.

Echium fastuosum.   Pride-of-Madeira.

Native of Canary Islands; cultivated in California as an ornamental for its unusual habit and tall spikes of blue flowers,
rarely naturalized along the coast of California. Shrubby perennial, 3 to 6 ft tall, usually branched at the base and with
grayish hairy rosettes of leaves; flowers blue, loosely arranged
on a tall spike standing above the leaves. On the Tiburon Peninsula, Marin County, in September 1972, a beagle dog developed severe liver damage from ingesting parts of this plant.

Heliotropium curassavicum.   Heliotrope, Seaside
Heliotrope, Alkali Heliotrope.

Native throughout California in alkaline or saline soils,
often in weedy situations; in other western states as well, and
in all other continents. Perennial from a very knotty rhizome,
stems and leaves somewhat fleshy, glabrous, glaucous; leaves
alternate, mostly lanceolate; flowers numerous, white with
yellow to purple center, in slender, curling, one-sided racemes.
Entire plant has pyrrolizidine alkaloids that cause liver damage
when used in herbal teas over a long period of time.

Heliotropium europaeum.   European Heliotrope.

Native of Europe; abundant weeds of roadsides and disturbed areas in California only in the northern Sacramento Valley and in Siskiyou and Modoc counties. Coarsely hairy annual plants 6 in. to $2\frac{1}{2}$ ft tall; ovate leaves; flowers white or light blue, $\frac{1}{8}$ in. across; fruit of four tuberculate nutlets. Plants contain two pyrrolizidine alkaloids that occur as more soluble oxides and are readily absorbed in the rumen. Livestock poisonings from European Heliotrope are not known in California, but extensive losses to sheep occur in Australia. Cattle are more susceptible but rarely poisoned.

Symphytum × *uplandicum* [*S. peregrinum*].   Russian Comfrey, Quaker Comfrey.

Hybrid of Prickly Comfrey, *Symphytum asperum,* and Common Comfrey, *S. officinale.* Cultivated in California for herbal teas and as a forage crop, but has never been recommended for such uses by university or government specialists. Rough hairy perennial with a thickened fleshy taproot to 2 in. across; leaves in a basal rosette, to 18 in. long, ovate-lanceolate, rounded at the base, stem leaves sessile, without petioles; flowering stems 4 to 5 ft tall, flowers somewhat pendulous, calyx lobes narrowly pointed or acuminate, tubular corolla with five lobes, to $\frac{3}{4}$ in. long, pink turning bright blue; sterile, forming no seeds.

Toxic part: Entire plant.

Toxin: Pyrrolizidine alkaloids; eight different kinds have been identified.

Symptoms: Typical of pyrrolizidine alkaloid poisoning described above under *Amsinckia.*

In California during 1982 there were two cases of poisoning of horses that had been fed Russian Comfrey by their owners.

Symphytum asperum, Prickly Comfrey, is established as a weed in the river bottomlands near Arcata, Humboldt County. It is distinguished from Russian Comfrey by its stem leaves with petioles and the bluntly rounded tips of the calyx lobes. The heavy large seeds formed by these plants make them abundant locally but not widely dispersed. In pastures, plants of Prickly Comfrey are not eaten by grazing animals.

Comfrey preparations are used medicinally in Europe as

ointments to relieve muscular pains and inflammations of the veins. Experimentally it has been shown that there is no measurable amount of absorption of the toxic alkaloids through the skin.

Brassicaceae, Mustard Family

This family, also called Cruciferae, is not an important source of poisonous plants.

Toxic part: Entire plant.

Toxin: Mustard oil glycosides that break down to form mustard oils (isothiocyanates) or by further breakdown to form thiocyanate ions (SCN⁻) or organic nitriles.

Symptoms: Severe irritation of the mucous membranes of the intestinal tract of humans and livestock.

Mustard oil glycosides or glucosinolates occur abundantly in plants of the Mustard Family. They have been found in all species that have been tested for them, particularly in Horseradish, *Armoracia rusticana;* various species of mustards, *Brassica;* and Winter Cress, *Barbarea vulgaris.* Mustard oil is not to be confused with the mustard gas of World War I, a synthetic compound so-named because of its odor.

The glucosinolates are hydrolyzed by the enzyme thioglucosidase to form a glucose sugar, a sulfate fraction, and either isothiocyanates (mustard oils) or organic nitriles. The isothiocyanates may further break down to release the thiocyanate ion, SCN⁻, which is combined to form an organic thiocyanate. The isothiocyanates, or mustard oils, have a pungent odor and taste and are irritating to the skin and mucous membranes, including the gastrointestinal tract. Liver and kidney damage may result from these substances as well.

Isothiocyanates are metabolized to the thiocyanate ion, and therefore some of their effects are probably the result of this ion. Isothiocyanates also inhibit the uptake of iodine in the thyroid, but 90% of the thyroid problems in humans are the result of a deficiency of iodine in the diet and not from ingestion of isothiocyanates in large quantities of cabbages and related plants. Investigations have shown that milk from cows on diets high in isothiocyanates may have a temporary drop in iodine content. Depending upon the amount of thiocyanate present, this deficient milk could cause goiters.

In San Joaquin County in September 1958, pigs refused to eat ground barley screenings that contained an unusually large amount of seeds of Wild Mustard, *Brassica kaber*.

Many members of the Mustard Family accumulate free nitrates and can be sources of nitrate poisoning of livestock; they are listed in Appendix C.

Cardaria draba.   Heart-podded Hoarycress.

Native of eastern Europe and western Asia; widespread serious agricultural pest in California. Creeping perennial with grayish green foliage, numerous white flowers in dense flat-topped clusters and heart-shaped pods. Has an undescribed toxin mildly poisonous to livestock; such toxicity is unknown in California but has been reported from other countries. The plants usually are not eaten by livestock.

Descurainia pinnata.   Tansymustard. Native to California; also widespread in the southern United States.
Descurainia sophia.   Flixweed. Native of Europe; established as a weed at scattered locations in California.

Both species of *Descurainia* are winter annual herbs, slightly pubescent; leaves pinnately divided, fern-like; yellow flowers less than $\frac{1}{8}$ in. long on slender flowering stalks in a raceme; fruits are slender pods of various lengths in the two species.

Toxic part: Tops of the plants when in flower. Young plants are apparently eaten by livestock without harm.

Toxin: Unknown.

Symptoms: A condition called "paralyzed tongue" develops with ingestion over a long period of time. Cattle show nervousness, blindness, muscular twitching, and inability to eat or drink; may be fatal.

Cases of paralyzed tongue have not been reported for California. These plants, however, form considerable forage in winter or early spring throughout California and should be avoided as livestock feed when in flower. A large corporation ranch based in California had this condition develop in cattle on their property on the rim of the Grand Canyon north of Seligman, Arizona. The species in this case was *Descurainia sophia*. The cattle had used this weed as their principal source of green feed and in addition were being given about 2 pounds of concentrate feed per head per day. The animals responded to treatment by intravenous therapy and by drenching with water introduced by a stomach tube to bypass the swollen tongue.

Stanleya elata, Panamint Prince's Plume; occurs in the Death Valley region and into adjacent Nevada and Arizona

Stanleya pinnata, Golden Prince's Plume; native in California from the Cuyama Valley, Santa Barbara County, through the Mojave Desert to Inyo County, but extending widely into the central part of the United States

Stanleya viridiflora, Green Prince's Plume; native from northeastern Nevada to Montana and Wyoming; once reported from Lassen County in California just west of the Nevada stateline but has not persisted

All three species are tall, erect perennials becoming woody at the base, unbranched or with a few branches toward the top of the plant, 6 to 9 ft tall; lower leaves pinnately lobed, upper leaves nearly entire; flowers bright yellow, small but very numerous, forming dense spikes to 2 ft tall. When in flower these are conspicuous plants in the desert.

Plants of *Stanleya* are indicators of selenium; California plants accumulate up to 46 ppm, which is above the toxic level. Under range conditions, however, animals have not been observed to eat these plants. Indians of the Death Valley region reportedly ate cooked young shoots of *Stanleya,* but they discarded the first water used for boiling the plants as it was capable of producing illness.

Buxaceae, Boxwood Family

Buxus is the only genus with toxic properties.

Buxus sempervirens. Box, Boxwood.

Europe, northern Africa, western Asia; commonly cultivated in California; has many cultivated varieties (cultivars). Evergreen shrubs or small trees; branchlets four-angled, lightly hairy; leaves opposite, simple, small, leathery; flowers inconspicuous.

Toxic part: Entire plant, especially the leaves and twigs.

Toxin: Buxine and other alkaloids, as well as a resin and an essential oil.

Symptoms: Nausea, vomiting, diarrhea, abdominal pains, convulsions, and respiratory failure.

This species can be fatal if large amounts are ingested. Small amounts are emetic and cathartic. Since livestock, including sheep, horses, cattle, and pigs, are affected by eating the foliage, clippings of these shrubs should never be placed within their reach. Poisoning in humans has not been reported.

Cactaceae, Cactus Family

The Cactus Family is not an important source of poisonous plants. Cactus plants are well known for the mechanical injuries that can be caused by the spines, many of which are easily broken off the stems and fruits. Even more injurious than the spines are the glochids, small barbed bristles occurring in the areoles of chollas and prickly pears, *Opuntia*. If they become embedded in the skin, glochids result in a very painful and troublesome mechanical injury. Often seen in cactus collections are Indian Fig, *Opuntia ficus-indicus;* Bunny Ears, *O. microdasys;* and Beavertail Cactus, *O. basilaris*. All have many glochids and few if any spines. Glochids blown about by wind currents in the enclosed space of a car have been known to cause injuries.

Lophophora williamsii.   Peyote, Mescal Button (pl. 7b).

Native to dry desert regions of southern Texas and northern Mexico; rarely cultivated. Plants solitary or in clumps with a carrot-like root and only the pincushion-like stem above-ground, stems to 3 in. tall and across, dome-shaped but depressed in the center, blue green, surface of the dome divided into seven to ten radiating sections or ribs, each with a central line of widely spaced tufts of hairs; flowers pink or rarely white, $\frac{1}{2}$ in. across, in the indented center of the stem. The carrot-like or turnip-shaped root is nearly as thick as the stem and tapers downward.

Toxic part: The stem is the peyote or mescal button.

Toxin: Mescaline and other alkaloids; up to thirty phenylethylamine and tetrahydroisoquinoline alkaloids of which mescaline, a form of phenylethylamine, is the most important.

Symptoms: When the peyote buttons are eaten or made into a tea, mescaline produces a paralyzing effect on the central nervous system, a disturbance of normal mental functions, and hallucinations. Nausea, vomiting, and chills are followed by heaviness in the arms and legs and then euphoria, color visions, flashing lights, and the appearance of strange figures; there is also an exaggerated perception of sounds. Different effects, often unpleasant, are produced by the several alkaloids.

Peyote, the sacred cactus of Mexico and one of the best known of the New World hallucinogens, has been used for more than 2,000 years by people in Mexico as a vision-inducing plant in ritualistic religious ceremonies. Known to the Aztecs as *Peyotl,* it was being used in religious rites at the time the Spaniards came to Mexico. With the establishment of the Catholic Church in Mexico unsuccessful attempts were made to suppress its use.

The use of Peyote spread northward from the native range of the plant. In 1880 the Plains Indians of the United States used it in their religious rites. This practice spread northward to the Great Lakes and west from there. In 1918 Indian groups in Oklahoma practicing what came to be known as the peyote religion were incorporated as the Native American Church. In the 1960s this denomination claimed about 200,000 members. In California possession of Peyote, *Lophophora williamsii,* is prohibited by state laws.

Lophophora diffusa, the only other species in this genus, has a yellowish green dome-shaped stem, irregularly segmented with a tuft of hairs on each segment; flowers in the depressed center of the stem are yellowish white to white in color. This cactus is native to a small area of central Mexico. The plant contains mostly the alkaloid pellotine and practically no mescaline.

A number of other species of cacti have been used in native medicine in Mexico and the Andean parts of South America. Species of *Ariocarpus, Coryphantha, Mammillaria,* and *Trichocereus* have been used for religious rites. Some of these "false peyotes" have very deleterious side effects and are even capable of causing permanent insanity.

Calycanthaceae, Calycanthus Family

Calycanthus occidentalis.   Western Spicebush.

Native of California in wet areas from Napa County north to Trinity and Shasta counties and south along the western slopes of the Sierra Nevada to Tulare County; sometimes cultivated. Shrubs 10 to 12 ft tall; leaves opposite, flowers fragrant, large to 2 in. across, sepals and petals reddish brown, similar, numerous, together with numerous stamens seated on top of a cup-shaped flower tube that becomes the woody fruit at maturity, containing many seeds.

Toxic part: Seeds.

Toxin: Calycanthine and related alkaloids.

Symptoms: May produce convulsions. The alkaloids calycanthine and calycanthidine exert an action similar to that of strychnine.

No human poisonings have been reported, but Western Spicebush has been listed in California as a livestock-poisoning plant of minor importance.

Cannabaceae, Hemp Family

Also called Cannabinaceae, this plant family is sometimes included in the Moraceae.

Cannabis sativa [regarded as a single species including such forms as have been named *C. indica*].   Hemp, Indian Hemp, Marijuana, Gallow Grass, Grass, Pot (fig. 13; pl. 8*a*).

There is evidence of human use of *Cannabis sativa* from China and Turkestan as far back as 3000 to 4000 B.C., which suggests that its original area may have been Central Asia. The plant has been cultivated for hundreds of years for its fiber, oil, and narcotic resin and is widely naturalized in many areas of the world. Annual, mostly 6 to 12 ft tall; leaves alternate, palmately divided, usually three to seven leaflets; male flowers in panicles to 15 in. long; female flowers in spikes about 1 in. long. The leaves are so characteristic in shape that the plants are readily recognized.

Treated stems yield strong, durable fibers. Fiber plants are usually grown in temperate areas, where they are planted very closely together. In warmer areas plants are grown widely

FIG. 13. *Cannabis sativa.* Marijuana. Upper left: Female flower subtended by a bract. Upper right: Achene and staminate inflorescence.

spaced for drug production. The resinous dried flower and fruiting branches produce the drugs marijuana or cannabis, hashish, charas, bhang, and ganja.

Toxic part: All parts of the plant, but highly variable in amounts. The largest amounts of resin are found in the female flowers, particularly when grown in a warm climate. Federal legislation prohibits the possession of any part of the plant.

Toxin: A resin, tetrahydrocannabinol, the principal compound causing psychoactivity; there are a number of additional substances with biological activity.

Symptoms: Humans vary greatly in response to this material. Elation, hallucinations, depression, confusion, and coma are all produced. The bitter plants are not eaten by livestock. Seeds (achenes) formerly were included in seed mixtures as feed for caged birds. Although the viability of the seed is killed by heat, in actual practice it could not be destroyed in all seeds when treated in bulk. Therefore the seed is no longer permitted in bird feed.

In July 1981, in San Bernardino County, nine persons were

affected by eating a zucchini cake in which marijuana had inadvertently been included. Symptoms varied with the individuals but included dizziness, confusion, drowsiness, blurred vision, dryness of the mouth, lapse of memory, anxiety, and tingling. The symptoms appeared 1 to 2 hours after eating the cake. Persons who repeatedly use marijuana may be less affected, and they may interpret the symptoms differently from those not using the drug. The active compounds of marijuana are three to five times more potent when inhaled than when eaten because they are insoluble in water. When fatty materials are used in baking, however, these active compounds are dissolved and their effects are more readily produced. The active substances of marijuana move quickly from the blood into the tissues of the body; complete elimination from the body may take as long as 30 days.

A case of commercial but illegal use of marijuana in baked goods was reported in San Francisco in early 1981 and again in December 1982. Cookies laced with marijuana were being made in a home bakery and sold at a handsome profit until police apprehended the baker.

Reportedly dogs have been poisoned from eating cookies containing marijuana or eating the dried green plant material. Depression of the central nervous system has been the most consistent sign of marijuana poisoning in dogs. In addition to being sleepy or unresponsive, dogs may show muscle tremors, vomiting, diarrhea, and lowered temperatures.

Humulus lupulus.   Hop, European Hop, Lupulo.

Cultivated and occasionally naturalized. Perennial, rapidly growing vigorous twiner; leaves opposite, three to seven lobed or entire; staminate flowers in conspicuous panicles; female flowers in cone-shaped heads. Hop pickers may develop vesicular dermatitis on the hands, face, and genitals. The rough hairs on the leaves may cause mechanical abrasion.

Capparaceae, Caper Family

Wislizenia refracta.   Jackassclover.

Native to southeastern California and the Owens Valley extending east to Texas and south into Mexico, introduced into the San Joaquin Valley and north to Sacramento; may be abundant on alkaline soils. Bad-smelling annual plants 1 to 6 ft tall, resembling sweetclover plants; compound leaves each with three leaflets about ½ in. long; small yellow flowers in dense racemes, petals about ⅛ in. long, six conspicuously exerted stamens; racemes elongate markedly in fruit, each fruit bearing two nutlets on recurved stalks. Livestock do not eat Jackass-clover plants, but a feeding trial based on suspicions of toxicity proved that these plants are lethally toxic at less than 1% of the animal's body weight. They are highly valued as honey plants.

Caprifoliaceae, Honeysuckle Family

Lonicera periclymenum.   Woodbine.
Lonicera tatarica.   Tatarian Honeysuckle.

Both native to the Old World; occasional ornamentals in California. Shrubs or partially climbing; leaves opposite; flowers in pairs with the corolla two-lipped, white or yellow white tinged pink or red; berries red. Berries contain an unknown toxin causing gastroenteric irritation and diarrhea; reported from Europe, not known as such in the United States. Children are likely to eat the berries.

Sambucus.   Elderberry, Elder.

Five native species of elderberries are distributed almost throughout California from coastal areas up to 10,000 ft elevation in the mountains, also in the desert mountains; the native species and others from Europe and eastern Asia are cultivated in California. These shrubs or small trees are readily recognized by the pithy stems; in cross section the pith comprises about 90% of the diameter; opposite, pinnately compound leaves; flowers small, white or yellow, in showy terminal clusters; fruits small red or blue berries.

Toxic part: Roots, bark, stems, and leaves.

Toxin: Uncharacterized toxic alkaloids, cyanogenic glycosides, and purgative substances.

Symptoms: Nausea and upset stomach.

Cases of poisoning from elderberries are rare but have been recorded for cattle, hogs, and humans. The roots and extensive rhizomes of elderberries have been postulated as the most toxic parts of the plant. Hogs should not be allowed to root in soil near elderberry bushes. Cattle rarely eat these shrubs. Ripe berries are not toxic, but they will produce nausea if too many are eaten. Cooked berries are not toxic. On removing the pith, the hollow stems have been used by children as blowguns or whistles and have resulted in cases of poisoning.

In August 1983, a group of twenty-five people in Monterey County were served a drink made from crushed berries, stems, and leaves of the native Southwestern Elderberry, *Sambucus mexicana* (fig. 14). Within 15 minutes after drinking the mixture several of the group experienced nausea and vomiting. The

FIG. 14. *Sambucus mexicana.* Southwestern Elderberry. Upper: Fruiting branch. Lower: Flowering branch. Upper right: Seed and berry. Lower right: Flower.

eight who became most ill also reported abdominal cramps, weakness, dizziness, and numbness. One who became stuporous was hospitalized. All recovered after treatment. Although ripe elderberries are edible, particularly if cooked, the leaves and stems, especially young leaves, can produce hydrocyanic acid and cause illness.

A composite sample of elderberry shrubs taken on March 15, 1974, from a small foothill native pasture in Orange County contained 4.4% nitrates as potassium nitrate, an extremely high percentage. Hence elderberries are nitrate accumulators.

Symphoricarpos albus.   Snowberry, Waxberry.

Native of eastern North America; cultivated to a limited extent in California. Deciduous shrub to 3 ft tall with small pink bell-shaped flowers $\frac{1}{4}$ in. long; the fruit is a waxy, snow-white berry about $\frac{1}{2}$ in. long, the conspicuous feature of the plant. The berry contains chelidonine, an isoquinoline alkaloid. A single case of poisoning has been recorded for the United States; the only symptom was vomiting. There are seven species of *Symphoricarpos* native to California. They have not been implicated in poisonings, but the berries remain on the plants and are not eaten by wildlife.

Caryophyllaceae, Pink Family

All poisonous plants in this family have essentially the same toxic properties. With present-day methods of cleaning grain and controlling weeds, the toxic seeds of members of this family no longer cause contamination of human or animal food. Animals have been poisoned by eating seeds or large quantities of the plants, but they usually refuse to eat food contaminated with such toxic material. The toxins are saponins, variously named as githagenin, saporubin, or saporubic acid. About $\frac{1}{10}$ ounce (3 to 4 grams) of seeds in wheat flour has produced gastrointestinal and circulatory disturbances in a human.

Agrostemma githago.   Corn Cockle, Purple Cockle.

Native of Europe; formerly a widespread weed in grain fields. Entire plant grayish hairy; winter annual or biennial; 1 to 3 ft tall; leaves linear to 4 in. long; one to several flowers on long stalks, petals separate, magenta purple or rarely white, the attenuate calyx lobes longer than the petals.

Saponaria officinalis.   Bouncing Bet, Soapwort, Fuller's Herb, Sweet Betty, Scourwort, Hedge Pink.

Native to Europe; a garden ornamental that has become naturalized along roadsides and ditchbanks. Stout, erect, clumped perennial; sessile opposite leaves; pink flowers in dense terminal clusters. In past times, as far back as ancient Greece, the stems and leaves were gathered, and after soaking or boiling in water, the liquid was used as a soap.

Celastraceae, Staff-Tree Family

Celastrus scandens.   Bittersweet, American Bittersweet, Climbing Bittersweet.

Native to eastern North America; occasionally ornamental in California; the stems with fruits are commonly used in dried floral arrangements. Vigorous climber; leaves alternate, mostly ovate; flowers inconspicuous; fruits yellow capsules that split open revealing the seeds covered by red arils. Toxin unknown but reputed to cause vomiting and diarrhea; no detailed human poisoning has been reported.

Euonymus europaea.   European Spindle Tree.

Europe, western Asia; occasionally cultivated as an ornamental in California. Deciduous shrub or small tree; leaves opposite, ovate; flowers yellow green in few flowered clusters; fruit a pink to red capsule; several seeds with an orange red covering.

Toxic part: Especially the fruits, but also the bark and leaves.

Toxin: Unknown.

Symptoms: Acts as a strong cathartic to humans and animals. Children, sheep, and goats are recorded as having suffered from its effects.

Chenopodiaceae, Goosefoot Family

Any member of this family must be regarded as capable of accumulating sufficient amounts of nitrates to be toxic to cattle. See Appendix C.

Atriplex canescens, Fourwing Saltbush; *A. nuttallii,* Nuttall's Saltbush; *A. rosea,* Red Orache, and perhaps other *Atriplex* species are secondary or facultative selenium absorbers.

Bassia hyssopifolia [*Echinopsilon hyssopifolium*]. Fivehook Bassia.

Native of eastern Europe and western Siberia; now widely spread as a weed of disturbed dry ground in most of California and other western states, often abundant in alkaline soils. Annual plants, softly hairy, to 5 ft tall with many lateral branches, the lowest prostrate on the surface of the soil; hairy calyx of five sepals each with a stiff hooked spine; abundant spines give the top of the plant a woolly look. Before spines are formed the plant may be mistaken for Summer Cypress, *Kochia scoparia,* also a member of the Chenopodiaceae.

Bassia hyssopifolia has been shown experimentally to contain acid potassium oxalate, a toxic salt that is not broken down to a nontoxic form in the rumen of animals. These plants are not usually eaten by livestock, but care should be taken that they are not forced to do so or that this weed is not included in hay. The oxalates remain toxic in dried plants. Experimental feedings have shown that Fivehook Bassia plants are about twice as toxic as those of Halogeton.

Beta vulgaris. Beet.

Native of Europe; widely cultivated as the garden beet, sugar beet, and chard. Biennial with a fleshy taproot. The leaves may accumulate oxalates in sufficient quantity to be toxic to livestock when eaten in large amounts. Cooked leaves are an edible potherb. In the winter of 1949, in Humboldt County, fourteen cattle died after eating frozen roots of the forage beet, mangel-wurzel.

Halogeton glomeratus. Halogeton (fig. 15; pl. 7c).

Native of Europe, western Asia, and northern Africa; introduced into western United States in the early 1930s and now covers millions of acres. It is established in California, in disturbed sandy soils, in southeastern Lassen County; new infestations appear sporadically in Inyo and San Bernardino

FIG. 15. *Halogeton glomeratus.* Halogeton. Upper left:
Flower with winged sepals, loosely enclosing the black
achene. Upper right: Flower with sepals without wings, tightly
enclosing the brown achene. Lower right: Mature fruits or
achenes with the spiral embryo of the seed visible.

counties and are eradicated. Annual plants; the first branches
at the base are prostrate on the ground and give rise to characteristic stiffly erect branches resembling a candelabrum to 2 ft
tall. Soft, very fleshy leaves to $\frac{3}{4}$ in. long with a hair-like bristle
at the tip of each leaf; flowers minute, in clusters among the
leaves; fruit is a single achene surrounded by five sepals that
become enlarged in some flowers to form showy white or red
papery wings, making the plant conspicuous.

Toxic part: Entire plant.

Toxin: Oxalates in large quantities.

Symptoms: Depression, coma, and death may result.

Sheep are poisoned by Halogeton more commonly than cattle, but this difference has been attributed to the methods in handling the animals. Cattle have a free run of the range and can select forage plants whereas sheep in large flocks are restricted where they can eat.

Signs of halogeton poisoning are incoordination of the posterior of the animal, increased salivation, and apprehensive behavior followed by belligerence. The incoordination may be followed by prostration and coma. Bloating and blueness of the skin and mucous membranes develop. Death can be very rapid.

The formation of calcium oxalate as insoluble crystals continues to remove calcium from the blood. Calcium crystals are deposited in the kidneys causing mechanical damage. Uremic poisoning from the lack of kidney function and the deficiency of calcium in the blood result in the poisonings seen. Affected animals have been saved with intravenous administration of calcium solutions. Transparent crystals of calcium oxalate found in the kidney tubules of animals that have died from Halogeton poisoning differentiate these animals from those dying from grass tetany, which is caused by deficiency of magnesium.

Halogeton plants are commonly distasteful to sheep and cattle, but these animals will eat them. Ranchers have learned to live with this abundant weed in the Great Basin of the western United States by being careful not to drive hungry animals into areas where the weed is growing. Losses of livestock are now rather rare.

Kochia scoparia.   Summer Cypress, Kochia, Burning Bush.

Native of Europe; abundant weed in northeastern California in alkaline waste ground where other vegetation has been drowned by winter rains; casual weed throughout the rest of the state. Summer annual plants to 6 ft tall, softly hairy throughout; flowers inconspicuous, sepals with a conspicuous midrib. This plant is considered an excellent source of protein. Toxicity has not been reported for California, but in other western states cattle have been poisoned from eating a diet of only *Kochia scoparia,* particularly with a water supply having a high sulfate content (2.3 grams per liter). Photosensitization with damaged tissues below the white areas of the skin, depression, continuous rolling of the eyeballs, loss of appetite, and prostration are all typical of this poisoning. Clinical findings reveal polioencephalomalacia and damage to the liver and kidneys. A substance causing the destruction of thiamine has been found.

Sarcobatus vermiculatus. Greasewood, Black Greasewood (pl. 7*d*).

Native in western United States, Canada, and Mexico on alkaline soils; found in California on the Colorado and Mojave deserts northward into Inyo, Lassen, and Modoc counties. Much branched, rounded, spiny shrub to 6 ft tall; leaves narrow, fleshy; numerous flowers small, of one sex, male flowers in slender terminal spikes, one to three female flowers in a leaf axil; fruit with a conspicuous rounded papery wing.

Toxic part: Entire plant.

Toxin: Soluble oxalates.

Symptoms: Depression, coma, and death.

Black Greasewood plants are considered good browse for animals in the late fall and winter. By providing additional nontoxic forage these plants can be used by the animals without harm. The plants are known to be toxic to sheep, especially when hungry animals are released into almost pure stands of the shrub. There have been no recent reports of deaths from Black Greasewood in California.

Convolvulaceae, Morning-glory Family

Cuscuta. **Dodder.**

California has some seventeen species, native or introduced. Annual leafless parasites, without chlorophyll; stems slender, twining, yellow or orange in color, fastened to the host plant by suckers; flowers small, usually clustered, corolla white, urn-shaped. May be placed in a segregate family, Cuscutaceae. Dodders are not toxic plants, but as serious parasites of alfalfa they cause the cut hay to be held in dense clumps that do not dry properly. This clumping results in moldy hay that can be distressful to livestock.

Ipomoea batatas. Sweet Potato, Yam.

Nativity uncertain; cultivated as a commercial crop. Perennial with large tuberous roots. Black rot of Sweet Potatoes is caused by infection by several species of fungi. Infected roots form ipomeamarone, a sesquiterpene, sufficiently bitter to be unpalatable to humans. Domestic animals have been poisoned from eating partially decayed Sweet Potatoes.

Ipomoea purga is cultivated in Mexico and the West Indies for its dried tuberous root, called jalap, which is used medicinally as a cathartic. Wild Jalap is *Podophyllum peltatum,* a member of the Barberry Family, Berberidaceae, native to eastern North America. *Mirabilis jalapa,* a member of the Four-o'clock Family, Nyctaginaceae, was so-named because formerly it was thought to be the source of jalap.

Ipomoea tricolor [*I. violacea* in some references].
Morning-glory (fig. 16).

Native of tropical America; used in southern Mexico as a hallucinogen in religious ceremonies; grown in California gardens as a summer annual ornamental vine. *Ipomoea tricolor* is variable in cultivation; several of its cultivars that differ in flower color are 'Heavenly Blue', 'Pearly Gates', 'Wedding Bells', and 'Flying Saucers'. Vigorous perennial vine without hairs; leaves heart-shaped to 10 in. across; flowers funnel-shaped to 4 in. across, blue, sometimes pale blue, pink, rose, or white.

Toxic part: Seeds.

Toxin: Peptide-type ergot alkaloids ergosine and ergosinine; only slightly different from the psychotomimetic lysergic acid diethylamide (LSD).

Symptoms: From ingestion of a small amount of seeds there is visual distortion, restlessness, relaxation, heightened awareness, and increased rapport with other persons. A medium amount results in hallucinations: visual, auditory, and spatial distortions lasting 1 to 4 hours. A higher dose gives the same results but is followed by a feeling of euphoria. Drowsiness, nausea, and a cold feeling in the extremities also develop.

Misuse of Morning-glory seeds to produce hallucinatory effects has caused very deleterious results, even death. Excessive use can result in complete dissociation from reality.

Ipomoea tricolor has a long history as a hallucinogen in

FIG. 16. *Ipomoea tricolor* [*I. violacea*]. Morning-glory. Upper left: Leaf. Lower left: Seeds. Lower right: Seed capsule.

southern Mexico where the seeds are used in the preparation of a drink. The seeds contain psychoactive ergot indole alkaloids, principally ergine and isoergine, which are amides of lysergic acid. These alkaloids have also been found in *Turbina corymbosa* [*Rivea corymbosa*], native of tropical America, and in *Argyreia nervosa,* native of India, called Woolly Morning-glory or Hawaiian Baby Wood Rose.

Coriariaceae, Coriaria Family

Coriaria ruscifolia, C. japonica, and *C. nepalensis.*

From tropical America, eastern Asia, or New Zealand; occasional ornamentals in California. Shrubs; leaves opposite or whorled, ovate, often three to seven veined from the base;

flowers small, in long racemes, sometimes hanging. At maturity petals enlarge, become fleshy and purple, and cover the single-seeded fruit. Entire plant, except the fleshy petals, contains toxic lactones called picrotoxins that act on the central nervous system causing convulsions, exhaustion, and coma; may be fatal.

Cornaceae, Dogwood Family

Aucuba japonica.   Japanese Laurel, Japanese Aucuba.

Native from the Himalayas to Japan; common ornamental in California. Evergreen dioecious shrub; leaves opposite, ovate, coarsely serrate; flowers small, unisexual; fruit showy, scarlet, fleshy, one-seeded. Fruit contains triterpene saponins and alkaloids. Although toxicity is not known in the United States, several children in Paris, France, experienced mild vomiting and fever after ingesting this plant.

Corynocarpaceae, Corynocarpus Family

Corynocarpus laevigatus.   New Zealand Laurel, Karaka.

From New Zealand; occasional ornamental in California. Small tree; ovate leaves leathery, glossy; flowers small, greenish yellow; fruit an oval drupe with a single large seed, orange-colored when mature. Fruits and seeds contain karakin, a glycoside, causing spasms and paralysis in humans and animals. Nectar is lethal to bees. The Maori people heat-treated the large starchy seeds to make them edible.

Crassulaceae, Stonecrop Family

Cotyledon orbiculata.   Pig's Ears.

Native of South Africa; cultivated in California as an ornamental for its showy red flowers. Stoutly branched, succulent plant, woody at the base, to 4 ft tall; leaves opposite, arching upward, oblong to obovate, to 4 in. long, sessile, grayish, mealy, sometimes margined with red; flowers red, to 1 in. long, drooping, on a stem standing above the leaves, five petals, reflexed above, forming a tube in their lower half.

Toxic part: Entire plant.

Toxin: Unknown.

Symptoms: In animals, the disease called cotyledonosis typically includes respiratory spasms and arching of the back and twisting of the head to rest on the flank. Bloating and paralysis may also occur.

There are two cases of sheep deaths recorded for California. The animals ingested prunings of *Cotyledon orbiculata*, which were thrown into the pastures. In both cases the plants had finished flowering. In October 1958, three sheep died from eating such plants in Novato, Marin County. Twenty-one sheep, two-thirds of the flock, died on a ranch west of Petaluma, Sonoma County, in September 1965 when fed plants of this species. The plants are regarded as most toxic from the time of flowering until the formation of the fruit.

In South Africa several species of *Cotyledon* and *Kalanchoë* have caused this poisoning in sheep and goats but rarely in horses, cattle, and poultry. The meat of animals killed by cotyledonosis remains toxic even after cooking and should not be eaten or fed to animals. Dogs are extremely susceptible to such secondary poisoning. For several weeks poisoned animals may show convulsions, muscular tremors, and paralysis. Animals recovering from cotyledonosis often have a peculiarly twisted neck for months.

Since many have not been tested, all species of *Cotyledon* and *Kalanchoë* should be regarded as toxic. Species in cultivation in California and known to be poisonous are *Cotyledon cacalioides* [*Tylecodon cacalioides*], *C. orbiculata, C. wallichii* [*Tylecodon wallichii*], *Kalanchoë lanceolata,* and *K. rotundifolia.*

Crassula arborescens.   Silver Jade Plant, Silver Dollar, Chinese Jade.
Crassula ovata [*C. argentea*].   Jade Tree, Jade Plant, Baby Jade, Cauliflower Ears, Dollar Plant.

Both species from South Africa; common ornamentals in California. Succulent shrubs; leaves opposite, rounded; flowers numerous, small, with spreading pink petals in showy clusters. Plants contain an unknown toxin that acts as an intestinal irritant causing colic and diarrhea.

Kalanchoë.   Palm-Beach-bells.

A number of species are listed as toxic in South Africa and in Australia where naturalized in pastures. Many species are

cultivated as succulents in California. The two species described below are propagated by small plants that form in
notches along the margins of the leaves or leaflets.

Kalanchoë daigremontiana. Devil's Backbone.

Native of Madagascar; common container-grown ornamental in California. Succulent herb; leaves lance-shaped, margins
toothed, mottled purple on the rounded lower surface; flowers tubular, purple. Entire plant contains an unknown toxin.
Household pets have been poisoned by eating small amounts.
Signs include convulsions, labored breathing, and paralysis;
shown experimentally to be highly toxic to chicks. Plants
should be kept away from small children.

Kalanchoë pinnata [*Bryophyllum calycinum, B. pinnatum*].
Air Plant, Life Plant, Miracle Leaf, Sprouting Leaf.

Nativity uncertain; cultivated and naturalized in tropical
areas; common container-grown ornamental in California.
Succulent herb; leaves pinnately compound of three or five leaflets with notched edges. Reported as poisonous in Australia.

Cucurbitaceae, Gourd Family

Bitter substances, cucurbitacins, occur in plants throughout the
family. Many edible members of the family have both bitter
(inedible) and nonbitter (edible) variants. The occurrence of
bitter and poisonous forms is not well understood, but commercially produced seeds do not yield fruits with bitter qualities. Zucchini Squash, *Cucurbita pepo* 'Zucchini', grew on a
compost pile as volunteer plants in San Rafael, Marin County.
They produced very bitter inedible squash that caused gastric
distress to members of the family who ate them.

Cucumis myriocarpus. Paddy Melon.

Native of South Africa; known as an introduction in California from three localities: the Santa Ynez Valley, Santa
Barbara County, and spot infestations near Lemoore, Kings
County, and near Kerman, Kern County. Plants are readily
eaten by cattle and deer in the Santa Ynez Valley and are not
toxic. Paddy Melon is well-known for its toxicity in its native

range in South Africa, however, where it has caused the death of humans and livestock.

Datiscaceae, Datisca Family

Datisca glomerata. Durango Root (fig. 17; pl. 8*b*).

Native in California from Shasta and Siskiyou counties at low to middle elevations of the Coast Ranges and the Sierra Nevada south into the mountains of Baja California; in wet areas or streambeds. Stout perennial herb with clusters of stems 3 to 6 ft tall from a cluster of dahlia-like roots; leaves alternate above but appearing opposite or even somewhat whorled below, leaf blades irregularly cut into unequal lobes; flowers in dense clusters in the axils of the leaves; fruit a capsule. Plants have been mistaken for Marijuana by the casual observer.

Toxic part: Entire plant in flower or fruit.

Toxin: Unknown.

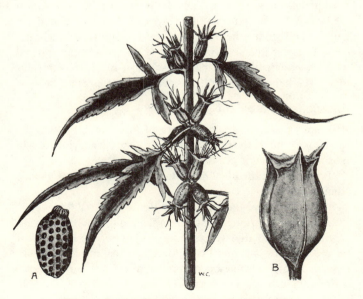

FIG. 17. *Datisca glomerata.* Durango Root. A. Seed. B. Fruit.

Symptoms: Depression, diarrhea, increased respiration rate, and death.

Durango Root was a well-known fish poison used by the California Indians. Experimentally it has been demonstrated to be highly toxic to cattle when the plants are in flower or in fruit. Despite the widespread distribution of the plant in California, there are very few records of livestock poisonings. The plants are obviously quite distasteful to livestock and are eaten only when other forage is lacking.

In July 1944, near Bridgeport, Mariposa County, five head of cattle died and others had severe diarrhea. The cattle had removed all the available forage and also had eaten Durango Root plants down to 2 ft in height. In Calaveras County a single cow, having eaten Durango Root plants almost to the base, retreated into the darkness of the barn and died there with the typical symptoms. In Tuolumne County at a middle elevation of the Sierra Nevada, a number of beef cattle died during a very dry year. There remained only the stumps of the Durango Root plants along the waterways. In September 1976, in a burn area on Mt. San Antonio, Los Angeles County, nine dairy heifers died from a herd of ninety. Although there was adequate forage available, Durango Root plants had been sprayed with a herbicide that gave them a salty taste relished by the animals. The heifers became listless and had watery diarrhea.

Ericaceae, Heath Family

Poisonous plants of several genera of this family contain the same toxic compounds that cause nearly identical symptoms in poisoned animals or persons.

Toxic part: Entire plant including the nectar and honey made from the nectar.

Toxin: Andromedotoxins, resins derived from diterpenes.

Symptoms: Burning sensation in the mouth shortly after ingestion, followed after several hours by excessive secretions of the mouth, eyes, and nose, vomiting, diarrhea, muscular weakness, lack of coordination, dimming of vision, slowed heart action, and possible coma and death. Most fatal poisonings in children appear to result from eating or chewing leaves.

Plants of *Kalmia,* laurels or American laurels, and *Rhododendron,* both rhododendrons and azaleas, have been found to be sources of toxic honey. The honey seems to produce sporadic, usually nonfatal poisoning. Bees normally do not work the flowers of these members of the Heath Family.

Kalmia.

Species in this genus are among the most poisonous members of the Heath Family, but because of their limited use as ornamentals or the native distribution in high elevations of the Sierra Nevada or wet habitats at lower elevations in the Coast Range, they seldom cause poisonings.

Kalmia angustifolia. Calfkill, Dwarf Laurel-wicky.

Native to eastern United States; occasionally cultivated in California. Evergreen shrub; leaves opposite or in threes, lanceolate; showy flowers in lateral clusters with a spreading bell-shaped corolla, $\frac{1}{4}$ to $\frac{1}{2}$ in. across, rose or red, anthers inserted and held in ten tiny pockets of the corolla; the terminal leafy shoot extends beyond the flower cluster.

Kalmia latifolia. Mountain Laurel, Mountain Ivy, Ivy Bush, Calico Bush.

Native of eastern North America; occasional ornamental in California. Evergreen shrub; leaves alternate or irregularly whorled, oval; flowers in terminal clusters, corolla white to pink about $\frac{3}{4}$ in. across; otherwise like *Kalmia angustifolia.*

Kalmia polifolia. Alpine Laurel, Bog Laurel, Pale Laurel, Bog Kalmia.

Native of northern California, widespread beyond; sometimes cultivated in California gardens. Somewhat straggling evergreen shrub; leaves lanceolate, sessile or nearly so, margins revolute, glaucous white beneath; flower clusters terminal, corolla rose to purple; otherwise like the two species described above.

Ledum glandulosum.   Western Labrador Tea, Trapper's Tea.

Native in northern California in wet meadows and bogs in the Sierra Nevada from 6,500 to 10,000 ft, lower elevations in northwestern California; widespread to the north and east. Erect, evergreen, rigid shrub, 2 to 5 ft tall; leaves alternate, ovate, resin-dotted beneath, fragrant when crushed; flowers yellowish white, numerous, in terminal clusters, five nearly separate petals, stamens longer than the petals. Only slightly toxic to livestock and rarely causes losses under range conditions.

Leucothoë axillaris [L. catesbaei] and *Leucothoë fontanesiana.*   Drooping Leucothoë.

Both natives of southeastern United States; occasional ornamentals in California. Evergreen shrubs; leaves alternate; flowers urn-shaped, in axillary racemes. Entire plants with andromedotoxins that can produce cardiovascular disturbances.

Leucothoë davisiae.   Sierra Laurel, Black Laurel.

Native to California from about 6,000 to 8,500 ft, northward to southern Oregon; in wet meadows and boggy places. Erect, evergreen shrub, 2 to 5 ft tall; leaves alternate, somewhat leathery, oval, 1 to 3 in. long, short stout petioles to $\frac{1}{4}$ in. long; flowers white and waxy, in terminal racemes, 2 to 4 in. long, corolla urn-shaped, $\frac{1}{4}$ to $\frac{1}{2}$ in. long, five lobes of the corolla longer than the tube, ten stamens, included in corolla tube.

Black Laurel is one of the most poisonous species of the Heath Family. As little as 1 to 2 ounces of leaves will kill a sheep. In August 1973, sheep were lost in an area of Placer County in the Sierra Nevada where Black Laurel and Labrador Tea were growing. The sheep were under the care of a newly hired individual, unfamiliar with the area, who allowed them to graze on the poisonous plants.

Menziesia ferruginea.   Rusty Leaf.

Restricted in California to redwood forests of Del Norte and Humboldt counties; extends north to Alaska, east to Idaho and Montana. Straggling deciduous shrub, 6 to 15 ft tall; leaves alternate, ovate, with scattered rusty hairs below; flowers yellow or greenish purple with short cylindrical corolla, in terminal clusters. Poisonous to cattle but does not cause problems in California because of its limited distribution.

Pernettya mucronata.
From Chile; occasional ornamental in California. Plants contain andromedotoxins; berries are reported as poisonous to humans, and cattle have been poisoned by the leaves.

Pieris floribunda.   Fetterbush, Mountain Pieris; native to southeastern United States; occasional ornamental in California.
Pieris japonica.   Lily-of-the-valley Shrub, Japanese Pieris, Andromeda; from Japan; common ornamental in California gardens, having several cultivars.
Both species of *Pieris* contain andromedotoxins and produce cardiovascular disturbances.

Rhododendron.
There are nearly 1,000 species of *Rhododendron;* azaleas, formerly considered to belong to their own genus, are now placed in *Rhododendron*. Presumably all species of this genus contain andromedotoxins, toxic diterpenoid resins. Four di-terpenes called rhodojaponins are found in *Rhododendron japonicum*.

Rhododendrons are largely evergreen shrubs or small trees; azaleas are mostly thin-leaved and deciduous; leaves alternate, short-petioled, entire; flowers showy and often brightly col-ored, corollas bell-shaped, tubular, or rotate with five or ten stamens. Many species and hybrids, too numerous to mention, are cultivated in California gardens. The following two species are native to California.

Rhododendron macrophyllum.   California Rose-bay, West Coast Rhododendron.
Native to the northwest coast of California, north to British Columbia; occasionally cultivated in California gardens but not so frequently as species from other parts of the world. Evergreen shrub to 8 ft or more; leaves leathery, oblong, 3 to 6

in. long; flowers rose purple or paler, corolla bell-shaped, to $1\frac{1}{2}$ in. long, ten stamens. Poisonous to sheep.

Rhododendron occidentale.   Western Azalea (pl. 8c).

Native to California in the North Coast Ranges, Sierra Nevada, and mountains of southern California, along streams and other wet places. Differs from *Rhododendron macrophyllum* in having thin-textured, deciduous leaves, hairy along the margins; flowers white or shades of pink, yellow, or salmon, corolla rotate, deeply five-lobed, with five stamens.

On land having many azalea plants, sheep losses are reported heavy when animals are pastured there for a considerable period of time. In August 1979, near Igo, Shasta County, 15 to 20 goats in a flock of 200 became ill, presumably from eating Western Azalea. The sick animals had vomiting and diarrhea. Hemorrhagic enteritis was observed in the two goats that died.

Euphorbiaceae, Spurge Family

Many plants in the family are poisonous, some extremely so, although in some species, as with the Poinsettia, reports are sometimes conflicting and exaggerated. Most species have milky juice that in some, including the cactus-like spiny shrubs, is caustic to the skin and particularly to tender areas. Phorbol, a diterpene alcohol, is the parent compound of toxins formed in plants of some species of *Euphorbia* and *Croton;* these toxins act as cocarcinogens causing tumors in skin as long as 1 year after a carcinogen has been applied.

Aleurites fordii.   Tung Oil Tree.

Native to western China and central Asia; occasionally cultivated in warm parts of California as an ornamental. The seeds are the source of tung oil used in paints and quick-drying varnishes. Deciduous small tree rarely to 40 ft tall, with milky juice; ovate leaves sometimes three-lobed, to 5 in. long; flowers to 1 in. long, white with red or orange veins, usually before the leaves; fruit rounded, 2 to 3 in. across, brown, rough-coated; seeds with white flesh. All parts of the tree, especially the seeds, are toxic. The toxin is a saponin (a toxalbumin or phytotoxin) that is derived from the irritant phorbol. After eating the nut-like seeds, humans suffer intense abdominal pain, nausea, vomiting, severe diarrhea, weakness, and finally exhaustion. In severe cases death may occur. In February 1957, three children were hospitalized in Riverside County after ingesting seeds of this plant.

Eremocarpus setigerus. Turkey Mullein, Dove Weed.

Widespread California native from coast to desert, up to 2,500 ft elevation, often abundant in dry areas. Low grayish annual forming prostrate mats as much as 8 to 10 in. across; leaves and stems covered with star-shaped hairs and even longer bristly unbranched hairs making the plant harsh to touch; oval thickish leaves, three-veined from the base, to 3 in. long; flowers small and inconspicuous; capsule with a single grayish mottled seed about ⅛ in. long.

Seeds of this species are eaten by wild birds, quail, doves, and turkeys. California Indians used these plants to stupefy fish. Following ingestion of the hairy leaves and stems, illness and occasional death of cattle, sheep, and hogs appear to be caused by the solid indigestible fibrous balls that accumulate in their digestive tracts.

An October 1928 report indicated that after ingestion of a particular lot of alfalfa hay forty dairy cows developed profuse diarrhea, a drop of 45% in milk production in 12 hours, and one abortion. The alfalfa hay, which had been grown in the Antelope Valley, Los Angeles County, was of excellent quality except for the presence of varying amounts of Turkey Mullein in the different bales. In experimental feedings a goat and a cow refused to eat this hay; another cow ate a small amount and refused to eat more after showing distress symptoms. In September 1979, near Ridgecrest, San Bernardino County, three 6-month-old geese died from eating these plants when turned into a yard for the first time. The Turkey Mullein had been left in the yard because the owners liked the sweet odor of the plant.

Euphorbia. **Spurge.**

There are more than 1,000 species in this genus, which is worldwide in distribution. A number of species are native in California, others are introduced weeds, and others are cultivated ornamentals, such as *Euphorbia pulcherrima*, Poinsettia. Several euphorbias are spiny and cactus-like but may be distinguished from cactus plants by the presence of milky juice. Herbs or shrubs of diverse habits, some spiny and cactus-like, with milky juice; leaves variable; flowers small, of one sex, grouped together into cyathia, cup-like structures with marginal glands and sometimes petal-like fringes on the glands; fruit a capsule with three seeds.

The milky juice of some species is strongly irritant to the skin and eyes, and it may cause temporary loss of sight. If ingested, the juice may produce nausea, vomiting, and gastrointestinal irritation.

Euphorbia cotinifolia. Red Spurge.

Native of Mexico to northern South America; occasionally cultivated in California. Shrub or small tree; stems reddish; leaves usually in threes, ovate to rounded, 2 to 5 in. long, purplish green to maroon; flower clusters whitish; capsules white. The abundant milky juice is exceptionally caustic and causes blistering and inflammation of the skin. It has been used by Indians of Central America as an arrow poison and for poisoning fish.

Euphorbia cyathophora [has been confused with
E. heterophylla]. Painted Leaf, Fire-on-the-mountain, Mexican Fire Plant, Fiddler's Spurge, Annual Poinsettia, Dwarf Poinsettia.

Native of eastern United States and Mexico; a self-seeding annual grown in California gardens. Plants resemble a poinsettia; floral bracts and upper leaves, twice as long as wide, sinuately lobed or fiddle-shaped, basal spot of white, pink, or red on the bracts. Besides causing dermatitis, plants of this species have been suspected of poisoning children and livestock.

Euphorbia cyparissias. Cypress Spurge.

Native of Europe; cultivated in California particularly in the colder areas, sometimes as a border plant. Glabrous creeping perennial with shoots about 1 ft tall; numerous linear leaves crowded on upper stems; cyathia in umbel-like clusters. Where naturalized in the eastern United States and Canada, plants have been poisonous to cattle when large amounts have contaminated hay; no record of poisoning from California.

Euphorbia esula.   Leafy Spurge, Wolf's Milk.

Native of Europe; a noxious weed in California where it is almost eradicated, an abundant weed in other parts of the United States and Canada. Deep-rooted creeping perennial; single stems 1 to 4 ft tall from buds on lateral roots; narrow leaves 1 to 4 in. long; umbel-like inflorescence terminates each stem and is subtended by a whorl of leaves, each branch forked once or more; involucral bracts heart-shaped and clasping the stem. The entire inflorescence including the involucral bracts turns a bright yellow and becomes showy.

Fresh Leafy Spurge is not eaten by animals, but the dried plants in hay are poisonous to livestock. Sheep, more resistant to these toxic plants, have been used to control them in pastures. On occasion, however, sheep have been poisoned by Leafy Spurge.

Euphorbia heterophylla.   Catalina.

From the southern tip of Texas southward, widely distributed in tropical America; California has a single infestation of this weed in a cultivated field near Yuba City, Sutter County. A summer annual; upper leaves and floral bracts are narrowly lanceolate, sharply pointed, four to five times as long as wide; floral bracts have a small white spot at the base. Cases of poisonings attributed to this species are referable to *Euphorbia cyathophora,* Mexican Fire Plant, because of an almost worldwide misuse of the name *Euphorbia heterophylla.*

Euphorbia lactea.   False Cactus, Candelabra Cactus, Candelabra Spurge, Hatrack Cactus, Dragon Bones, Mottled Spurge.

Native of the East Indies; cultivated in collections of succulents in California. Tall cactus-like succulent with milky juice; branches thick, three or four-angled, angles undulate with paired spines; rarely with clustered tiny yellow flowers.

Milky juice is highly irritant and can produce severe eye inflammation, temporary blindness, and blistered skin.

Euphorbia lathyris.   Caper Spurge, Mole Plant, Gopher
Plant.

Native of Europe; cultivated in California and somewhat
naturalized, especially along the coast. Winter annual or biennial to 3 ft tall, gray-glaucous in color; sessile leaves mostly
opposite and appearing in four vertical rows, lanceolate to 5
in. long; cyathia in umbels; rounded spongy-inflated capsules
about $\frac{1}{2}$ in. across.

The common name, Caper Spurge, alludes to the resemblance of the spongy capsules to capers, which are young
flower buds from quite a different plant, *Capparis*. Caper
Spurge fruits were mistakenly pickled by several women and
when ingested caused severe illness. Children have been poisoned from eating the capsules and seeds.

The milky juice is an irritant to humans. When ingested the
plants are poisonous to cattle. After goats eat this plant their
milk is said to have the same toxic properties as the fresh plant.
The name Gopher Plant or Mole Plant refers to the practice of
growing these poisonous plants to repel gophers or moles in
the garden. Experimentally this practice has not succeeded.

Euphorbia marginata.   Snow-on-the-mountain, Ghost Weed,
Mountain Snow (fig. 18).

Native of Mexico and northward into central United States;
grown in California as a summer ornamental. Erect annual to 3
ft or more, single stem from the base and branched repeatedly
above; ovate leaves 1 to 3 in. long, nearly sessile; lower leaf
blades narrowly margined with white, upper leaves smaller,
crowded, their broader white margins giving them an almost
petal-like appearance; cyathia with conspicuous white margins, in the axils of the uppermost leaves.

Caustic milky juice causes swelling and blistering of the
skin much like that caused by Pacific Poison Oak. At one time
a few cattlemen in Texas used the juice of this plant to brand
cattle because they considered the scars to heal more satisfac-
torily than those made by a hot iron. One human fatality has

FIG. 18. *Euphorbia marginata.* Snow-on-the-mountain.
Upper left: Mature capsule and seed. Upper right: Cyathium
containing one female flower and many male flowers. Center
right: White-margined bract below the cyathia. Lower right:
Enlarged cyathium with petal-like appendages on the glands.

been recorded following ingestion of a decoction of this plant
to induce abortion. In areas where it is abundant, this plant has
caused honey to have a disagreeable taste and sufficient toxicity
to cause vomiting and diarrhea.

Euphorbia milii.   Crown-of-thorns, Christ Plant, Christ
Thorn (pl. 8*d*).

Native of Madagascar; popularly cultivated in California as
a container plant and outdoors in protected areas. Sprawling
shrub, irregularly branched, 1 to 4 ft tall; stems with numerous
sharp spines to $\frac{1}{2}$ in. long; leaves mostly on young growth, en-

tire, obovate, to $2\frac{1}{2}$ in. long; bright, showy red bracts surround
two or more cyathia in a cluster and have often been mistaken
for a single flower. Milky juice is caustic and irritating to the
skin and eyes.

Euphorbia peplus.   Petty Spurge.

Native of Europe; a common garden weed in California,
often in moist shady places and in greenhouses. Erect annual to
10 in. tall; lower leaves broadly ovate $\frac{1}{2}$ to 1 in. long, upper
leaves not as broad; cyathia with large horned glands. Milky
juice is irritating to skin and eyes; if swallowed it causes vomiting, severe gastroenteritis, and diarrhea.

Euphorbia pulcherrima.   Poinsettia, Christmas Flower,
Painted Leaf, Mexican Flame Leaf, Easter Flower, Christmas
Star, Lobster Plant.

Native of Mexico and Central America; cultivated commercially on a large scale in California and used as a container
plant or in gardens. Shrub, multistemmed from the base, to 15
ft tall; leaves elliptic or ovate, entire or more or less sinuately
lobed, to 6 in. or as much as 10 in. long; cyathia yellow and red
in openly branched clusters at the ends of branches, subtended
by bright red, or sometimes white or pink, very showy, leaf-
like bracts that are the conspicuous parts of the plant.

The copious milky juice appears less irritating than that of
other members of the genus *Euphorbia*. But perhaps because
of the report of a fatality in Hawaii in 1919 and because of the
plant's popularity during the Christmas season, considerable
controversy has centered around the toxicity of this plant.

In 1973, of 228 cases of poinsettia ingestion registered with
the National Clearinghouse for Poison Control Centers, Be-
thesda, Maryland, the "most severe" symptoms reported in
only fourteen cases were a feeling of being unwell and vomiting. Chemical studies have shown no irritant diterpenes in the
plant. Handling the plant is said to produce local irritation.
Therefore the same care should be exercised in handling it as
with other plants having milky juice, but it is not known to be
fatal.

Euphorbia tirucalli.   Milkbush, Malabar Tree, Indian Tree Spurge, Finger Tree, Pencil Tree, Rubber Euphorbia.

Native of tropical and southern Africa; occasionally cultivated in California in collections of succulents and outside in protected places. Succulent shrub with milky juice and numerous slender cylindrical pencil-like stems, variously branched, spineless, 6 to 8 ft or eventually taller; very small leaves fall early; flowers rarely seen.

Copious milky juice is highly irritating to the skin and eyes, often causing the loss of sight for several days. In East Africa it has been used as a fish poison and a medicinal plant. Ingestion results in severe irritation to mouth, throat, and stomach. Injudicious medicinal use in East Africa has caused fatalities.

Jatropha curcas.   Barbados Nut, Physic Nut, Poison Nut, Purging Nut.

Native of tropical America; rare ornamental in California. Shrub or small tree; leaves broadly oval and lobed; flowers yellow; fruit a globular or oval capsule to 1 in. long, seeds $\frac{3}{4}$ in. long, oblong, brownish black, sweetish to taste. Seeds and other plant parts contain curcin (a toxalbumin), a dermatitis-producing resin, a purgative oil, and a glycoside. Symptoms of intense abdominal pain, nausea, vomiting, and diarrhea may appear within an hour after ingestion. In severe poisonings, muscular spasms and collapse may follow. May be fatal. Children are especially vulnerable because of the pleasant-tasting seeds. As few as three seeds may cause poisoning. All species of *Jatropha* should be regarded as potentially toxic.

Mercurialis annua.   Annual Mercury.

Native of Europe; established in coastal San Francisco and San Mateo counties since 1919; has a record of toxicity to livestock in Europe and Israel but not in California. Plants are dioecious, herbaceous, branched from the base, with opposite leaves having toothed margins. In grazed areas animals will not eat the plants, but they have been poisoned by fresh or dried plants mixed with edible fodder. Blood-stained urine, discolored body fat, and dark brown or blackened kidneys are found in poisoned animals.

Pedilanthus tithymaloides.   Slipper Flower, Redbird Cactus, Redbird Flower, Devil's Backbone, Japanese Poinsettia, Slipper Plant, Ribbon Cactus.

Native of tropical America; occasional ornamental, may be grown as a houseplant. Succulent shrub with milky juice; stems zigzag; leaves ovate, falling early; small flowers grouped into terminal red irregularly shaped slipper-like involucres.

Toxic part: Copious milky juice.

Toxin: Euphorbol and other terpenes.

Symptoms: Dermatitis with occasional blistering and conjunctivitis. Mild gastritis if eaten, severe poisoning with larger amounts.

Ricinus communis.   Castor Bean, Castor Oil Plant, Wonder Plant, Palma Christi (pl. 9*b*, *d*).

Probably native in tropical Africa; now widely distributed after having been planted as an ornamental for its large leaves and tropical effect and for the commercial oil obtained from its seeds; introduced into California and now frequently naturalized in warmer coastal areas. Tree-like shrub to 15 ft or more; soft herbaceous stems that become woody with age, or may be grown as an annual; peltate leaves glossy reddish brown to green, leaf blades palmately five to eleven-lobed, 6 to 12 in. across, or reported to be as much as 3 ft across; small inconspicuous flowers in dense panicles; fruit with a smooth surface or covered by soft spines, separating into three parts when mature; three seeds in each fruit, oval, mottled, $\frac{1}{2}$ to 1 in. long. Sometimes the entire plant is a purplish brown color.

Toxic part: Seeds and, to a lesser extent, the leaves.

Toxin: Ricin, a toxalbumin (plant lectin); one of the most toxic compounds known. It inhibits protein synthesis in the intestinal wall and agglutinates red blood cells.

Symptoms: On ingestion of the seeds, after several hours, burning of the mouth, nausea, vomiting, diarrhea, great thirst, prostration, sometimes convulsions, and circulatory collapse. The plant may cause contact dermatitis and pollen allergies, as well.

The number of seeds that will produce serious illness or death varies. The seed must be broken and chewed to release the ricin. As few as two or three seeds have been fatal to children; two to four may cause severe illness or death in an adult; in one instance eight seeds were fatal to an adult. The severity of the poisoning probably depends on the amount of toxic material ingested and absorbed. A case was reported in 1982 of a 24-year-old male who partly chewed and swallowed twelve seeds in a suicide attempt. After 1 hour he became nauseated, vomited, and developed diarrhea. Seed fragments were found in the vomitus and stool. The patient was admitted to the hospital for treatment and after 6 days was discharged. Animal poisonings from Castor Bean seeds occur; horses, sheep, cattle, and poultry have been affected.

Deaths and allergic reactions attributed to Castor Bean were examined thoroughly by the Division of Chemistry, California Department of Food and Agriculture, particularly for the period 1951–1956 when Castor Bean was grown extensively as a crop plant. In this time twenty-three serious cases of allergy from Castor Bean were documented, frequently from using the pressed cake of Castor Bean as a lawn fertilizer. Deaths of three children from eating Castor Bean seeds were documented, one in 1958 and two in 1961. In 1958, five men, thinking them to be edible, ate Castor Bean seeds that were on the floor of a van they were loading. By evening all the men had developed bad cases of vomiting and diarrhea that persisted until the next morning. All five men recovered without any apparent ill effects.

Castor oil obtained from the seeds is a strong cathartic but does not contain the toxic ricin. The oil is used commercially as a lubricant, in the manufacture of soaps, and in medicine.

Castor Bean plants can reduce populations of plant-parasitic nematodes in the soil. The root-knot nematode *Meloidogyne incognita* and the root-lesion nematode *Pratylenchus alleni* were the species investigated. The compound responsible for this damage to nematodes is not known.

Fabaceae, Pea Family

Also called Leguminosae, the Pea Family has almost 100 genera listed as containing various toxins. Only those known to cause distress to humans or to animals are mentioned here.

Proteins called lectins, found in the seeds of some species of legumes, can agglutinate red blood cells of animals. Certain lectins known as mitogens change lymphocytes, the smallest white blood cells, from a resting condition to an active metabolic state. These changes result in cell division and may affect the total immune responses in the blood. Clinically, however, no practical use has been made of extracts of these seed proteins. Mitogens have been found in:

Abrus precatorius, Rosary Pea
Canavalia ensiformis, Jack Bean
Lens culinaris, Lentil
Phaseolus vulgaris, Kidney Bean
Pisum sativum, Garden Pea
Vicia faba, Broad Bean
Wisteria floribunda, Japanese Wisteria

Abrus precatorius. Rosary Pea, Jequirity Pea, Precor Bean, Lucky Bean, Prayer Bean, Love Bean, Crab's Eye, Gidee-gidee, Indian Bead, Coral-bead Plant, Red-bean Vine, Indian Licorice, Seminole Bead (pl. 9*a*).

Native of the warmer parts of India from where it has spread to most tropical and subtropical regions; naturalized in Florida and Hawaii. It does not grow in California, but the colorful seeds have been brought into the state in seed-bead necklaces and other kinds of beadwork. Slender twining vine, more or less woody at the base; leaves alternate, pinnately compound, many small leaflets; flowers lavender or white, inconspicuous; legume oblong, 1 to 1½ in. long, in clusters, each containing four to eight shining hard seeds, half-scarlet and half-black in color; the point of attachment to the pod is in the black half of the seed.

Toxic part: Seeds.

Toxin: Abrin, a lectin that is a phytohemagglutinin, a seed protein that agglutinates red blood cells; very poisonous.

Symptoms: Severe gastroenteritis may develop within a few hours to several days after ingestion; abdominal pain, nausea, vomiting, and diarrhea, as well as low blood pressure, dilation of the pupils, liver damage, circulatory collapse, coma, and convulsions. Can be fatal.

One well-chewed seed can cause the death of a child. Because of the hard seed coat, a seed might be harmless if swallowed whole. Seeds drilled to make strings of beads are therefore a distinct hazard.

Acacia greggii.   Catclaw, Gregg's Catclaw, Catclaw Acacia, Devil's Catclaw.

Native in creosote-bush scrub and pinyon-juniper woodland; in southeastern California on the Colorado and southern Mojave deserts and eastward. Spiny shrub or small tree; leaves twice pinnately compound with small leaflets; numerous tiny pale yellow flowers in dense spikes; legumes flat, constricted between seeds.

Toxic part: Twigs and leaves.

Toxin: A cyanogenic glycoside.

Symptoms: Typical of hydrocyanic acid poisoning; rapid respiratory rate, convulsions, paralysis, coma, and death.

There are no records of poisoning from this species in California. In Arizona, cattle have been poisoned in the fall at the time of frosts. When there is no other forage available, cattle will eat heavily on *Acacia* foliage. The dried leaves may retain high amounts of the cyanogenic glycoside for some time.

Astragalus.   **Locoweed, Rattleweed, Crazyweed** (fig. 19).

A large genus in North America and the Old World with about 370 species in North America, mostly in the West. The name Rattleweed comes from a distinctive feature of many species: conspicuously inflated indehiscent pods in which the loose seeds rattle when the mature pods are shaken.

Poisonous species are found throughout the range of the genus; the largest concentrations are in western North America where livestock losses occur each year in the important grazing areas. Many species of *Astragalus* are not toxic and are good forage plants, but variations in toxicity in plants of the same species have been found. Unless definitely known to be edible, all species of *Astragalus* should be regarded as potentially dangerous to livestock.

The poisonous species are divided into three groups: (1) those that accumulate selenium, (2) the locoweeds that

FIG. 19. Upper left: *Astragalus nuttallii*. Gray Loco. Upper right: *Astragalus curvicarpus*. Sickle Milkvetch. Center left: *Astragalus crotalariae*. Rattle-box Milkvetch. Center right: *Astragalus hornii*. Sheep Loco. Lower left: *Astragalus filipes*. Threadleaf Locoweed.

form an unknown toxin, locoine, and (3) those that synthesize aliphatic nitro compounds. Species of all three groups are found in California. Identification of the many species of *Astragalus* is specialized and requires a good deal of study.

1. Selenium Accumulators

Toxic part: Aboveground parts of the plants.

Toxin: Selenium accumulated in the plants.

Symptoms: As discussed under selenium poisoning (see pp. 307–310), acute poisoning is rare but causes depression, rapid breathing, coma, and death. Chronic poisoning is characterized by wandering ("blind staggers"), excitability, eventual depression, faulty respiration, and death or by emaciation and the loss of hooves ("alkali disease") followed by death from thirst or starvation.

Selenium poisoning of livestock has not been reported from California. Plants vary exceedingly in the amounts of selenium they accumulate from the soil, often only in very localized spots. Apparently these high-selenium plants have never occurred in California where livestock have been forced to eat them. The following California species are known to accumulate selenium.

Astragalus crotalariae.   Rattle-box Milkvetch (fig. 19).

Ill-smelling plants with conspicuous bright purple flowers 1 in. long; found in the Coachella Valley, Riverside County, south along the eastern edge of San Diego County and into the Imperial Valley with a few localities in adjacent Arizona and Baja California. Following two or three wet winters in succession, Rattle-box Milkvetch may become extremely abundant on the Colorado Desert, but these areas are not used as rangeland. The fetid odor of the plants is typical of selenium accumulators.

Astragalus preussii var. *preussii.*   Preuss' Milkvetch.

Fetid-smelling plants with pinkish purple flowers almost 1 in. long, fading to bluish purple when dried; native in extreme southern Nevada and extending into adjacent Arizona and Utah; occurs in California only in the northeastern corner of

San Bernardino County and in a variant form (var. *laxiflorus*) near Lancaster, Los Angeles County.

2. Locoweeds

Toxic part: Aboveground parts of the plants.

Toxin: Unknown toxic compound or compounds called locoine.

Symptoms: In livestock, the nervous system is affected; depression, lowering of the head, tremors, incoordination, poor vision, excitability, inability to eat or drink, paralysis, and death. Birth defects and abortions develop in pregnant animals.

Cattle, sheep, and horses are affected. Under usual grazing conditions these toxic species are left untouched by the animals even though the forage plants are used heavily. Animals can, however, become addicted to eating these plants. Despite the name "loco," most animals become depressed.

The following species of locoweeds in the genus *Astragalus* are found in California. Plants of the genus *Oxytropis* also cause locoweed poisoning.

Astragalus agnicidus. Lambkill Milkvetch.

Stout plants to 3 ft tall with numerous, up to forty, small pendulous white flowers to $\frac{1}{2}$ in. long and papery inflated incurved pods to $\frac{3}{4}$ in. long. A poorly understood species known from a single ranch near Miranda, southern Humboldt County, where it became abundant in disturbed cleared areas but has not been collected since 1954. At one time this species caused heavy losses of lambs. The toxin has not been identified; the plants tested had only a trace of nitro compounds.

Astragalus asymmetricus [*A. leucophyllus*]. Horse Loco (fig. 20; pl. 9c).

Conspicuous plants to 3 ft tall with inflated pods each extending from the dried calyx on a narrow stipe to 2 in. long; native in the inner Coast Ranges from western Kern County northward to Solano County and on the floor of the Salinas Valley and into the Great Valley south to Madera County. The Horse Loco is quite common or even abundant locally and is a threat to livestock, especially horses.

FIG. 20. *Astragalus asymmetricus.* Horse Loco. Each pod exerted from the calyx on a long stipe. A. Flower. B. Seeds.

Astragalus douglasii. Douglas' Loco.

The variant form *douglasii* occurs in the lower Sacramento Valley and the inner Coast Ranges from Yolo County south to Kern County; var. *parishii* is found at scattered locations through the coastal mountains of southern California from the San Bernardino Mountains south into Baja California. Sprawling stems forming a mat to 4 ft across; flowers ten to thirty-five at the ends of somewhat erect stems, white, to $\frac{1}{2}$ in. long and forming spreading inflated pods to 2 in. long and 1 in. wide.

Douglas' Loco is the most common locoweed of the inner South Coast Ranges and can be locally abundant. It is dangerous to livestock, particularly horses. Two separate instances of poisonings of horses have been ascribed to var. *parishii* although in each instance the plants were removed from the rangeland before proper identification; one case on Mt. San Jacinto, Riverside County, the other in the mountains east of San Diego. The horses showed excitability when disturbed, a peculiar high-stepping gait, and impaired vision. Bees have been poisoned from this species in San Benito County.

Astragalus hornii.   Sheep Loco (fig. 19).

Annual plants with somewhat procumbent branching stems to 4 ft long; small white flowers less than $\frac{1}{2}$ in. long in a dense head at the stem tip, inflated pods $\frac{3}{4}$ in. long in a similar dense head; growing at the margins of lakes or pools where winter water remains; native of the floor of the San Joaquin Valley in Kings, Kern, and Tulare counties, probably introduced at one location on the western edge of the Mojave Desert in Los Angeles County and on the western shore of Pyramid Lake in Washoe County, northwestern Nevada. Two other populations are recorded for southern California, but these have been extinct for many years.

At one time Sheep Loco was considered the most dangerous species of *Astragalus* in causing livestock losses in California. Its occurrence is now quite rare in the San Joaquin Valley. Leveling of the land for agriculture has removed many of the habitats, and purposeful eradication from grazing lands has destroyed other populations.

Astragalus lentiginosus.   Freckled Milkvetch, Spotted Loco.

This widespread species is a complex of a large number of varieties; sixteen or more varieties are recognized for California plants with many intergrading forms. Plants are mostly perennials; flowers of different varieties from $\frac{1}{4}$ in. to almost 1 in. long, white, yellow, or purple in color; pods strongly inflated and usually divided into two chambers by a septum down most of the length of the pod; the silvery hairy var. *borreganus* of the southeastern deserts has erect uninflated pods scarcely divided into two sections.

The varieties of Freckled Milkvetch native in California are found from the southern part of the state north through the South Coast Ranges and the San Joaquin Valley and along the eastern slopes and crests of the Sierra Nevada, the White and Inyo mountains, and north into Siskiyou and Modoc counties and eastward into other western states.

This species has been implicated in livestock poisonings, but apparently some of the accounts of toxicity are actually poisonings from *Astragalus douglasii*.

Signs typical of poisoning from Freckled Milkvetch or Spotted Loco include depression, disorientation, excitement when stressed, a dry rough coat, and a peculiar pacing gait with the head held high. The gradual decrease in the number of white blood cells may indicate effects on immune responses. Animals having access to this species appear to have more cases of pneumonia, foot rot, and pinkeye.

Sheep experimentally fed this species during gestation manifested a decreased blood supply to the uterus and developing embryos. Small weak lambs and abortions were attributed to this poisoning. A few newborn lambs were deformed, as well, but the causes were difficult to determine.

Astragalus nuttallii [*A. menziesii*].   Gray Loco (fig. 19).

Large leafy clumps of plants to 3 ft tall or prostrate when subjected to the prevailing coastal winds; whitish flowers tipped with light purple, $\frac{1}{2}$ in. long; conspicuous rounded bladdery pods 1 to 2 in. long. Native to the meadows and sandy fields along the coast from western Santa Barbara County to the Monterey Peninsula and from San Francisco to Mendocino County.

Gray Loco formerly was a serious toxic species. Before 1875, along coastal California thousands of horses, cattle, and sheep were poisoned by this *Astragalus*. Most of the habitats for this species have been destroyed.

3.   Aliphatic Nitro Compounds

Toxic part: Tops of plants, particularly the leaves.

Toxin: Nitro-bearing compounds: 3-nitro-1-propanol or 3-nitropropionic acid. The name miserotoxin has been given to the compound that breaks down to form 3-nitro-1-propanol; it was first found in *Astragalus miser*. The 3-nitro-1-propanol is readily absorbed in the stomach of ruminants. The 3-

nitropropionic acid is not absorbed so easily but is produced in greater quantities in the plants.

Symptoms: In cattle, acute poisoning results in death 4 to 20 hours after ingestion. Labored breathing from emphysema of the lungs, cyanosis of the mouth and nose, weakness and swaying in the hindquarters, and lowered extended head are all characteristic of this acute toxicity. Chronic poisoning may occur after several days of eating lesser amounts of nitro-containing plants or larger amounts of those with lower concentrations. Degeneration of the spinal cord results in loss of motor control and weakness in the hindquarters, "cracker heels" and goose-stepping, knocking over of the fetlocks, and the hind feet knocking together when walking. Moreover, from the emphysema there is labored breathing that has a wheezing or roaring quality. Impaired vision and running into objects are also characteristic of chronic poisoning. Animals poisoned by these nitro compounds may collapse and die when forced to move. Animals so affected do not usually recover.

The following California species and varieties of *Astragalus* have been found to contain significant amounts of aliphatic nitro compounds and must be considered distinct hazards to livestock. Because the greatest toxicity is in the leaves, some of the risk of livestock loss may be avoided by using rangelands after these species have flowered; then most of the leaves are dried and the nitro content drops rapidly.

Astragalus accidens var. *hendersonii*.　　Rogue River Milkvetch.

Sprawling plants to 3 ft across; flowers white or yellow, ½ to ¾ in. long; oval fleshy or leathery pods to 1 in. long on a thickened stipe ½ in. long. On grassy slopes and openings in forests at widely scattered locations from Tehama and Butte counties northward to Siskiyou County along the Klamath River and into adjacent Oregon where it is more abundant.

Astragalus agrestis [*A. dasyglottis*].　　Field Milkvetch.

Weak-stemmed mounds of plants 18 to 20 in. across, supported by grasses; flowers almost 1 in. long, reddish purple petals sometimes paler at the tips or white; erect pods to $\frac{1}{2}$ in. long, covered with silky hairs, grooved, and completely divided into two sections. Native in places that are wet for most of the growing season; in California only in the Madelaine Plains, Lassen County, but widespread from the north central states through the western United States and Canada (not in Alaska) and into eastern and central Asia.

Astragalus andersonii.   Anderson Milkvetch.

Plants grayish hairy with prostrate stems erect at the ends forming plants to 18 in. across; flowers $\frac{1}{2}$ in. long, white or yellowish, often with pale lavender veins; pendulous flattened pods slightly incurved, to $\frac{3}{4}$ in. long, two-celled or almost completely so. In sagebrush vegetation, forming colonies and often weedy following clearing of the land; from Honey Lake, Lassen County, south along the edge of the Sierra Nevada to the Truckee Valley, in adjacent Nevada, and again in California north of Mono Lake, Mono County.

Although a plant from Washoe County, Nevada, had a significant amount of nitro compounds when tested, plants of *Astragalus andersonii* are nevertheless routinely eaten to ground level.

Astragalus bicristatus.   Two-keeled Milkvetch.

Stems to 18 in. tall, somewhat decumbent; flowers $\frac{3}{4}$ in. long, yellowish and tinged with purple or green; pods curved to almost one-third of a circle and prominently crested especially toward the beak. Sometimes locally common in open areas of the desert valleys; very restricted in distribution in the transverse ranges of eastern Los Angeles County and western San Bernardino County.

Astragalus bolanderi.   Bolander Milkvetch.

Plants to 18 in. tall, finely hairy; flowers to almost $\frac{1}{2}$ in. long, white, sometimes with light lavender, drying yellowish; pods almost solid, curved, woody when mature. In large colonies in openings in the forests; the only *Astragalus* at high elevations on the west slope of the Sierra Nevada above 5,000 ft in the north, above 6,500 ft to the south, and growing to 10,000 ft; Plumas County to Kern County, also on Mt. Rose in adjacent Nevada.

Astragalus californicus. Klamath Milkvetch.

Stout plants to 2 ft tall, stems and leaves covered with spreading or curly grayish hairs; flowers about ½ in. long, yellowish white; pods pendulous and on a noticeable stipe hanging down from the dried calyx. Forming large colonies in the woodlands of the inner North Coast Range from Mt. Shasta region, Shasta County, into the middle and upper Klamath River drainage of Siskiyou County and adjacent Oregon, also extending infrequently into adjacent sagebrush areas.

Astragalus calycosus var. *calycosus.* Torrey Milkvetch.

A very variable species; in California plants are small tufts of several crowns on a single taproot (in central Nevada these tufts are compounded and much larger), the stems completely covered by the leaf bases, leaves 1 to 3 in. long, one to five leaflets, covered with straight appressed hairs; flowers less than ½ in. long, white or purplish tinged; papery pods less than 1 in. long, straight or slightly curved, erect at maturity. Often, however, the entire plant may be prostrate. In sagebrush or open hillslopes; locally common in California from Death Valley, Inyo County, to Mono Lake, Mono County; widespread eastward in the Great Basin.

Astragalus canadensis var. *brevidens* [*A. mortoni* and *A. mortonii* of California references]. Short-toothed Canada Milkvetch, Morton's Loco.

Stout plants with stems to 2 ft tall arising singly or in clusters from underground rhizomes, with axe-shaped hairs (dolabriform); flowers ½ in. long, yellowish or greenish white, rarely flushed with purple; pods erect, ½ in. long. Locally abundant in winter-wet places in sagebrush and openings in the drier forests; native east of the Sierra Nevada in Mono County, eastern Placer County north of Lake Tahoe, to eastern Siskiyou and Modoc counties, widespread in the Great Basin and the Columbia Basin. Short-toothed Canada Milkvetch was considered very poisonous by early ranchers in California. There are no recent reports of such poisonings.

Astragalus clevelandii.   Cleveland Milkvetch.

Slender plants 3 ft tall; flowers white, ¼ in. long, reflexed downward, in loose very open clusters; pods ¼ in. long, somewhat curved and hanging downward, completely two-celled with two or three seeds in each side. Rare in wet serpentine areas, soils containing unusually large amounts of magnesium, iron, nickel, cobalt, and chromium and only minute amounts of potassium, calcium, and sodium; northern Napa County and adjacent Lake County and in southern San Benito County.

Astragalus curvicarpus var. *curvicarpus* [*A. whitedii*].
Sickle Milkvetch (fig. 19).

Robust plants to 15 in. tall, usually hairy throughout; flowers ¾ in. long, pendulous, white or yellowish; pods 1 to 1½ in. long, markedly recurved (sickle-shaped) or even coiled into a spiral ring. In sagebrush areas or fixed sand dunes, forming large patches in plains and foothills; central Siskiyou County through Modoc and Lassen counties and in Mono County to over 9,000 ft elevation; into the Great Basin.

Astragalus filipes.   Basalt Milkvetch, Threadleaf Locoweed (fig. 19).

Erect stems to 6 ft tall or more; linear or linear oblong leaflets, lower stipules fused or connate into a sheath around the stem; flower ½ in. long or less, white or with a yellow or green tinge; pods papery and shining, flattened and one-celled, 1 in. long on a stipe ½ in. long. Central Siskiyou County and northeastern Shasta County through Modoc and Lassen counties and in southern California on Mt. Piños, Ventura County, the desert slopes of the San Bernardino Mountains, and the coastal slopes of the San Jacinto and Santa Rosa mountains; widespread in the central and northern Great Basin and also in Baja California.

Astragalus gibbsii.   Gibbs Milkvetch.

Like *A. curvicarpus* but more prostrate; stems to 14 in. long, lower stipules connate as a sheath around the stem; flowers deflexed, $\frac{1}{2}$ in. long, definite yellow drying brownish; pods incurved to as much as a half-circle, over 1 in. long, on a stipe. Often abundant in meadows, sagebrush, and yellow pine forests of the eastern slope of the Sierra Nevada in Lassen, Plumas, Sierra, Alpine, and Mono counties; the Mono County record needs confirmation.

Astragalus layneae.   Layne Milkvetch.

Common and forming extensive colonies from underground rhizomes that form grayish hairy buds and separate plants; stems about 6 in. tall; flowers $\frac{1}{2}$ in. long, white with shades of purple on the tips of the petals; pods to $2\frac{1}{2}$ in. long, incurved from one-third circle to a complete spiral. In desert flats and foothills in creosote bush scrub, rarely in sagebrush scrub; central San Bernardino County, eastern Kern County, southern Inyo County, and adjacent parts of Nevada and Arizona.

Astragalus pachypus var. *pachypus.*   Bush Milkvetch.

Bushy plants 18 to 20 in. tall; striate stems and upper leaf surfaces silvery white with parallel appressed hairs; flowers $\frac{3}{4}$ in. long, white, sometimes light pink at the base, drying yellowish; pods formed vertically, slightly incurved, to 1 in. long on a stipe. Often abundant in Digger Pines about the head of the San Joaquin Valley from Kern County to Santa Barbara County and in the Mojave Desert near Ricardo, Kern County, and in the Antelope Valley, Los Angeles County; also in the southeastern part of San Benito County. Var. *jaegeri,* with yellow flowers, occurs in the foothills of Riverside County from Banning south to the Temecula River.

Astragalus serenoi var. *serenoi.*   Naked Milkvetch.

Bushy plants to 2 ft tall having a naked rush-like appearance, with wiry branched stems and few leaves; flowers 1 in. long, purple with white tips on some petals; woody inflated pods 1 in. long, somewhat erect. Often forming colonies of plants in sagebrush and pinyon-juniper woodland at low elevations of the Cottonwood and Inyo mountains, Inyo County, extending eastward in Nevada.

Astragalus umbracticus. Sylvan Milkvetch.

Slender plants to 2 ft tall, nearly lacking hairs; flowers $\frac{1}{4}$ in. long, yellowish white; pods 1 in. long, horizontal, evenly incurved to as much as a half-circle. Rare in California in the lower Redwood Creek and Trinity River areas, Humboldt County; more frequent in southwestern Oregon. This species has the potential of becoming abundant in disturbed ground as from logging operations.

Astragalus webberi. Webber Milkvetch.

Somewhat prostrate stems 18 in. long, covered with fine satiny hairs more readily seen in dried foliage; flowers $\frac{3}{4}$ in. long, arching or somewhat erect; pods 1 to $1\frac{1}{2}$ in. long, flattened partially, green or red spotted, becoming woody with maturity. Only in Plumas County west of the crest of the Sierra Nevada; specimens reported from Sierra Valley, in Sierra County, have not been confirmed.

Astragalus whitneyi. Balloon Milkvetch.

An extremely variable species; stems 1 ft tall but becoming more vigorous plants in northeastern California; flowers variable up to $\frac{1}{2}$ in. long, yellowish white, pink, or light purple, wing petals often lighter in color at the tips; pods $\frac{3}{4}$ to $2\frac{1}{4}$ in. long, markedly inflated, balloon-shaped, on a small stipe often concealed by the dried calyx. The named varieties have many intermediate forms. Var. *whitneyi* occurs on the summits of Mt. Piños, Ventura County, and the Piute Mountains, Kern and Tulare counties, and along the crest and eastern slopes of the Sierra Nevada in Inyo, Mono, and Alpine counties and the White Mountains of Inyo County. Var. *lenophyllus* is on the crest of the Sierra Nevada in Placer and Nevada counties. Var. *siskiyouensis* extends from parts of Shasta and Siskiyou counties to the southern foothills of the Trinity Alps, Trinity County; some Shasta County and Lassen County plants are intermediate between this variety and var. *confusus* of Modoc County. This variable species extends into Nevada and the Pacific Northwest.

An additional twenty-eight species of *Astragalus* have tested positive for appreciable or trace amounts of nitro compounds, but the plants are sufficiently small or the nitro compounds are in such slight amounts that they cannot be considered dangerous to livestock.

Caesalpinia gilliesii [*Poinciana gilliesii*].   Bird-of-paradise Bush, Bird-of-paradise Flower, Paradise Poinciana (fig. 21).

Native of South America; cultivated in California especially in the hot interior valleys. Straggling shrub, sparsely branched; leaves alternate, twice pinnately compound with numerous small leaflets, $\frac{1}{4}$ to $\frac{1}{2}$ in. long; showy, densely glandular, yellow flowers in racemes, stamens bright red, long protruding; legume flat, 4 in. long, $\frac{3}{4}$ in. wide.

Toxic part: Pods and seeds.

Toxin: Unknown.

Symptoms: Severe gastroenteritis.

FIG. 21. *Caesalpinia gilliesii*. Bird-of-paradise Bush. Top: Pod and seed.

Both humans and cattle have been poisoned from these seed pods but usually recover. In 1962 at Lodi, San Joaquin County, six children became ill from eating seeds of this plant; all recovered.

Caesalpinia pulcherrima [*Poinciana pulcherrima*].
Barbados Pride, Dwarf Poinciana, Flower Fence.

Native to the West Indies; in California only occasionally cultivated in warm areas. Spiny shrub; leaves twice pinnately compound with numerous leaflets ½ to ¾ in. long; flowers red and yellow in a raceme; stamens red; legume flat, thin, 4 in. or less long.

Toxic part: Seeds.

Toxin: Not known.

Symptoms: Severe gastroenteritis develops within ½ hour after ingestion; nausea, profuse vomiting, and diarrhea; recovery usually within 24 hours. The leaves of this species are reportedly used as a fish poison in Guatemala and Panama.

Cassia. Senna.

The dried leaves of certain species, commonly called senna, have been used medicinally as a cathartic. There are no records of toxicity in California for these species.

Cassia didymobotrya [*C. nairobensis*].

Native to tropical Africa; commonly cultivated in California; naturalized in the Santa Barbara area. Smelly shrub with showy flowers having yellow petals and protruding red stamens. In central and eastern Africa it has been toxic to livestock and the pounded leaves and stems have been used as a fish poison; an infusion of the roots and leaves is said to be a drastic purgative.

Cassia fistulosa. Golden Shower.

Native of India; occasionally cultivated in California. Leaves, bark, and the sticky pulp around the seeds are toxic, containing emodin anthraquinone glycosides. Recorded as causing nausea, vomiting, dizziness, abdominal pain, and catharsis.

Castanospermum australe.   Moreton Bay Chestnut, Black Bean.

Native of Australia; occasionally cultivated in warmer parts of southern California. Evergreen tree with alternate compound leaves of eleven to fifteen leaflets; flowers in racemes to 6 in. long, red to yellow, formed on the older wood of branches; rounded legume to 9 in. long with chestnut-like seeds. Toxicity not known for California. The toxin is castanospermine, an indolizidine alkaloid, similar to pyrrolizidine and quinolizidine alkaloids. It inhibits the breakdown of sugars. In Australia, ingestion of seeds by cattle caused numerous cases of gastroenteritis, with some fatalities. Horses have also been affected. Ingestion of seeds by early European settlers, either raw or roasted, caused severe, painful diarrhea.

Cytisus.   Broom.

Several species are grown in California; of these, some have escaped and become invasive weeds along roads and in disturbed areas, including *Cytisus monspessulanus,* French Broom, and *C. scoparius,* Scotch Broom. Evergreen shrubs; alternate leaves, each with three small leaflets; flowers bright yellow, sweet-pea-like; legume flat, containing several seeds.

Toxic part: Probably the entire plant.

Toxin: Two poisonous alkaloids, cytisine and sparteine, are present in small amounts.

Symptoms: Cytisine is rapidly absorbed in the mouth, stomach, and intestine. Symptoms appear very soon after ingestion; nausea, vomiting, but little abdominal pain or diarrhea. Livestock may develop staggering followed by paralysis.

Children have developed nausea and vomiting from sucking on the flowers of French Broom. In April 1962, four boys were hospitalized in Eureka, Humboldt County, after they had eaten flowers of this species.

Erythrina.   Coral Tree.

About twenty-five species are cultivated in California including *Erythrina caffra, E. coralloides, E. crista-galli,* and *E. lysistemon.* Trees or, rarely, shrubs; leaves alternate, divided into three leaflets; flowers red, numerous in showy clusters; legumes with black, brown, or red seeds. Seeds and bark of some species contain toxic alkaloids having a curare-like action, producing a curare poisoning that causes death from paralysis of the respiratory system. Crushed stems and leaves have been used as a fish poison.

Gymnocladus dioica.   Kentucky Coffee Tree, American Coffee Berry.

Native to eastern North America; rarely cultivated in California. Large tree; deciduous leaves alternate, twice pinnately compound, to 3 ft long; flowers small, whitish, in racemes at ends of branches; legume hard, to 6 in. long and nearly 2 in. wide, with several seeds. Seeds and surrounding pulp are toxic, containing cytisine, an alkaloid. Symptoms are gastrointestinal disorder, irregular pulse, nausea, diarrhea, and coma. Cytisine content of the seeds is so low that chewing on a seed or two probably will not produce toxic effects. The roasted seeds have been used as a coffee substitute.

Laburnum anagyroides.   Golden Chain, Laburnum.

Native of Europe; cultivated in California for its showy yellow flowers. Shrub or small tree; leaves alternate, divided into threes; flowers sweet-pea-shaped, bright yellow, in long hanging racemes; legume flat with about eight seeds.

Toxic part: Flowers, leaves, seeds, bark.

Toxin: Cytisine, a toxic alkaloid.

Symptoms: Because cytisine is absorbed rapidly through the mucous membranes of the mouth, stomach, and intestine, symptoms usually appear within $\frac{1}{2}$ hour; nausea, vomiting, sometimes dilation of the pupils, irregular pulse, difficulty in breathing, weakness, drowsiness, with dizziness and muscular incoordination in the most severe cases. Can be fatal if ingested in large quantities.

Seeds may be mistaken for peas especially by children. Horses are more susceptible to Laburnum poisoning than cattle or goats. In Europe toxic honey has reportedly been traced to this species.

Lathyrus.   Vetchling, Wild Pea.

Native to the northern hemisphere and mountains of tropical America and Africa. Several species are cultivated for fodder and ornamentals; perhaps the best known is the Sweet Pea, *Lathyrus odoratus*, of southern Europe, popular in gardens for its fragrant colorful flowers. The Tangier Pea, *Lathyrus tingitanus*, and the Everlasting Pea, *L. latifolius*, are also cultivated. All three species are naturalized occasionally in California. Annual or perennial herbs, usually tendril-bearing climbers; stems winged or angled; leaves alternate, pinnately compound; several to many flowers in clusters, showy, sweet-pea-like; legume usually flat, containing several seeds.

Toxic part: Seeds.

Toxin: Toxic amino acids.

Symptoms: The toxins affect the nervous system with a disease called lathyrism manifested by paralysis, irregular pulse, shallow breathing, and convulsions. They may also produce enlarged joints, degeneration of cartilage plates, enlarged calluses at fractures, and hemorrhages.

Human beings and livestock, particularly horses, are affected after ingesting large amounts of seeds. Moderate amounts in the diet are not harmful. *Lathyrus sativus*, Grass Pea, Indian Pea, or Green Vetch, and *L. clymenum*, Spanish Vetch, are two European species that have caused lathyrism in Europe, North Africa, and India and often with much loss of human life. Young men are more affected than women or older men.

Lathyrus hirsutus, Rough Pea, Singletary Pea, Austrian Winter Pea, made up about 60% of some bales of oat-vetch hay grown in Sonoma County in the winter of 1969. When cut for hay the plants had many green pods with maturing seeds. This hay caused severe lathyrism in dairy cattle and in horses on two different ranches. The animals showed stiff gaits and excitability when disturbed. Some animals apparently had impaired vision.

Lotus corniculatus.   Birdsfoot Trefoil.
Lotus tenuis.   Narrow Birdsfoot Trefoil.

Both native to Europe and Asia; grown for forage and occasionally escaped in California in uncultivated fields and along roadsides. Decumbent perennials to 2 ft tall; leaves alternate with three leaflets; showy yellow to red-tinged pea-shaped flowers, ½ in. long; slender rounded legumes in flat clusters resembling a bird's foot. The entire plant is presumably toxic; contains linamarin and lotaustralin, cyanogenic glycosides. Natural populations of these plants contain both cyanogenic and noncyanogenic plants; the difference is genetically determined. For this reason some plants test positive for hydrocyanic acid while the toxicity of others is doubtful. Cases have been suspected in England and in Australia and have been confirmed in the USSR. None are recorded for California. Most of the trefoil grown or naturalized in California is *Lotus tenuis*.

Lupinus. Lupine.

About 200 species of western North America and the Mediterranean region; a few species are cultivated in gardens in California. Annual or perennial herbs or perennials somewhat woody at the base; leaves alternate, palmately compound with usually a number of leaflets; pea-like flowers blue, yellow, or whitish, in showy clusters; legume flat, sometimes constricted between the seeds. Plants of some species of lupines are not toxic and are eaten with the other forage; those of other species are not eaten by livestock. Unless they are specifically known to be harmless, it is best to regard all species of lupines as toxic. The problem of identifying the toxic species of *Lupinus* is complicated further by different interpretations regarding the limits of a particular species. The following lupines are known to be toxic to livestock:

> *Lupinus caudatus*, Tailcup Lupine; in California in Inyo County north to Shasta and Modoc counties, extending further north and east into the Great Basin
> *Lupinus latifolius*, Broadleaf Lupine (pl. 9*e*); native in varied forms from the coast to above 10,000 ft elevation of the Sierra Nevada, almost throughout California from San Diego County to Del Norte and Modoc counties and north
> *Lupinus laxiflorus* [*L. arbustus* and *L. argenteus*], Velvet Lupine, Grassland Lupine; found in California in different varieties from the north slope of the San Gabriel

Mountains, Los Angeles County, through the Sierra Nevada at middle elevations to Siskiyou and Modoc counties and northeast into the Great Basin

Lupinus leucophyllus, Woolly-leaf Lupine, Western Lupine; in the Coast Ranges from Colusa and Trinity counties and in the Sierra Nevada from Plumas County north to Modoc and Siskiyou counties and into Montana and Washington

Lupinus onustus, Plumas Lupine, Woodland Lupine; Plumas and Lassen counties to Trinity and Del Norte counties and adjacent Oregon

Lupinus pusillus subsp. *intermontanus,* Intermontane Low Lupine; east of the Sierra Nevada in Inyo and Modoc counties in sagebrush areas; restricted distribution in California but extending into adjacent states where it has been a serious problem even though these are small plants

Lupinus × alpestris [*L. macounii*], a hybrid of *Lupinus caudatus × L. laxiflorus,* is found above 6,500 ft elevation in the bristlecone pine forest of the Panamint Mountains, Inyo County, extending to Colorado and Montana. Plants have tested positive for anagyrine, but livestock poisonings from this hybrid are not known in California

Lupinus sericeus, Pursh's Silky Lupine; reported in California from a single location south of Mono Lake where apparently it has not persisted but widespread in the northwestern states from Utah, Wyoming, and Montana westward to British Columbia; a serious threat to livestock in these areas

Toxic part: Especially seeds and fruits; also stems and leaves when very young.

Toxin: A series of lupine alkaloids.

Symptoms: Nervousness and excitability; then depression, reluctance to move about, difficulty in breathing, twitching of leg muscles, loss of muscular control, frothing at the mouth, convulsions, and coma.

Sheep are most susceptible to lupine poisoning but cattle, horses, deer, goats, and hogs have also been affected by it. The amount of lupine material that is fatal to an animal depends on the species and the amount ingested. Large amounts eaten in a short period of time are lethal, but the same quantity taken over a longer period will not be harmful because the poisons are excreted from the body of the animal. Dried plants are as toxic as fresh ones, particularly when fruits and seeds are present since these parts retain their toxicity.

Lupinus caudatus, L. latifolius, L. laxiflorus, and *L. sericeus* contain the very toxic alkaloid anagyrine. This alkaloid is most abundant in young plants in the spring and in the maturing seeds still retained in the pods that shatter at maturity. Cows that eat these plants with high concentrations of the alkaloid while in the first 40 to 70 days of pregnancy develop deformed calves, a condition called "crooked calf disease." The numbers vary from year to year in a herd, but as many as 30% of the calves may show deformities. These calves have crooked legs with permanently bent or flexed joints, arched backs, and perhaps cleft palates that leave them unable to nurse. Some calves die at birth; others may live to be sold as yearlings, but the legs and back never straighten. The obvious way to avoid crooked calf disease is to change the breeding cycle so that the cows are not in that stage of pregnancy when the lupines are most toxic. However, the best use of forage on the rangeland coincides with the period of greatest toxicity. Cows bred in July will be in the second to third month of pregnancy at the end of August when lupine plants are forming seeds in the green pods. Cows will search out these plants and eat them at this time.

A case in which deformities of a human were linked to lupine toxin passing through the milk of an animal was reported in 1981. A baby born in Trinity County in 1980 showed bone deformities. The mother drank goat's milk regularly during her pregnancy. During and after this time the goats had stillborn or deformed offspring. Moreover, deformed puppies were born to a dog fed goat's milk during this period. *Lupinus latifolius* formed the main forage of areas where the goats fed regularly during the early months of the mother's pregnancy. Analysis by gas chromatography indicated that seeds and dried leaves of this lupine have a very high content of anagyrine.

When lupine seeds were fed to a milk goat experimentally, almost immediately the anagyrine and other alkaloids were detected in the milk.

Since toxic alkaloids are not stored in the body, animals may eat lupines so long as there is not enough alkaloid at one time to cause poisoning. Some of these species are widespread and constitute part of the native grasslands that are regularly grazed by livestock. Presumably the anagyrine content is not sufficient to cause deformed offspring in cattle. Native hay containing 10% or more dried plants of *Lupinus nanus* is reported to cause livestock poisoning. Although they have not been involved in livestock poisoning, plants of the following lupine species have been found to contain anagyrine:

Lupinus densiflorus, Dense-flowered Lupine; from Sonoma and Sacramento counties south to San Diego County

Lupinus longifolius, Watson's Bush Lupine; from San Diego County to San Bernardino and Los Angeles counties

Lupinus nanus, Douglas' Annual Lupine; widespread in coastal valleys from Mendocino County to Los Angeles County and in foothills of the Sierra Nevada

Lupinus polycarpus, Small-flowered Annual Lupine; throughout the length of California west of the mountains and north to British Columbia

Lupinus sparsiflorus, Loosely-flowered Annual Lupine; in the Mojave Desert and extending into adjacent parts of Mexico and Arizona

Medicago sativa. Alfalfa, Lucerne.

Native of southwestern Asia but widely cultivated for centuries; one of the most important crop plants in California. Deeply rooted perennial herb branching from the base, 3 to 5 ft tall; leaves of three leaflets, leaflets wedge-shaped and toothed across the tip; flowers to $\frac{1}{2}$ in. long in loose clusters, bluish purple in color, sometimes white in naturalized plants; fruit a loosely spiraled or coiled pod.

Toxic part: Entire plant.

Toxin: Bloat-causing proteins, saponins, photosensitization compounds, and estrogenic substances.

Symptoms: Bloat in cattle and sheep is related to the amount and type of proteins in the cytoplasm of the cells in the leaf blades; saponins may or may not be involved in bloat; photosensitization rarely occurs; infertility occurs, but rarely, from the estrogenic compounds.

Alfalfa has a high capacity for producing bloat in cattle because of the proteins in the cells of the leaves. When animals break into an Alfalfa field and eat only the young tips of the plants, they ingest an excessive amount of proteins that cause bloat. If a cow has an empty rumen without other food to dilute these proteins, the possibility of bloat is further increased. Tall plants cut for green-chop feed do not produce this condition as readily. Moreover, high stocking rates reduce the chances of bloat because a large number of animals in a small area will be forced to eat entire plants and not just the tender tips with higher amounts of proteins. Bloat is discussed further under "Plant Toxins" (see pp. 365–366).

In Alfalfa there is a temporary compound reported to cause photosensitization in all species of animals, a toxicity that is seldom seen. This photosensitization is the primary kind and occurs without liver damage. California Burclover, *Medicago polymorpha* [*M. hispida*], has caused photosensitization, as well.

Estrogenic compounds, coumestrol and coumestan, which cause infertility, have been found in Alfalfa, but infertility in cattle and sheep from this source has occurred only rarely, mostly experimentally. Alfalfa meal, selected for high coumestrol content, can increase weight gains in wether lambs. Infections by fungi are known to increase the amounts of coumestrol, the principal estrogen found in Alfalfa.

Cultivated varieties of Alfalfa have been determined to contain as many as thirty saponins of various structures, each having different sugars and different aglycones. The aglycone formed from medicagenic acid is considered to be the most troublesome.

Saponins in the diet of nonruminant animals cause different reactions. Poultry have proved to be the most susceptible to poisoning from saponins. A diet of alfalfa meal at 10% of their weight will depress the growth of chicks. Laying hens have lowered egg production from similar quantities of Alfalfa in their diet. Using alfalfa meal low in saponins and particularly eliminating the saponins containing medicagenic acid permits the inclusion of certain amounts of Alfalfa in poultry diets without adverse effects. The bitter taste of the saponins may cause some of the effects merely by lowering the intake of feed. Alfalfa without the bitter saponins causes an increase in weight in poultry. Breeders of Alfalfa continue to select newer cultivated varieties with lesser amounts of toxic compounds.

Melilotus. Sweetclover.

About twenty species of herbs native of the Old World; three species have been planted as cover crops or as forage plants in past years and are widely naturalized in California as weeds of disturbed ground. Annual or biennial herbs, sometimes perennial under California conditions; leaves of three leaflets but the central leaflet on a longer stalk and therefore the leaflets pinnately arranged, leaflets about 1 in. long, somewhat succulent, finely toothed on the margins; flowers pea-shaped in dense racemes with the leaves at the ends of the branches; pods barely exceeding the calyx, rarely opening.

Melilotus albus [*M. alba*]. White Sweetclover.

Commonly 5 to 6 ft tall, branched above; numerous flowers, white, less than $\frac{1}{4}$ in. long, in dense racemes 2 to 4 in. long.

Melilotus indicus [*M. indica*]. Annual Yellow Sweetclover.

Shorter than the other two species, to 3 ft tall; flowers yellow, smaller, about $\frac{1}{8}$ in. long, in loose racemes to 2 in. long.

Melilotus officinalis. Yellow Sweetclover.

Plants about 4 ft tall; leaflets more ovate in shape; flowers yellow, $\frac{1}{4}$ in. long, in dense racemes to 4 in. long.

Toxic part: Aboveground parts of plants.

Toxin: Dicoumarol, 3,3'-methylenebis (4-hydroxycoumarin), produced in moldy hay, derived from coumarol, the coumarin glycoside found in these three species of *Melilotus*.

Symptoms: Lack of blood coagulation, the result of toxin

interfering with the formation of prothrombin in the liver; probably related to vitamin K production since it is an effective antidote for poisoning from Sweetclover; most commonly seen in cattle, rarely in sheep, but can occur in other animals as well. Internal hemorrhages may cause stiffness; animals tend to remain stationary but are otherwise apparently normal. Swellings full of blood may develop, particularly along the flanks and back. Death may result from bleeding following injuries or from surgery such as dehorning or castration. If pregnant cows are fed moldy Sweetclover hay, their newborn calves may die even though the dams themselves have shown no signs of Sweetclover poisoning.

It is very difficult to make these succulent plants into dried hay without spoilage by molds. Care must be taken on the rangeland when feeding hay containing dried Sweetclover plants that no excess is left to decay from the moisture of the winter rains. Three different fungi capable of producing toxicity have been isolated from this moldy hay, one species each of the genera *Aspergillus, Mucor,* and *Penicillium*. Many other species of fungi can produce the intermediate toxic compound, 4-hydroxycoumarin, in the laboratory, but these fungi have not been found in decaying hay.

The coumarin glycoside, coumarol, and the enzyme for its breakdown occur naturally in plants of these three species of *Melilotus*. The glycoside hydrolyzes to glucose and coumarin. Coumarin is present as coumaric acid, which the fungi change into 4-hydroxycoumarin in improperly dried hay. Formaldehyde from the decaying plant tissue reacts with the 4-hydroxycoumarin to form dicoumarol, 3,3′-methylenebis(4-hydroxycoumarin), the lethal toxin, in the moldy hay.

Oxytropis. Locoweed, Crazyweed.

Four species of this genus are native in California, but all occur above 9,000 ft elevation in Inyo, Mono, and San Bernardino counties in areas not used for grazing livestock. The locoweeds of this genus are serious problems to grazing livestock in other western states and adjacent Canada. The toxic locoweeds in California are species of *Astragalus*. Incoordination and an erratic gait are typical signs of poisoning from plants of *Oxytropis*. The toxin is cumulative and may not show up for 6 to 10 weeks after grazing these plants. Animals develop a preference for eating these poisonous plants and must to be removed from areas where they are growing.

Phaseolus limensis.   Lima Bean.

Native of tropical South America; an important edible bean, widely cultivated. To this species belong many wild forms as well as named cultivars; perhaps this species could be included as a form of *Phaseolus lunatus,* Sieva Bean. Robust twining annual; leaves pinnately divided into leaflets to 5 in. long; flowers white or yellowish to ⅜ in. long, in racemes; pods flat, thick-edged; seeds flat, green to white.

Toxic part: Aboveground parts of plants, especially the seeds.

Toxin: Phaseolunatin (a cyanogenic glycoside) and an enzyme that releases hydrocyanic acid from the glycoside.

Symptoms: Typical for hydrocyanic acid poisoning. (See pp. 225, 319.)

Normally Lima Beans are not toxic, but plants of *P. lunatus* are known to be sufficiently high in the cyanogenic glycoside to be a threat to livestock. The cyanogenic content is quite variable and depends upon many environmental factors. The seeds of *Phaseolus lunatus* also contain phytohemagglutinins (lectins).

Vines and pods with some seeds of the Lima Bean, *Phaseolus limensis,* caused the death of fifty-eight hogs in Stanislaus County in 1958. The plant material tested positive for hydrocyanic acid. The vines had been obtained from a commercial freezer plant as had been done many times before without any difficulties. This crop had just been irrigated because it was expected to be kept in the field for some time, but a change in plans required the immediate harvest of the plants in a wet condition.

Rhynchosia pyramidalis.

Vine native of Mexico; the colorful red and black seeds have been imported into California strung as beads. The seeds are about the size and shape of those of *Abrus precatorius,* and seeds of the two vines might be mistaken for one another. The seeds of *Rhynchosia pyramidalis* show the black portion greater than the red, however, with the hilum scar in the red portion; seeds of *Abrus precatorius* differ in having the red portion greater than the black with the hilum scar in the black portion. Although not so dangerously toxic as those of *Abrus,* seeds of *Rhynchosia* contain an alkaloid with hallucinogenic properties.

Robinia pseudoacacia. Black Locust (pl. 9*f*).

Eastern and central North America; widely cultivated in California, commonly persisting from cultivation and naturalized. Deciduous tree with stout unbranched thorns on the stems below the leaves; leaves alternate, pinnately compound; flowers whitish, pea-shaped, in pendant showy clusters; legume flat, reddish brown, containing several seeds, persisting on the tree through the winter.

Toxic part: Seeds, bark, leaves. Children are often poisoned by chewing or eating these parts. Flowers are not usually poisonous and are used by bees.

Toxin: Robin and phasin, plant lectins (phytohemagglutinins, proteins that agglutinate red blood cells); robatin, a glycoside.

Symptoms: In humans, symptoms may occur after about an hour: nausea, vomiting, diarrhea, reduced heart action, coldness of the extremities, and stupor. Can be fatal but only rarely so. In smaller doses there is inflammation of the intestine, hemorrhages of the lymphatic tissues, and liver damage. Cattle and horses may also display paralysis of the hindquarters.

Black Locust is well known to be toxic to all classes of livestock and to human beings. The toxicity is variable in occurrence. In June 1981, a valuable horse was lost from eating Black Locust near Hidden Hills, Los Angeles County.

Bees routinely visit Black Locust flowers. Very rarely large numbers of dead bees have been found beneath the trees when in flower in late winter or early spring. Such a loss of bees occurred in 1955 in Nevada City, Nevada County. These deaths are not caused by a toxin in the plant but are the result of a sudden drop in temperature causing the bees to become paralyzed. This same situation is found with winter-flowering species of *Eucalyptus*.

Sesbania punicea [*Daubentonia punicea, Sesbania tripetii*]. Rattlebox, Scarlet Wisteria Tree, Glory Pea (pl. 10*a*).

Native of South America; cultivated widely in warm areas of California. Shrub or small tree; leaves alternate, pinnately compound, each leaflet 1 in. long or less; flowers in hanging showy clusters, pea-shaped, orange to red; legumes about 4 in. long, rounded with four wings running the length of the pod; seeds brown.

Toxic part: Seeds especially; also flowers.

Toxin: Saponins and sesbanine, an alkaloid; actual cause of toxicity is still not determined.

Symptoms: Diarrhea, depression, rapid pulse, weakness, and difficult breathing. Can be fatal. Symptoms may not appear for 24 to 48 hours after ingestion. Cases of poisoning have been reported for sheep, goats, cattle, and poultry. Since the seed pods hang on the small trees all year long, even while the following year's pods are being formed, toxic seeds are continuously present on these plants.

In October 1965, a child in Sacramento became ill apparently from eating pods and seeds of this plant. The child vomited immediately after eating the seeds (the same reaction seen in several other cases as well) with no apparent aftereffects.

Some species of *Sesbania* containing pyrrolizidine alkaloids are being used in herbal teas. Other plants used in this way, species of *Crotalaria, Heliotropium,* and *Senecio,* are discussed under pyrrolizidine alkaloids (see pp. 332–334).

Sophora secundiflora. Mescal Bean, Texas Mountain Laurel.

Native of Texas, New Mexico, and adjacent Mexico; occasionally cultivated as an ornamental tree in California; sometimes the attractive seeds are strung into necklaces. Evergreen shrub or small tree; leaves alternate, pinnately compound with seven to nine leaflets; sweet-pea-like blue violet flowers about 1 in. long in racemes; cylindrical legume brown with a silvery cast, hard and woody, somewhat constricted between hard red seeds about ½ in. long. Seeds contain anagyrine and other toxic alkaloids related to those found in *Lupinus,* and cytisine, a toxic quinolizidine alkaloid, a strong hallucinogen. Symptoms are nausea, vomiting, diarrhea, excitability, delirium, and coma with death as a result of respiratory failure. Children have been fatally poisoned, and one seed, when chewed, is considered to be sufficient to cause death of a human being. The hallucinogenic seeds were used in visionary rites by people in the native range of the species. The drink made from the red beans is highly toxic and has often resulted in death from overdoses. Its use has been abandoned in favor of Peyote.

Trifolium.   Clover.

The following species are native to Europe and the Mediterranean region; they are grown in California as forage crops and are also naturalized:

> *Trifolium hybridum,* Alsike Clover
> *Trifolium pratense,* Red Clover
> *Trifolium repens,* White Clover, and its giant form, Ladino Clover
> *Trifolium subterraneum,* Subterranean Clover, Subclover

Toxic part: Entire plant.
Toxin: Phytoestrogens: isoflavones and coumarins.
Symptoms: Infertility and an effect on the growth rate of animals; perhaps increased mastitis in cows. Unknown toxins cause photosensitization or "bighead."
Infertility of ewes was first demonstrated in those feeding almost exclusively on *Trifolium subterraneum.* This species solidly occupies vast areas in Australia and is the only source of food for these animals. Subclover is not widely adapted to California conditions; sparingly naturalized mostly along the immediate central and northern coast.

The phytoestrogen content of the Subterranean Clover cultivar 'Yarloop' may be 200 times greater than that of other cultivars. This amount exceeds the level that is considered the cause of infertility in ewes. Phytoestrogens have been found in the sperm of males feeding on plants of Subterranean Clover.

Genistein, from Subterranean Clover, has been found to be the most potent of the isoflavones. Coumestrol, however, is 30 to 100 times as potent as the isoflavones. Although California has no records of infertility in animals feeding on legumes with phytoestrogens, animals fed exclusively on such legumes should be watched for infertility. These estrogens also occur in Alfalfa, *Medicago sativa,* and Berseem Clover, *Trifolium alexandrinum,* the latter rarely grown in California.

Vicia faba.   Fava Bean, Broad Bean, Horse Bean, Windsor Bean, European Bean, Field Bean.

Native range unknown; perhaps from North Africa or southwestern Asia or a cultivated form of the southern European bean, *V. narbonensis;* cultivated in California and occasionally escaped and naturalized. This species is the bean of antiquity, long grown for the seeds or immature pods as food, for forage, or for "green manure." When ingested, the seeds cause a condition known as favism in certain individuals.

Toxic part: Seeds, raw or cooked, and pollen when inhaled.

Toxin: An antimetabolite, 2,6-diaminopurine; a glycoside, vicine.

Symptoms: In susceptible humans, within a few minutes of inhaling pollen or several hours after eating the beans, an allergic reaction occurs with dizziness, diarrhea, nausea and vomiting, abdominal pain, and severe prostration. Blood appears in the urine, which turns reddish brown to black. Anemia develops within a few hours or a day.

Male children are the most frequently affected, and apparently all fatalities have been children. Favism, the severe hemolytic anemia, occurs only in susceptible individuals who have inherited a deficiency of an enzyme, glucose-6-phosphate dehydrogenase. This genetic trait occurs among people of the Mediterranean region and among black Africans. Most individuals have this enzyme and hence are not affected.

Wisteria. **Wisteria.**

Native in eastern United States and eastern Asia. The species commonly cultivated in California for their attractive spring flowers are:

Wisteria floribunda, Japanese Wisteria (fig. 22)
Wisteria sinensis, Chinese Wisteria
Wisteria venusta, Silky Wisteria

Toxic part: Entire plant, including the flowers, but especially the seeds.

Toxin: An uncharacterized glycoside, wistarin, and a poisonous resin.

Symptoms: Nausea, abdominal pain, repeated vomiting,

FIG. 22. *Wisteria floribunda.* Japanese Wisteria. Lower left: Seed pod, seed, and flower. Upper left: Seed (enlarged).

followed by collapse. Two seeds are sufficient to cause serious illness in a child. Fatalities have not been reported, however, and usually the patient recovers within 1 or 2 days.

Fagaceae, Beech Family

Fagus sylvatica. European Beech.

Native to Europe; cultivated in California. Handsome deciduous tree with smooth very dark gray bark; leaves glossy green, ovate, toothed or sometimes cleft; fruit three-angled nut, enclosed in a prickly covering that splits open into four parts. Gastrointestinal distress is caused by a saponin-like substance in the raw nuts or the cake after the oil has been removed from the seeds. Raw nuts should not be eaten in quantity, but in Europe they are eaten after roasting. The oil is not toxic. Nuts of the American Beech, *Fagus grandiflora,* native in eastern North America and sometimes cultivated in California, are said to be edible.

Quercus. Oak.

California has sixteen native species but in addition many species from other regions are cultivated for ornament. Evergreen or deciduous trees; leaves variously cut or lobed; fruit an acorn, a one-seeded nut, partly enclosed in a cup. The identification and taxonomy of the species are not important to this discussion because their toxic properties are similar.

Toxic part: Leaves and acorns.

Toxin: Tannin.

Symptoms: In cattle, principally kidney damage accompanied by loss of appetite, thirst, edema, frequent urination, gastrointestinal pain, passing of compacted dark brown feces, and finally diarrhea. Kidney damage is permanent and usually results in death. The toxin causes lesions and perforations of the alimentary tract, permitting the entry of bacteria that affect the lungs, liver, and other internal organs. In acute cases, fluids accumulate in the body cavities in large quantities.

Poisoning of cattle and occasionally sheep by browsing on oaks, particularly buds and young leaves, causes annual economic losses in the Southwest. Such losses are rare in California, but serious poisonings of cattle have occurred.

In late April 1960, in the lower elevations of the northeastern side of the Sacramento Valley, following a very dry period, a snowfall of 6 to 18 in. covered the little available forage. About 2 weeks later from Shasta Lake south to Mineral, Tehama County, there were losses of range cattle because of heavy feeding on young sprouts of California Black Oak, *Quercus kelloggii.* Deaths were especially common in fields that had been burned or bulldozed in the past year or so and where California Black Oak sprouts were abundant. Large quantities of twigs, leaves, and buds were eaten. At the same time during these unusual weather conditions but in other locations, deaths of cattle occurred from eating Deathcamas, *Zigadenus* sp., and California Buckeye, *Aesculus californica.*

In 1985, unusually warm and dry winter weather had brought up the forage to its best stage of utilization and cattle were dispersed on the rangeland. On March 26 there was a heavy snowfall on the northwestern side of the head of the Sacramento Valley; 6 to 9 in. were recorded in Redding. For 3 days cattle had nothing to eat but the young sprouts of Blue Oak, *Quercus douglasii.* The weight of the snow broke down large branches of the older trees that became readily available to the animals. On sixty ranches at the lower elevations of the inner Coast Range in Shasta and Tehama counties, more than 2,500 cattle were lost. The immediate cause of death of one calf was found to be collapse of the lungs from accumulation of fluid in the chest cavity.

Acorn poisoning of cattle is less common in North America than in Europe. Acorns are not a source of poisoning in humans. Although the raw nuts would be toxic if eaten in quantity, their bitterness precludes overeating. Children who might gnaw on a raw bitter acorn or two probably would not get a harmful amount of tannin. In California and other parts of North America acorns were a staple food of the Indians, who leached them under running water to remove the tannin and render them palatable.

Fumariaceae, Fumitory Family

Corydalis caseana. Fitweed (fig. 23).

Endemic native found only in California; middle to higher elevations of the Sierra Nevada from Lassen and Shasta counties south to Placer County and in Tulare County. Perennial

FIG. 23. *Corydalis caseana.* Fitweed.

herbs to 2½ ft tall; leaves pinnately dissected, 6 to 18 in. long and across, ultimate leaflets with a prominent tip; flowers in a small panicle, each flower horizontal, spurred, petals white or pinkish with purple tips.

Toxic part: Top of the plants; stems, leaves, or flowers.

Toxin: Isoquinoline alkaloids.

Symptoms: Within minutes to a few hours after ingestion sheep and cattle show depression, convulsions, excitability, muscular rigidity, pawing of the ground, and grasping objects with the mouth. Death may follow.

Fitweed grows in running water or on stream banks in shaded conditions. Animals may seek out these plants as they are succulent and initially palatable. Although plants are rare throughout its range, this species has been regarded as the most serious cause of livestock loss in the Plumas National Forest. Fitweed plants initially are palatable, but almost immediately after ingestion they produce a distinct, long-lasting, acrid aftertaste. Cattle accustomed to the area usually will not eat these plants.

Dicentra.

Various species of *Dicentra* have a number of toxic isoquinoline alkaloids that are structurally related to those of *Papaver. Dicentra chrysantha,* Golden Ear-drops, is native from Baja California north to Calaveras and Lake counties. Plants are covered with a gray waxy bloom; flowering stems to 5 ft tall with scattered twice-pinnately dissected leaves; flowers golden yellow, to $\frac{1}{2}$ in. long. Plants may be abundant after burns and have been suspected in cases of livestock poisoning. Toxic alkaloids are present in plants of *Dicentra formosa,* Western Bleedingheart, but this species is so rare in California that it does not present a threat to livestock.

Gentianaceae, Gentian Family

Centaurium floribundum.   June Centaury.

This species has caused distress to livestock in California. These summer annual plants, 6 to 18 in. tall, covered with cymes of pink flowers, are abundant in wet places in the foothills of the Sierra Nevada from Butte County to Fresno County and the inner Coast Ranges from Marin County to Humboldt County. The last of several reports received by the California Department of Food and Agriculture was a case in San Joaquin County in 1954. June Centaury plants, eaten only where there is no palatable forage, are rarely a problem for livestock. The entire plant contains an unknown toxin that causes sluggishness and lack of appetite, gastroenteritis, diarrhea, and frequent urination; can be fatal. *Centaurium beyrichii* and *C. calycosum* have caused trouble to livestock, particularly sheep, in Texas and Mexico. *Centaurium calycosum* occurs along the Colorado River in California.

Geraniaceae, Geranium Family

Erodium.   Filaree.

Seven species of *Erodium* are native or naturalized in California. Several of these winter annual species form valuable forage plants that have not been recorded as toxic to livestock in California. In southwestern New Mexico in November and December 1983, under unusually good growing conditions followed by frost, *Erodium* plants produced hydrocyanic acid in sufficient quantity to be lethal to sixty cattle.

Hippocastanaceae, Buckeye Family

Aesculus.   Buckeye, Horse Chestnut, Conquerors, Conkers, Fish Poison.

Thirteen species of trees or large shrubs; opposite, deciduous leaves palmately compound; flowers in showy panicles terminal on branches; fruit a smooth or spiny capsule.

> *Aesculus californica*, California Buckeye; widespread native at low elevations of the Sierra Nevada and Coast Ranges from Kern and Los Angeles counties to Siskiyou County; flowers white, in narrow panicles
>
> *Aesculus hippocastanum*, Horse Chestnut; from southern Europe, occasionally cultivated in California; flowers white with red or yellow markings, in large panicles to 1 ft long
>
> *Aesculus × carnea*, Red Horse Chestnut; of hybrid origin, cultivated in California; flowers light to dark red in color, in panicles 6 to 8 in. long

Toxic part: Leaves, flowers, twigs, bark, seeds.

Toxin: Aesculin, a coumarin glycoside.

Symptoms: In humans, restlessness, circulatory disturbances, involuntary urination and defecation, vomiting, diarrhea, muscular twitching, lack of coordination, dilated pupils, weakness, paralysis, and stupor. Only rare fatalities have been known in Europe, and no deaths have been reported for the United States. Persons have become ill from eating the seeds.

Buckeye plants are poisonous to cattle and are reported to cause abortion. Honeybees are poisoned by the nectar and pollen.

California Indians used ground-up seeds of the California Buckeye as a fish poison. They also rendered the seeds edible by roasting or leaching, but the seeds were used only as an emergency food supply.

In late April 1960, near Manton, Tehama County, a heavy snowfall covered the forage plants on the ground. A number of cattle died from heavily browsing both the broken and un-broken branches and leaves of the California Buckeye, *Aesculus californica*. Autopsy revealed that pneumonia caused the death of one cow, but the condition was precipitated by foreign material in the lungs. It was postulated that the Buckeye plant material had been regurgitated and became lodged in the lungs.

Hydrophyllaceae, Waterleaf Family

Phacelia.

California has eighty-seven species; hairy, sometimes viscid-glandular, annuals or perennials; leaves usually alternate, simple or pinnately divided; flowers blue or white to pink, some large and showy, scattered or in densely curled racemes. Plants of many species, at least those having viscid-glandular hairs, cause an allergic contact dermatitis in sensitive individuals resembling that produced by Pacific Poison Oak, *Toxicodendron diversilobum*.

Turricula parryi [*Nama parryi*].   Poodle-dog Bush.

Southern half of California, Sierra Nevada, inner Coast Ranges, Panamint Mountains, Tehachapi Mountains, south to Baja California; dry areas, especially abundant after fires. Stout, sticky glandular, and ill-scented perennial, woody at base, to 6 ft tall; leaves alternate, lanceolate, sessile; flowers purple, in a coiled raceme, corolla $\frac{1}{2}$ to $\frac{3}{4}$ in. long.

Turricula parryi is known to cause severe dermatitis. In 1941, in the San Gabriel Mountains northeast of Glendora, Los Angeles County, three persons were affected after exposure to old flowering stems of the previous year. This was the third year following a fire, and plants had become abundant. When plants were in flower during the previous year no one was affected. Hairs on the old flowering stalks, easily broken from the stems, were the cause of a rash with edema, intense itching, and burning.

Hypericaceae, St. Johnswort Family

Hypericum perforatum.   Klamathweed, Common St. Johnswort (pl. 10*b,d*).

From Europe; naturalized as a weed in northern California

and extending south along the coast into Mendocino County and along the western slopes of the Sierra Nevada south to Tuolumne County. The form of this species established as a widespread weed in California is more vigorous than any native form found in Europe. Taprooted perennial with short sterile basal shoots and many upright stems from the base, to 3 ft tall; leaves opposite, sessile, several-nerved from the base, with translucent dots in the blades; flowers to 1 in. across, yellow, in flat-topped clusters at the ends of the branches; sepals and petals black-dotted; stamens in three clusters; seeds minute, easily spread in the wind.

Toxic part: Entire plant, especially leaves and flowers.

Toxin: Hypericin, a fluorescent pigment derived from dianthrone. It remains unchanged through digestion and is absorbed into the bloodstream of the animal.

Symptoms: Photosensitization in cattle, horses, sheep, and goats, causing areas of white skin to slough. Dark areas of an animal's skin are not affected. Animals sometimes suffer aftereffects with refusal to eat, loss of weight, and blindness. Death may occur from starvation.

Biological control of this weed species has restored more than a million acres of California rangeland to economic use. The primary cause of the reduction of the weed has been the Klamathweed flea beetle, *Chrysolina quadrigemina*. It is estimated that less than 1% of the former population of Klamathweed is left. Small populations of the weed still exist in shady areas, boggy situations, north-facing slopes, and roadsides where the bettle is less active. A root-boring beetle, *Agrilus hyperici,* has also been released.

Juglandaceae, Walnut Family

Juglans. **Walnuts.**

Deciduous trees; leaves alternate, pinnately compound; male and female flowers on the same tree; fruit an edible nut contained in a husk. Black walnuts are native in the United States; the cultivated English or Persian Walnut is native in southeastern Europe and western Asia.

Toxic part: Moldy nuts.

Toxin: Penitrem A, the mycotoxin produced by the fungus, *Penicillium crustosum,* growing on decaying walnuts.

Symptoms: Gastroenteritis followed by generalized convulsions, 5 to 30 minutes after ingestion of the moldy walnuts, only in dogs; other signs are raised temperature, panting, enlarged pupils, urination, and defecation.

In California, poisonings of dogs are associated with English Walnut, *Juglans regia,* and black walnuts. Dogs readily eat English Walnuts, cracking and eating the nutmeats that are usually harmless to them. Toxicity occurs in late summer and again in winter. English Walnuts that have been on the ground during the winter rains and have become heavily infested with fungi are the cause of the winter poisonings. In late summer, English Walnuts in the hulls cause the same poisoning of dogs. Portions of the green hulls damaged by larvae of the Walnut Husk Fly are blackened, and the growth of fungi can be seen in these areas. At maturity, nuts of black walnut trees commonly have blackened hulls.

The wood of Eastern Black Walnut, *Juglans nigra,* and Butternut, *Juglans cineraria,* contains juglone, 5-hydroxy-1,4-naphthoquinone, which has proved toxic to plants and animals. Alfalfa, tomato, and other plants do not grow in the root area of trees of these two species in central and eastern United States. Lameness and edema of legs of horses occur when wood shavings from these two species are included in the bedding. This toxicity has not been demonstrated in California.

Lamiaceae, Mint Family

Also known as Labiatae, this family is of minor importance as a source of toxic plants.

Glechoma hederacea [*Nepeta glechoma, N. hederacea*].
Ground Ivy, Gill-over-the-ground, Creeping Charlie,
Runaway Robin, Field Balm.

Native of Europe; cultivated in California as a garden plant, has escaped and become naturalized, in some places a serious lawn and garden weed. Creeping perennial herb making a ground cover; leaves ½ to 1 in. long, rounded; flowers ½ in. long, blue. Entire plant contains an irritant oil toxic only to horses. Signs are labored breathing, sweating, and salivation; death may occur. Causes difficulty only from eating large quantities of either fresh or dried plants. Poisonings have occurred in Europe; reported only once in North America.

Mentha pulegium. Pennyroyal.

Native in Europe and western Asia; occasionally cultivated since its strong mint flavor and fragrance make it a useful flavoring herb; escaped and naturalized as an unpalatable rangeland weed of the North Coast Ranges of California. Creeping, more or less decumbent stems to 12 in. tall; small leaves ovate to nearly round; flowers lilac, in densely flowered, spaced whorls.

Toxic part: Entire plant.

Toxin: Pulegone, a ketone volatile oil that is toxic when used injudiciously in large amounts. It is marketed as an insect repellent.

Symptoms: Retching, delirium, unconsciousness, and respiratory depression. Causes liver damage.

Although pulegone is an irritant, it has no specific action on the uterus. In Colorado in late 1979 three cases were reported of women who had ingested concentrated extracts of the plant, two in unsuccessful attempts to induce abortion. Two women recovered, but the third woman who died was known to have taken Pennyroyal as well as other herbal preparations for self-treatment of minor ailments unrelated to abortion.

Lauraceae, Laurel Family

Cinnamomum camphora. Camphor Tree.

Native of China and Japan; common in cultivation in California. Implicated in the deaths of forty-nine budgies or parakeets in a 24-hour period in November 1982. As reported by the Fresno Veterinary Laboratory of the California Department of Food and Agriculture, no new birds had been introduced onto the premises, and there were no management changes except the introduction of Camphor Tree foliage into the cages. Small caged birds are known to be extremely sensitive to noxious fumes. The leaves of Camphor Tree contain oil of camphor, a volatile oil. Camphor is a crystalline ketone obtained from distillation of the wood; it has caused poisoning of humans, especially children, and death may result from respiratory failure. Leaves, stems, flowers, and fruits of Camphor Tree are listed as containing hydrocyanic acid.

Persea americana. Avocado.

Native of tropical America; two races are grown in California. The toxic cultivars belong to the Guatemalan race, var. *americana*, leaves not anise-scented; the nontoxic cultivars make up the Mexican race, var. *drymifolia*, leaves anise-scented. A few cultivars are hybrids between the races. Widely spreading, many-branched evergreen tree 30 to 40 ft tall; leaves alternate, elliptic or ovate, to 8 in. long; flowers minute in large terminal panicles formed during the winter; fruit fleshy, edible, pear-shaped to globose with a tough skin and a single large seed.

Toxic part: Leaves, branches, seeds.

Toxin: Unknown.

Symptoms: Congestion of the lungs and inflammation of the udder; can be fatal.

All animals are apparently poisoned by cultivars of the Guatemalan race. Cows and goats have shown a marked drop in milk production. Goats apparently are more susceptible; some died after eating large amounts of Avocado leaves and branches.

In a 1939 report from Los Angeles County, thirty canaries in an aviary were given fresh branches of Avocado that were eaten readily. Twenty birds died within a few hours. Canaries have also been reported as dying from eating Avocado fruits, presumably with the seeds also present.

In June 1982, ten milk goats were allowed to feed for an hour on the leaves of 'Anaheim', a Guatemalan cultivar. Twenty-four hours later they fed again for an hour on these leaves. When milked after this feeding, production was about one-quarter that of the previous day, the udders were edematous, and the milk curdled. Two days later the curdled milk could not be forced from the udders. Later the animals developed edema of the neck and brisket. After 10 days, the older goats had dried up. This toxicity was repeated experimentally even though the Avocado foliage of the same cultivar had been kept under refrigeration for 2 weeks. A similar case of milk goats drying up entirely was reported from Los Angeles

County in 1939. The same report also described the drying up of female rabbits when leaves and immature fruits of Avocado were blown into the pen during a severe windstorm.

Later a feeding trial resulted in death of rabbits fed leaves of 'Fuerte', a Guatemalan cultivar, and 'Nabal', a cultivar hybrid of the two races. The rabbit fed leaves of 'Mexicola', a Mexican cultivar, survived unharmed. Fish died in a pond where Avocado leaves fell into the water; the deaths stopped after the toxic leaves were removed and kept out of the pond. Mice died when experimentally fed Avocado seeds.

Sassafras albidum [*S. variicolor*].   Sassafras.

Native of eastern and central United States; rarely cultivated as an ornamental in California. Deciduous tree to 60 ft tall; leaves alternate, 3 to 5 in. long, ovate, strongly veined from the base, entire or with one or two lobes in leaves on the same tree; fruit a drupe about $\frac{1}{2}$ in. across, dark blue with a fleshy red stem. Safrole, the major constituent (80%) of oil of sassafras, is obtained from the root. It is considered by the Food and Drug Administration to cause cancer. The use of safrole as a flavoring in foods is now prohibited. For a long time sassafras oil has had a medicinal use topically as an antiseptic and internally as a carminative and stimulant. The root bark is used in making sassafras tea.

Umbellularia californica.   California Laurel, California Bay, California Olive, Oregon Myrtle, Pepperwood.

Native to southern Oregon and California; occasionally cultivated as an ornamental. Evergreen tree to 50 ft tall; leaves alternate, entire, lanceolate to ovate-oblong, 2 to 5 in. long, aromatic; flowers small, about $\frac{1}{4}$ in. long, in clustered umbels to 2 in. across; fruit a drupe to 1 in. long, yellow green to purple, the single stone occupying most of the inside of the fruit.

The pungently aromatic leaves contain an irritating oil of which the major constituent is umbellulone. Contact with the leaves has caused skin irritations in some persons. Violent headaches may be produced by inhaling the crushed leaves, an old scouting prank that has led to the common name Headache Tree. Irritation of the mucous membranes may cause pronounced sneezing. If leaves of the California Laurel are desired in cooking, they should be used more sparingly than those of the Grecian Laurel or Bay Leaf, *Laurus nobilis*.

Linaceae, Flax Family

Linum usitatissimum.　Flax.

May have originated in Asia; cultivated for its fibers. Grown rarely as a field crop in California, occasionally escaping in disturbed areas. Annual about 2 ft tall; leaves alternate, narrowly linear to lanceolate; flowers bright blue, five separate sepals and petals, the petals soon falling, five stamens; fruit a segmented capsule with several seeds.

Flax fiber from the stems is the source of linen, one of the oldest textiles used for clothing. Linseed oil is obtained from flax seeds, and the cake remaining after pressing oil from the seeds is used as a food for domestic animals.

Toxic part: Leaves and seeds.

Toxin: Linamarin, a cyanogenic glycoside. Linamarase, an enzyme present in the plant material, acts on the linamarin to release hydrocyanic acid.

Symptoms: Typical of hydrocyanic acid poisoning (see p. 225).

The amount of hydrocyanic acid present is usually not sufficient to affect adult animals, but it is enough to harm calves and on occasion to cause sudden death. Linseed cake can be made harmless by heat treatment. Hay with old flax plants containing seeds is potentially hazardous.

Plants of *Linum lewisii* [*L. perenne* subsp. *lewisii*], Prairie Flax, a blue-flowered native perennial similar to the cultivated Flax, have been found to have the same potential production of hydrocyanic acid. They are not known to cause livestock poisoning in California but were implicated in a case in Colorado.

Lobeliaceae, Lobelia Family

This family is sometimes placed in the Campanulaceae, the Bellflower Family.

Lobelia cardinalis subsp. *graminea* [*L. splendens*].　Scarlet Lobelia, Cardinal Flower, Indian Pink.

Native in southern California from Los Angeles and San Bernardino counties south to San Diego County, widely distributed elsewhere; cultivated for its bright red flowers. Erect herbaceous perennial; stems sometimes reddish purple; leaves narrowly lanceolate; flowers scarlet, rarely pink, showy, in clusters at the ends of the stems, irregular tubular corolla with three large lobes and two smaller ones.

Toxic part: Entire plant when green, less toxic when dried; leaves and seeds especially toxic.

Toxin: Lobeline and related pyridine alkaloids; similar in structure and activity to nicotine from tobacco.

Symptoms: Dryness of throat, nausea, vomiting, headache, dizziness, exhaustion, lowered temperature, abdominal pain, burning in the urinary tract, pupils constricted and not reactive to light, tremors, convulsions, respiratory failure, and coma. May be fatal. Poisoning is rare except for injudicious use or overdose when used in home-medicines.

Many alkaloids are found in plants of different species of *Lobelia,* the largest genus of the Lobelia Family. Lobeline, the most important alkaloid, has been isolated from American species of this genus including *Lobelia inflata,* Indian Tobacco, a native of the eastern and southeastern United States. Early settlers knew that Indians dried and smoked leaves of the plant. It was soon found that plants of other species of *Lobelia* native to the eastern United States had similar properties and were used for medicinal and other purposes. Overdoses of the medicinal preparations resulted in deaths, however, and these plants acquired a reputation of being poisonous. Other species cultivated in California, such as *Lobelia laxiflora* with red flowers and *L. siphilitica* with blue flowers, should be considered potentially toxic.

Loganiaceae, Logania Family

Several genera in the family yield drugs and poisonous substances, the best known of which is strychnine from *Strychnos nux-vomica* from tropical Asia.

Gelsemium sempervirens.   Carolina Jessamine, Yellow Jessamine, Carolina Jasmine, Yellow False Jasmine, Carolina Wild Woodbine, Wood Vine (fig. 24).

Native of southeastern United States; cultivated in California, often as a climber on fences. Woody evergreen vine; leaves

FIG. 24. *Gelsemium sempervirens.* Carolina Jessamine.
Flowering branch and seed capsule.

opposite; flowers yellow, showy, corolla tubular with five lobes;
fruit a capsule.

Toxic part: Entire plant.

Toxin: Gelsemine, gelsemicine, and related indole alkaloids.

Symptoms: Muscular weakness, dizziness, visual disturbances, dryness of the mouth, slowed pulse, impaired respiration, great anxiety, and convulsions. Can be fatal.

Domestic livestock and poultry have been fatally poisoned. Human poisonings have resulted from crude medicinal preparations of the root. Children in the southern states have been poisoned by eating flowers or sucking their nectar. In 1979 in Texas a 3-year-old girl became ill after eating an estimated five flowers. She was given prompt medical treatment and recovered in 24 hours.

The gelsemine alkaloid has been used in medicine as an antispasmodic and antineuralgic agent. Its chief effects are paralysis of the motor nerve endings, leading to respiratory arrest and a strychnine-like action on the spinal cord causing muscular rigidity and convulsions.

Gelsemium sempervirens is listed among the ten most poisonous plants native to North Carolina. In an experiment the leaves and stems of Carolina Jessamine only retarded the growth of poultry, but the roots were found to be lethally toxic to turkeys and chickens. Such toxicity is not recorded for California. Honeybees are poisoned by the plant, and their honey is toxic.

Lythraceae, Loosestrife Family

Heimia salicifolia.

Native of Mexico and south to Argentina; occasional ornamental in warm areas of California. Shrub to 5 ft tall or more; leaves linear to lanceolate, to 3 in. long; calyx with horn-like appendages alternating with the calyx lobes, petals yellow, $\frac{1}{2}$ to $\frac{3}{4}$ in. long. Plants contain several toxic quinolizidine alkaloids; cryogenine (vertine) is the most active, producing hallucinations. Symptoms include giddiness; surroundings become dark and appear to shrink; eventually deafness or auditory hallucinations may result with distorted sounds apparently coming from a distance. Its excessive use can be harmful.

Malvaceae, Mallow Family

Gossypium. **Cotton.**

Commercial cotton comes mainly from four species of *Gossypium,* two from the Old World and two from the New World.

Perennials but usually grown as an annual crop; leaves dotted with black oil glands; flowers variously colored; seeds covered with long fibers, the cotton.

Cotton, one of the oldest textile fibers, was used as cloth at least 2,500 years ago in India. Cottonseed, after removal of the fibers, is a source of an edible oil used as a substitute for olive oil and in the manufacture of margarine and soap. After the cottonseed has been processed to remove the oil, the residual cottonseed cake can be used as food for livestock because of its oil and protein content.

Toxic part: Pigment glands seen as black dots on the seeds.

Toxin: Gossypol, an unusual sesquiterpene-like, polyphenolic plant pigment, usually destroyed in making cottonseed meal.

Symptoms: In swine, respiratory problems, occasional appearance of a bloody froth at the mouth, emaciation and weakness despite normal appetite, convulsions, and cyanosis. Cattle and sheep show the same signs but are much more resistant to this toxicity.

Gossypol may be rapidly tied up with soluble proteins and decomposed in the rumen. In pigs gossypol may form a complex with iron in the liver that is then lost through the bile secretion. Therefore, it would appear that by this means gossypol reduces the oxygen-carrying capacity of the blood, causing respiratory distress. May eventually be fatal.

Malva parviflora.   Cheeseweed, Small-flowered Mallow.

Native of Europe and Asia; widespread weed in many parts of the world; in California, a weed in disturbed ground almost throughout the state, in flower most of the year. Branched annual, 1 to 2 ft tall, slightly hairy with umbrella-like hairs; leaves rounded, lobed, on long petioles; flowers small, whitish to pinkish; fruit composed of flattened pie-shaped segments that split apart at maturity. The shape of the fruit suggests the common name Cheeseweed.

Toxic part: Entire plant.

Toxin: Unknown.

Symptoms: Severe muscular tremors called "shivers" or "staggers," intensified by exercise. Prostration and death may follow. Mild cases recover.

This toxicity was reported for California many years ago but has not been fully investigated; it has also occurred in Australia and Africa. Horses, sheep, and cattle have been affected.

Ingestion of Cheeseweed and other mallows by chickens causes egg whites to turn pink during storage, the result of leakage of iron from the yolks. Two unsaturated fatty acids, malvalic acid and sterculic acid, are considered the cause of this toxicity.

Modiola caroliniana. Wheel Mallow, Ground Ivy, Red-flowered Mallow, Bristly Mallow.

Native of tropical America; a widely naturalized weed of turf and waste places. Low-growing perennial with many spreading branches; leaves rounded, irregularly toothed, and palmately lobed; flowers about ½ in. across, petals dull brick red; pie-shaped fruit splits into triangular one-seeded segments. *Modiola* has been implicated circumstantially as causing nervous disturbances in goats, sheep, and cattle. Animals showed incoordination and prostration; goats also developed paralysis of hindquarters; cattle and sheep had convulsions prior to death. The plant was suspected of causing two cases of staggers in sheep in Australia.

Sida leprosa var. *hederacea.* Alkali Mallow.

Widespread native of California and along the Pacific Coast, extending to Oklahoma and Mexico; often abundant in alkaline soils. Creeping perennial plants covered with a felty mass of stellate hairs; stems somewhat prostrate along the ground, 12 to 18 in. long; flowers about ¾ in. across, pale yellow. A report in 1939 from Los Angeles County described the death of more than fifty sheep pastured in an area where Alkali Mallow was abundant. Hair balls "by the hatful" were recovered from the dead sheep.

Meliaceae, Mahogany Family

Commercial woods are obtained from *Swietenia,* Mahogany, and from *Cedrella,* Cedar. Woods from a number of tropical species of this family produce dermatitis and respiratory symptoms in woodworkers.

FIG. 25. *Melia azedarach.* Chinaberry Tree. Lower left: Flower. Lower right: Cluster of yellow berries with stone above.

Melia azedarach.   Chinaberry Tree, China Ball Tree, China Tree, White Cedar, Cape Lilac, Pride-of-India (fig. 25).

Native of southwestern Asia, northern Australia, and eastern Africa; widely cultivated in warm areas as an ornamental, especially the cultivar 'Umbraculiformis', Texas Umbrella Tree, with a dense, spreading crown; has become naturalized in California. Deciduous, broad-crowned tree to 50 ft tall; large pinnately compound leaves; flowers small, numerous, in panicles, petals lavender, tube of stamens dark purple; fruit a rounded fleshy yellow berry about ½ in. across.

Toxic part: Entire plant, but the fruits most frequently cause poisonings; the toxin is in the fruit pulp.

Toxin: Tetranortriterpenoid neurotoxins and unidentified gastroenteric toxins.

Symptoms: In humans, symptoms begin within an hour or up to several hours after ingestion; nausea, vomiting, and diarrhea may be followed by faintness, mental confusion, and stupor; labored respiration, convulsions, and partial to complete paralysis have also been observed. The plant can be fatal if ingested in quantity.

In animals the symptoms are much the same: rapid weak heartbeat, cyanosis, excitability, and respiratory difficulties followed by paralysis. Poisoned animals may eventually develop fetid diarrhea. Asphyxiation is the cause of death.

Because the amount of toxin is relatively small, a fairly large amount of plant material must be ingested to produce significant poisoning.

Melianthaceae, Melianthus Family

Melianthus. Honeybush.

A genus of six species native of South Africa and India. Perennial herbs, woody at the base, tender to frost, odd-pinnately compound leaves with large stipules having an unpleasant odor when injured.

Melianthus comosus. Cape Honey Flower.

Native of South Africa, rarely cultivated in California. Greenish flowers orange yellow inside.

Melianthus major. Tall Cape Honey Flower.

Native of South Africa and India, occasional ornamental in California. Reddish brown flowers.

Plants of both species are shrubs 10 to 12 ft tall with large alternate pinnately compound leaves, grayish green; many flowers, slightly irregular, on a tall stalk; fruit a four-lobed inflated capsule. Entire plant is toxic, especially the root. It contains toxic bufadienolides (cardiac glycosides) that produce, in humans and animals, increased salivation, vomiting, bloody diarrhea, cyanosis of the mucous membranes, rapid weak pulse, and extreme exhaustion. Dead animals show hemorrhage and edema of the lungs, pericardial hemorrhage, general cyanosis, and congestion of the liver and kidney.

The plants are extremely toxic, requiring only a small amount to produce poisoning. They are utilized medicinally in South Africa, where human fatalities have resulted from their improper use. Domestic animals will not eat these offensively scented plants unless there is no other feed available. The toxin remains in the dried plant material. Both species have black nectar and are attractive to bees. The honey is quite black and is considered toxic.

Menispermaceae, Moonseed Family

About 65 genera and 350 species of warm, mostly tropical areas of the world. Plants of several species of the family contain alkaloids, including chondoinine and berberine, saponins, and toxic bitter substances including picrotoxin. *Cocculus laurifolius*, native of eastern Asia from the Himalayas to Japan and south to Java, is occasionally cultivated as an ornamental in California. Evergreen shrub or small tree; leaves ovate, usually 2 to 4 in. long, glossy green above, prominently three-nerved from the base; flowers small and inconspicuous. The bark contains the alkaloids cocculine and coclaurine, which have the properties of curare, a muscle relaxant. Indians of South America used the word "curare" for a group of related arrow poisons made from several alkaloids (chondoinine and others). Curare is used medicinally as a muscle relaxant but in large amounts can cause respiratory depression or arrest.

Moraceae, Mulberry Family

Ficus carica. Fig.

Probably native to Asia Minor, but it has been cultivated around the Mediterranean region and western Asia for its edible fruits for so many centuries that its geographic origin is uncertain; cultivated and occasionally spontaneous in warmer parts of California. Deciduous tree with milky juice, thickish, lobed, rough-hairy leaves, and edible fruits. As in all species of *Ficus* the small flowers are contained within a hollow structure that develops into the fig. The fig fruit is edible in *F. carica* but inedible in most other species.

Toxic part: Milky juice, particularly from the unripe figs or crushed leaves.

Toxin: 8-Methoxypsoralen, a photosensitizer; ficin, a proteolytic enzyme.

Symptoms: On latex-exposed areas, photosensitization, reddening, and blistering of the skin similar to sunburn occurs; a deep purple color may follow and sometimes even a prolonged disfiguration.

Protein-digesting enzymes are found in the latex. The milky juice of plants of many species of *Ficus* can cause irritant dermatitis; the best known is that from *F. carica* because of its wide use as a cultivated tree. This species is highly variable with many cultivars, and its irritant properties are variable as well. Fig gatherers may experience a direct irritation from the latex. Sometimes workers who dry, pack, or cook figs develop a chronic eczema on the hands. Young children who pick figs may acquire a dermatitis from the rough hairs on the leaves. Figs are not associated with systemic poisoning.

Myoporaceae, Myoporum Family

Myoporum laetum. No common name in California; called *Ngaio* in New Zealand.

Native of New Zealand; cultivated along the coast of California, rarely naturalized. Shrub or small evergreen tree 10 to 30 ft tall with broadly spreading crown; bark furrowed on old trees; young branches and leaves glandular-sticky and brownish green; leaves alternate, 2 to 4 in. long, shiny, glabrous, glandular-punctate; two to five small flowers in clusters, each flower ½ in. across, corolla white with short tube, hairs and purple spots inside tube; fruit a fleshy, oval, reddish purple drupe about ⅓ in. long. The prominent glandular punctations in the leaf are a characteristic feature of this tree.

Toxic part: Leaves most toxic, fruits less so.

Toxin: Ngaione, a furanoid sesquiterpene ketone that constitutes 70% to 80% of oil of ngaio, an essential oil.

Symptoms: In animals, dullness, loss of appetite, severe constipation with a small amount of bloodstained feces, and photosensitization from liver damage.

A dosage of 7.5 grams of dried leaves per kilogram of body weight killed a sheep in 18 hours. Ingestion of leaves of *Myoporum laetum* interferes with the excretion of bile salts and causes susceptibility to sunburn in animals. This photosensitization causes losses of livestock each year in New Zealand; cattle are the most commonly poisoned animals. The poisonings occur mostly from branches or trees felled during storms.

In July 1958, at Cypress, Orange County, one cow died and ten to twelve showed distress from eating shrubs of this species that were planted outside the fence of a dairy feedlot. The 145 cows in the herd were in excellent condition and being fed high-quality rations. When given a new lot of inferior hay, the cows overnight stripped the shrubs outside the fence.

Myristicaceae, Nutmeg Family

Myristica fragrans.   Nutmeg.

Two well-known spices are obtained from the fruit of the Nutmeg, a tropical tree not grown in California. Nutmeg comes from the large woody brown seed; mace comes from the thin net-like reddish covering of the seed. Large quantities of powdered nutmeg or mace affect the central nervous system, producing hallucinations and unpleasant side effects. In early stages of intoxication there may be headaches, dizziness, drowsiness and pronounced nausea, rapid pulse, delirium, coma. May be followed by hangover.

Myrsinaceae, Myrsine Family

Myrsine africana.   African Boxwood, Cape Myrtle.

Native to the Azores, South Africa, the Himalayas, China, and Taiwan; occasional ornamental in California. Evergreen shrub to about 8 ft tall, compact and densely leafy; leaves alternate, oval, $\frac{1}{4}$ to $\frac{3}{4}$ in. long, finely toothed in upper half, glabrous, lustrous green on upper surface; flowers tiny, of one sex only, clustered at leaf bases; fruit rounded pale blue berries about $\frac{1}{4}$ in. or less across, single-seeded. Fruits contain embelin, also called embelic acid, a mild cathartic.

Myrtaceae, Myrtle Family

This family is of very minor importance as a source of poisonous plants.

Eucalyptus.   Eucalypt, Eucalyptus, Gum Tree.

A large genus in Australia with more than 500 species. Many species, perhaps 200 to 300, have been introduced into

California at one time or another; perhaps 100 species are grown here as ornamentals; a few species have become naturalized. Small to large trees, flowers usually three or more in umbels or heads; its most distinguishing character is the fusion of the sepals and petals into a cap, or calyptra, that falls when the flower opens; numerous stamens, the conspicuous part of the flower; fruit a capsule, variable in shape and size, which opens at the top to release many small seeds.

Some species including *Eucalyptus viminalis,* Manna Gum, and *E. cladocalyx,* Sugar Gum, are cyanogenic and yield hydrocyanic acid in quantities large enough to be lethal. In Australia the Sugar Gum has caused poisoning of livestock. Nearly all species yield essential oils.

Dead bees have been found by the thousands beneath trees of *Eucalyptus.* Such losses are not caused by a toxin in the trees but by a sudden drop in temperature. On warm days, bees visit winter-flowering species of *Eucalyptus* in large numbers. A sudden cold front, however, can cause an immediate drop in temperature, paralyzing and killing the bees.

Eucalyptus globulus.   Blue Gum, Tasmanian Blue Gum.

Native of Victoria and Tasmania; cultivated widely in California and frequently naturalized along the coast. Large tree with smooth deciduous bark; juvenile leaves subopposite, lanceolate; adult leaves alternate, sickle-shaped, covered with a bluish waxy bloom; flowers showy, borne singly, numerous whitish stamens, cap very wrinkled. The only toxicity of Blue Gum plants is caused by the refined oil. Eucalyptol, a terpene, is the main constituent of oil of eucalyptus. In humans, ingestion of the oil may cause vomiting, vertigo, confusion, amnesia, and coma. In hypersensitive persons, watery blisters may result from handling.

Melaleuca quinquenervia.   Cajeput, Paper-bark Tree, Punk Tree, Tea Tree [has been confused with *M. leucadendron*].

From eastern coastal Australia, New Guinea, New Caledonia; cultivated in California as an ornamental. Tree with white spongy peeling bark, mostly pendulous branches; leaves alternate, lanceolate, to 3 in. long, usually five-nerved; flowers

white in dense hanging bottlebrush-like spikes with conspicuous stamens. The tree is a respiratory irritant from the wind-
borne volatile substances given off at the time of flowering (not
from the pollen, which is not windblown). Sensitive people in
the vicinity of a flowering tree may develop contact dermatitis.

Nyctaginaceae, Four-o'clock Family

Mirabilis jalapa.   Four-o'clock, Marvel-of-Peru, Beauty-of-
the-night (fig. 26).

Native of tropical America; commonly grown as a summer
annual, often spontaneous from self-sown seed. In warm areas
a perennial with a tuberous root weighing as much as 40

FIG. 26. *Mirabilis jalapa.* Four-o'clock. Lower center: Fruit.
Lower right: Fruit attached to calyx-like bract.

pounds, to 3 ft.tall and across; leaves opposite; calyx trumpet-shaped, 1 to 2 in. long, 2 in. across, showy, in shades of pink, red, yellow, or white, sometimes striped.

The seeds and roots contain the alkaloid trigonelline, which is an irritant to the skin and the lining of the stomach and intestine. The root was formerly used as a cathartic similar to that of Jalap, *Ipomoea purga*.

Oleaceae, Olive Family

> *Ligustrum japonicum* from Japan (fig. 27; pl. 10*c*)
> *Ligustrum lucidum* from China and Korea
> *Ligustrum ovalifolium* from Japan
> *Ligustrum vulgare* from Europe, North Africa, and western Asia

FIG. 27. *Ligustrum japonicum*. Japanese Privet. Upper right: Flower (enlarged). Lower left: Leaf. Lower center: Smooth black berry and seed (both enlarged).

These and several other species are cultivated in California as individual small trees, shrubs, or hedges. Woody evergreens; leaves opposite; small white flowers in terminal panicles, petals united into a tube; fruits nearly black berries.

Toxic part: Berries, leaves, and perhaps other parts.

Toxin: Uncertain; syringin (ligustrin), an irritant glycoside, has been identified.

Symptoms: In humans, severe gastric irritation, nausea, and vomiting develop shortly after ingestion of large quantities of berries; watery yellowish diarrhea, weak pulse, lowered body temperature, muscular twitching, and convulsions follow. May be fatal.

One patient became restless, collapsed, and died a few hours after eating the fruit. Deaths of horses and cows following ingestion of the foliage are recorded, especially from New Zealand and England. Despite their toxicity, birds readily eat the berries and regurgitate or pass the toxic seeds. Clippings of any privet should be kept away from children and livestock.

Oxalidaceae, Woodsorrel Family

Oxalis. **Woodsorrel, Lady's Sorrel.**

A large cosmopolitan genus of about 850 species of herbaceous or somewhat woody plants; species most numerous in South Africa and South America. In California a number of species are natives, ornamentals, or weeds. Small herbs, annual or perennial, many with bulbs, tubers, or rhizomes, rarely shrubby; leaves compound with three leaflets, often clover-like.

Oxalis corniculata. Creeping Woodsorrel.

A yellow-flowered woodsorrel spreading rapidly by stolons on the surface of the ground and by seeds formed even on very young plants; it can develop perennial roots. Creeping Woodsorrel is a widespread pernicious garden weed in California but never has been reported to cause livestock poisoning here; it has, however, been implicated in a few cases of sheep poisoning in Australia.

Oxalis pes-caprae [*O. cernua*]. Bermuda Buttercup.

Perennial with a fleshy thickened root and numerous scaly bulblets; leaves basal, petioles to 10 in. tall, each leaf with three leaflets, deeply notched, heart-shaped, and usually dotted with blackish spots; flowers showy, bright yellow, to 1½ in. across, in clusters or cymes of five to twenty flowers on stems 12 in. tall, above the leaves; flowers are sterile, no fruits or seeds formed, reproducing by underground bulblets.

Bermuda Buttercup is a widespread winter-growing garden plant becoming weedy wherever planted. It has spread as an agricultural weed only in a few areas as in citrus orchards in Tulare County and in artichoke fields in the Carmel Valley, Monterey County.

Poisonings from Bermuda Buttercup are not known in California, but because it is such a widespread garden weed care should be taken that plants are not accessible to livestock in rural garden areas. It is not found as a pasture weed in California.

In Australia feeding on Bermuda Buttercup plants has resulted in many fatalities to sheep grazing Subterranean Clover pastures. Subclover is alternated with crops of wheat, but cultivation of the wheat spreads the bulblets of the Bermuda Buttercup, which then becomes a weed of the Subclover pasture the following year.

The oxalate in Bermuda Buttercup is the acid potassium oxalate occurring in plant juices at a pH around 2. Presumably other species of *Oxalis* have the same toxin.

In acute poisonings from Bermuda Buttercup there is lowering of the calcium in the blood and then death without kidney involvement. Less rapid poisonings result in mild kidney complications and damage to the central nervous system with paralysis in the hindquarters or forequarters, muscular rigidity, trembling, and a very stiff gait. Chronic toxicity damages the kidney, and death results from its failure.

Papaveraceae, Poppy Family

About 25 genera and 250 species, mostly annual or perennial herbs; a few shrubs occur mainly in the Northern Hemisphere. Plants with milky or watery latex; leaves alternate, entire, lobed or deeply dissected; flowers usually large, brightly colored and showy, solitary or in few to many flowered inflorescences; petals separate, many stamens; fruit a capsule opening by pores at the top or elongated and splitting open at maturity.

The family is best known for opium obtained from the Opium Poppy. Alkaloids are found in all members of this family, particularly Opium Poppy, *Papaver somniferum;* Mexican Prickly Poppy, *Argemone mexicana;* Celandine Poppy, *Chelidonium majus.*

Argemone mexicana.　Mexican Poppy, Mexican Prickly Poppy, Thorn Poppy, Flowering Thistle, Jamaica Thistle.

Native of the West Indies and perhaps Central America and Florida; has become a widespread weed around the world; sometimes cultivated in California but also becomes weedy.

Argemone munita [*A. platyceras* in some California references].　Prickly Poppy, Chicalote (pl. 11*c*).

Native to California and adjacent states; becomes weedy especially on roadsides.

Both species are coarse prickly annuals, but some forms of *Argemone munita* are perennials; large white flowers in *A. munita,* large yellow flowers in *A. mexicana;* prickly elongated capsules with many brown or black seeds.

Toxic part: Entire plant, especially the seeds.

Toxin: Berberine, protopine, and other active isoquinoline alkaloids.

Symptoms: Use of argemone oil has caused vomiting, diarrhea, dilation of the capillaries of the skin with permanent damage, visual disturbances, and damage to the eye.

Argemone oil has been used as an emetic and a cathartic, but its action is less powerful than that of castor oil. Argemone seeds may be used as an adulterant of mustard seed, resulting in a mixture of oils.

Plants are not generally eaten by livestock because of their extremely spiny nature, but they have been suspected of poisoning cattle and horses in Australia. Domestic poultry are poisoned readily by the seeds.

Eschscholzia californica.　California Poppy.

Native of California; also widely cultivated as an ornamental; has become weedy in many countries of the world. Annual or short-lived perennial herb to 2 ft tall, grayish waxy; leaves much dissected; flowers showy in shades of yellow to deep orange, cultivated forms may be white, pink, or maroon.

Toxic part: Entire plant.

Toxin: A number of isoquinoline alkaloids.

Symptoms: Slightly narcotic with depressant effects on respiration.

Thirteen different alkaloids have been found in plants of this species. There is no morphine present; earlier reports have not proved justified. In western Australia the plant was suspected of poisoning livestock, although it is not usually eaten by animals.

Papaver somniferum.   Opium Poppy, Garden Poppy (fig. 28; pl. 11*d*).

Native of southeastern Europe and western Asia; formerly cultivated in California as a garden plant, commonly spontaneous in well-watered situations; possession prohibited by federal law. Erect waxy-covered annual to 5 ft tall; leaves coarsely toothed, the lower ones with a short petiole, the upper leaves markedly clasping the stem; flowers large, to 4 in. across, red, purple, pink, or white, petals sometimes with a fringed margin or with a dark spot at the base, flowers often double; fruit a round capsule capped with a lobed disk; many small dark seeds are released at the top through pores in the capsule when it is mature and dry.

The plant has a milky latex. Opium is the air-dried latex that is collected from the green fruits that have been cut. The maximum flow of the milky sap occurs in 24 hours. Opium has been known since ancient times as a drug for relieving pain and inducing sleep. Certainly it is one of the world's most frequently used drugs for the relief of pain, but it also has become one of the most dangerous and misused drugs.

Toxic part: Entire plant, but the dried sap as opium is the most toxic.

Toxin: Opium contains some twenty-five alkaloids, including morphine. Codeine and heroin are, in turn, derived from morphine.

Symptoms: Chronic poisoning from therapeutic use of opium results in sweating, particularly of the head; blistering

FIG. 28. *Papaver somniferum.* Opium Poppy. Right: Capsules. Upper left: Seed (enlarged).

and eczema of the skin may erupt in localized patches or cover the entire body and may be followed by peeling; headaches and fever are common; excitement, vertigo, and muscular twitching may result in lowered cardiac and respiratory rates.

Acute opium poisoning may result from accidental or suicidal taking of very large amounts of opium; the affected person becomes sleepy and finally passes into stupor and coma leading to respiratory and cardiac arrest.

Morphine is a narcotic drug, one that is analgesic, relieving pain, and also hypnotic, producing sleep. Morphine produces a feeling of well-being in the human body, but it is followed by depression and hence a desire for more morphine. Thus it is a strongly addictive drug. Codeine is a slightly modified mole-

cule of morphine, and there is good evidence that it is converted in the body to morphine. Some of its effects are produced only after this change. Codeine is only about one-tenth
as strong as morphine but has about one-fourth of morphine's
narcotic effect on the respiratory center of the brain. Therefore
codeine can be used more safely than morphine in slowing respiration and relieving respiratory distress and coughs. Codeine
is equally addictive, however, and subject to the same regulations as morphine. Addiction to opium or its derived alkaloidal
drugs leads to both physical and mental deterioration.

Cultivated varieties of *Papaver somniferum* bred for opium
content may yield 10% or even 14% of the dried weight of the
plant as opium. Bred as oilseed varieties, plants still have 1%
to 2% opium; the seeds yield an edible oil. Opium Poppy seeds
used in baking, sprinkled on the tops of breads and rolls, are
not harmful as the toxins are destroyed by heat.

Pedaliaceae, Pedalium Family

Sesamum indicum [*S. orientale*].   Sesame.

Native of the tropics; cultivated in California as an experimental crop or as a garden specialty. Annual herb to 3 ft tall;
leaves opposite, to 5 in. long, sometimes lobed; flowers borne
singly with the upper leaves, corolla tubular, white or rose,
two-lipped; fruit a capsule 1 in. long. In humans, with large
doses, the oil is a cathartic. Livestock fed exclusively on the
pressed seedcake develop colic, tremors, coughing, depression, eczema, and loss of hair.

The seed and its oil are reported as producing human abortions, but considering its widespread use in India where
abortions from this cause are unknown, the report must be in
error. The root is regarded as toxic in Africa. Sesamin, which
is found in the oil, has a synergistic effect, enhancing the toxicity of pyrethrin when used against houseflies.

Phytolaccaceae, Pokeweed Family

Phytolacca americana [*P. decandra*].   Pokeweed, Poke,
Pokeberry, Inkberry, Pigeonberry, Skoke (pl. 11*a*).

Native of the eastern half of the United States; widespread
casual weed of gardens and wet roadsides in California. Peren-

nial with a large carrot-like root often 4 to 6 in. in diameter, plant ill-smelling; succulent stems branching above, 6 to 9 ft tall, bright purple with maturity; leaves alternate, ovate; small whitish flowers to $\frac{1}{2}$ in. across in arching clusters at the ends of the shoots; fruit a berry, shiny purplish black, soft and juicy.

Toxic part: All parts are highly toxic, the root, leaves, and bark especially so. The ripe berries in small amounts are not toxic, but the seeds inside the berries are poisonous.

Toxin: Phytolaccatoxin, a resin; phytolaccigenin, a triterpene saponin. A toxic alkaloid, phytolaccine, has also been reported. Additional substances are unidentified mitogens that can be absorbed through abrasions in the skin. Mitogens are substances that increase the activity and divisions of the white blood cells and sometimes damage the red blood cells.

Symptoms: An immediate burning and bitter taste in the mouth; followed in 1 or 2 hours by soreness of the mouth and throat, salivation and thirst, lassitude, coughing, and persistent vomiting and diarrhea leading to severe dehydration and shock. Recovery usually occurs in 24 hours except for weakness and lassitude; rarely fatal.

Pokeweed was at one time widely used medicinally for a great number of purposes. The young shoots with leaves may safely be eaten as a boiled vegetable if they are boiled twice and the water is discarded after each cooking. Inadequate preparation or inclusion of any part of the root, however, can lead to poisoning. Despite these precautions, poisonings were reported in a day camp in New Jersey in 1981. Although Pokeweed was prepared in the usual way to ensure its edibility, after $\frac{1}{2}$ to 5 hours twenty-one of the campers became ill with nausea, headache, dizziness, burning in the mouth and throat, vomiting, stomach cramps, and diarrhea. Apparently there was a wide variation in individual responses. Following medical treatment, all recovered after 48 hours.

Poisoning has also occurred when young raw shoots were included in a salad or when the root was mistaken for horseradish. The berries are edible when cooked and have been used in pies. Children have been poisoned from eating raw berries, however, and these can even be fatal if large amounts are ingested. A 5-year-old girl died after drinking the juice prepared

by crushing the berries and adding sugar to make a drink resembling grape juice. Anyone handling Pokeweed should do so
with caution and wear gloves to prevent absorption of the toxins through breaks in the skin.

Rivina humilis.   Rouge Plant, Bloodberry, Baby Pepper.

Native in southeastern United States and American tropics;
cultivated, but rarely, in warm parts of southern California as
an ornamental, outside and in greenhouses. Perennial, sometimes woody at base, 1 to 2 ft tall; leaves thin-textured, ovate to
oblong, 2 to 4 in. long; flowers small, greenish white to pink,
in hanging racemes as long as the leaves; fruit a bright red
berry nearly ¼ in. in diameter, yielding a red dye. An unknown
toxin in the plant causes tainting of milk; not known to be
poisonous.

Pittosporaceae, Pittosporum Family

Pittosporum.   **Pittosporum.**

About 100 species native to the Himalayas, China, Japan,
Australia, New Zealand, South Africa, Lord Howe Island, Hawaii; about ten species are cultivated in California as ornamentals. Leaves alternate, simple; flowers creamy to greenish
white or pale to dark purple, in few to many-flowered clusters,
sometimes very fragrant; fruit a capsule that splits in halves or
thirds exposing several sticky seeds. The name *Pittosporum* is
taken from two Greek words meaning pitch and seeds in reference to the resinous coating of the seeds. The following species are the most commonly cultivated in California:

> *Pittosporum crassifolium;* from New Zealand; tree; leaves
> obovate, leathery, gray hairy beneath; flowers dark
> purple
>
> *Pittosporum erioloma;* from Lord Howe Island; shrub;
> leaves obovate to oblong, revolute margins; flowers
> creamy white to reddish purple mostly in clusters of two
> to seven.
>
> *Pittosporum eugenioides;* from New Zealand; tree; leaves
> elliptic-ovate, wavy margins; many small greenish yel-
> low flowers

Pittosporum phillyraeoides, Willow Pittosporum; shrub or small tree, branches pendulous; leaves linear-lanceolate, to 4 in. long; flowers yellow, to ½ in. long, solitary or few in a cluster

Pittosporum tenuifolium; from New Zealand; tree; leaves elliptic, about 4 in. long, wavy margins; flowers solitary or few in a cluster, dark purple

Pittosporum tobira, Japanese Pittosporum, Tobira, Mock Orange; from Japan; shrub or small tree; leaves obovate, about 4 in. long, revolute margins; flowers creamy white, fragrant, in many-flowered clusters

Pittosporum undulatum, Victorian Box, Mock Orange; from Australia; tree; leaves ovate, about 4 in. long, wavy margins; flowers creamy white to yellow, very fragrant, in clusters of twelve to fifteen

Toxic part: Entire plant.

Toxin: Saponins.

Symptoms: Extreme lethargy in fowl and in a horse has been reported in Australia; gastrointestinal distress in humans.

Plants of the genus *Pittosporum* usually are not reported as toxic. Several species native to New Zealand evidently have not caused problems there. A number of species, including *Pittosporum phillyraeoides* and the other species listed as cultivated in California, have varying amounts of saponins in their leaves and fruits. In Java and Fiji native species of *Pittosporum* are used as fish poisons.

A case was reported recently in San Francisco of a child who had eaten the fruits and some leaves of a pittosporum. The child experienced nausea, vomiting, and abdominal cramps for several days and then recovered.

Two separate deaths of horses were observed in California, near Woodland, Yolo County, and near Auburn, Placer County, after prunings of *Pittosporum tobira* were thrown into the pens. In each case the animal was confined to a small area, and both animals apparently had empty stomachs. In each instance there were many green seedpods on the branches. Saponins are not readily absorbed from the intestinal tract unless it has been injured, although the saponins themselves may cause such injury. Moreover, if there is a large quantity of food already present in the stomach, the saponins are absorbed by this material and will cause no harm. Thus it seems that saponin-bearing plants are a threat to livestock only if eaten when the stomach is empty.

Plumbaginaceae, Plumbago Family

Plumbago auriculata [*P. capensis*].   Cape Plumbago.

Native of South Africa; a popular ornamental in California. Evergreen shrub, somewhat sprawling, 5 to 6 ft or more tall; leaves alternate with a short petiole; flowers azure blue, in terminal clusters, calyx with stalked glands, slender corolla tube about two times as long as the calyx, five spreading corolla lobes. Entire plant, especially the root, contains plumbagin, a toxic naphthoquinone derivative, and oil of plumbago. The plant can cause severe skin irritation; in some people it may blister the skin. In South Africa the foliage is eaten by livestock, especially sheep, and by poultry, but under certain undetermined conditions it becomes poisonous to animals.

Polygonaceae, Buckwheat Family

Only a few members of this family are toxic.

Fagopyrum esculentum [*F. sagittatum*].   Buckwheat.

Native of central Asia; cultivated crop; at one time reported as escaped from cultivation in southern California. Annual about 3 ft tall; leaves spear-shaped; small white flowers in elongate erect or drooping clusters. Ingestion of foliage and seeds by domestic livestock followed by intense sunlight can cause photosensitization.

Rheum rhabarbarum.   Rhubarb, Garden Rhubarb, Pie Plant, Wine Plant (fig. 29). [The name *Rheum rhaponticum* has been misapplied to this species.]

Native to Manchuria; widely cultivated for its edible leaf stalks. Perennial from a large taproot; large leaves with blades to $1\frac{1}{2}$ ft long, wavy margins, on a stout, thick petiole of about equal length; flower stalks are removed to keep the leaves growing.

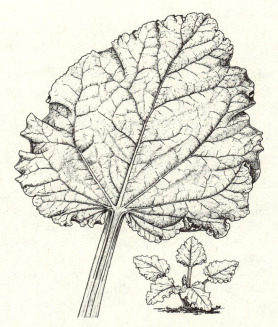

FIG. 29. *Rheum rhabarbarum.* Rhubarb.

Toxic part: Leaf blades.

Toxin: Oxalic acid and soluble oxalates of potassium and calcium.

Symptoms: First nausea and stupor, followed 1 to 2 days later by severe gastrointestinal disturbance, vomiting, sometimes bloody diarrhea, headache, and drowsiness. Coma and convulsions occur only in extreme cases. Fatal poisoning is rare.

Although the thickened leaf stalk or petiole is edible, the toxins in the leaf blade are occasionally strong enough to be potentially lethal when eaten by humans or animals. In England during World War I, in the interest of food conservation, it was recommended that the leaf blades of Rhubarb should be eaten. Several cases of poisoning and death resulted. Cattle and hogs have been poisoned from these leaf blades with some deaths recorded.

Rumex. **Dock, Sorrel.**

Mostly native in North America and Eurasia; about twenty-four species in California, both native and introduced weeds; some are grown for vegetable greens. Perennials or a few annuals; leaves various; flowers small, greenish or reddish, numerous, often in terminal panicles; fruit a three-angled achene or nutlet.

Plants contain soluble oxalates, toxic in large amounts to livestock; animals usually will not eat them.

Portulacaceae, Purslane Family

Portulaca oleracea. **Purslane.**

Native of Europe; widespread weed in California. Low mat-like succulent summer annual with fleshy wedge-shaped leaves scattered along the stems; clusters of small yellow flowers less than ¼ in. across, sessile, at the base of the leaves; fruit a capsule that splits around the middle, the upper half coming off as a lid. Sometimes used as a potherb, it is not reported as toxic in the United States. May accumulate toxic amounts of oxalates; has been suspected in Australia of poisoning sheep and cattle.

Primulaceae, Primrose Family

Toxic species of plants are rare in this family.

Anagallis arvensis. **Pimpernel, Scarlet Pimpernel, Shepherd's Clock, Poorman's-weatherglass.**

Native of Europe; common weed in California in gardens, roadsides, and fields. Annual; low-growing, slender stems spreading and branched; leaves opposite, sessile, oval, pointed, about ½ in. long; flowers borne singly with the leaves on long slender stalks; corolla salmon to brick red, rarely white or blue; rounded capsule splits horizontally with the upper half coming off like a lid to release the many seeds.

Toxic part: Entire plant.

Toxin: A triterpenoid saponin; a glycoside, cyclamin; and a volatile oil with a peculiar pungent acrid odor.

Symptoms: When eaten by humans, it causes intense headache and nausea with body pains for 24 hours. In quantity it can be narcotic. In sheep, feeding experiments have produced a staggering gait, gastroenteritis, constipation followed by bloody diarrhea, congestion of mucous membranes, and hemorrhages of the heart, rumen, and kidneys.

Scarlet Pimpernel has been more involved in poisoning of sheep, cattle, and horses in New Zealand and Australia than in the United States. It was suspected some years ago of having caused the death of horses in southern California. In July 1966, a veterinarian attributed the deaths of two horses to eating Scarlet Pimpernel in a new permanent pasture in Monterey County; it was an abundant weed in the small enclosure.

Cyclamen purpurascens [*C. europaeum*]. Common Cyclamen.

Native of central and southern Europe; a common florists' plant, planted outdoors in areas of mild climate. Perennial with tubers; basal leaves often mottled green above and reddish beneath; showy flowers fragrant, rose pink or crimson, borne singly on naked stems, nodding, and the petals reflexed and twisted. Not usually listed among poisonous plants but it contains a toxic glycoside, cyclamin. Tubers were formerly used as a cathartic.

Proteaceae, Protea Family

This family is of very minor importance as a source of poisonous plants.

Grevillea banksii. Red-flowered Silky Oak, Dwarf Silky Oak.

Native of Queensland, Australia; sometimes cultivated as an ornamental in southern California. Shrub or small tree to 20 ft tall with pinnately divided leaves and red flowers in showy one-sided clusters.

Grevillea robusta. Silky Oak, Silk Oak.

From northeastern Australia; cultivated in California as a street tree; in cold areas as an indoor potted plant. Tall tree to 150 ft with pinnately dissected leaves and orange flowers in large showy one-sided clusters.

Both species produce hydrocyanic acid in the flowers and seed pods, but no cases of poisonings are known from them. In

California the flowers of *Grevillea robusta* appear just after the flowering of orange trees. A very small amount of *Grevillea* pollen will turn an excellent orange blossom honey a dirty brown color, making it unmarketable. Keepers therefore remove bees from orange groves before the *Grevillea* trees come into bloom.

Ranunculaceae, Buttercup Family

Many members of this family are poisonous. Several species contain various toxic alkaloids or cardiac glycosides. In Europe bees are known to make toxic honey from plants of this family.

The following members of the Buttercup Family contain a glycoside, ranunculin, which breaks down by enzymatic action to form the aglycone protoanemonin, a highly irritant oil:

Actaea rubra, Red Baneberry
Actaea spicata, Herb Christopher
Anemone blanda and other species, Anemone
Caltha palustris, Marsh Marigold
Clematis, presumably all species, Clematis
Ranunculus, all species, Buttercup

Toxic part: Entire plant.

Toxin: Protoanemonin, an unstable compound that is soon converted to the harmless anemonin.

Symptoms: Painful gastrointestinal irritation; painful inflammation in the mouth and throat, sometimes with blisters and ulceration, profuse salivation; vomiting, bloody diarrhea, abdominal cramps, dizziness; and in severe cases convulsions.

Fresh buttercup plants, *Ranunculus,* growing in pastures are harmful to livestock, but the dry plants in hay are not. Livestock poisoning from buttercups is rare, however; their acrid, bitter taste usually causes animals to reject them. For the same reason buttercup poisoning in humans is rare, although poisonings are reported from using buttercup leaves in salads.

Aconitum. **Monkshood, Aconite.**

Plants of all *Aconitum* species are poisonous; the tuberous roots contain alkaloids of the aconitum group sometimes used in medicine. This genus is closely related to the larkspurs, *Delphinium,* and plants of the two genera are sometimes confused.

Aconitum columbianum.   Western Monkshood (fig. 30; pl. 12*a*).

Native in California throughout the Sierra Nevada. Although a well-known stock-poisoning plant, it seldom causes deaths of animals.

Aconitum napellus.   Garden Monkshood, Aconite, Garden Wolfbane, Helmet Flower, Friar's Cap, Soldier's Cap.

From Europe; cultivated in California gardens.

Both species are perennial herbs with tuberous roots and pithy stems; leaves alternate, much divided into segments,

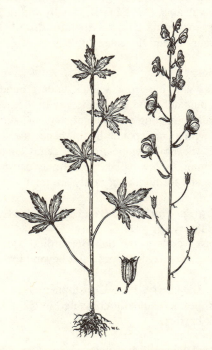

FIG. 30. *Aconitum columbianum.* Western Monkshood.
A. Mature fruits of five separate follicles.

upper leaves with very short petioles; flowers attractive and showy, blue or purple in clusters at the ends of the stems; five sepals, petal-like, colored, uppermost sepal large and hood-shaped, actual petals small.

From the dried roots of *Aconitum napellus* the drug aconite, a heart stimulant, is obtained. Formerly the roots were collected in Central Europe and used in preparing a liniment to treat neuralgia, sciatica, and rheumatism.

Toxic part: Entire plant but especially the leaves and roots; even after drying the toxic alkaloid remains.

Toxin: Aconitine, a highly toxic alkaloid, one of the most potent of plant toxins.

Symptoms: Cardiovascular disturbances and severe oppression of the chest are characteristic of aconite poisoning; tingling of the mouth and skin, nausea, lowering of the blood pressure, and convulsions. Can be fatal, usually within 2 hours but ranging from ½ to 6 hours. Various rhythmic disturbances of the heart may occur and are the cause of death.

Actaea. Baneberry.

Perennial herbs growing from a thick underground stem; leaves alternate, pinnately divided, leaflets with toothed margins; flowers small, ⅛ in. long with longer stamens; shiny berries red or white with a dark spot at the tip, hence the common name Doll's Eyes. Several members of this genus contain the toxin protoanemonin. As few as six berries may cause severe symptoms that persist for hours but rarely are fatal.

Actaea rubra subsp. *arguta.*   Red Baneberry, Doll's Eyes, Snakeberry, Coralberry (fig. 31).

Native in moist woods in the San Bernardino Mountains, the Sierra Nevada, the Coast Ranges, and beyond California.

Actaea spicata.   Baneberry, Herb Christopher.

Native of Europe; occasionally cultivated in California for its purplish black berries.

Adonis aestivalis.   Summer Pheasant's-eye.

Native of Europe; established in California as a weed along the Pitt River at Canby, Modoc County, and in a meadow pas-

PLATE 1

a. *Amanita pantherina*.
Panther Mushroom.

b. *Amanita muscaria*.
Fly Agaric.

d. *Agaricus xanthodermis*.

c. *Amanita phalloides*.
Death Cap.

PLATE 2

a. *Pteridium aquilinum*.
Bracken Fern.

b. *Taxus baccata* 'Stricta'.
Irish Yew.

c. *Toxicodendron diversilobum*.
Pacific Poison Oak. Leaves.

d. *Toxicodendron diversilobum*.
Pacific Poison Oak. Dry-season
leaves and fruits.

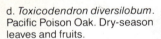

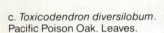

PLATE 3

a. *Conium maculatum*.
Poison Hemlock. Plant.

b. *Conium maculatum*.
Poison Hemlock. Stem and
base of petiole.

c. *Cicuta douglasii*.
Western Waterhemlock.

d. *Cicuta douglasii*.
Western Waterhemlock.
Longitudinal section of tuber showing
characteristic orange color and
chambered pith.

PLATE 4

a. *Ammi visnaga*.
Toothpick Ammi.

b. *Catharanthus roseus*.
Periwinkle.

c. *Nerium oleander*.
Oleander.

d. *Thevetia peruviana*.
 Yellow Oleander.

PLATE 5

a. *Asclepias fascicularis*.
Whorled Milkweed.

b. *Asclepias speciosa*.
Showy Milkweed.

c. *Centaurea solstitialis*.
Yellow Starthistle.

d. *Helenium hoopesii*.
Orange Sneezeweed.

PLATE 6

a. *Senecio jacobaea*.
Tansy Ragwort.

b. *Xanthium strumarium*.
Cocklebur.

c. *Tetradymia canescens*.
Spineless Horsebrush.

d. *Tetradymia glabrata*.
Littleleaf Horsebrush.

PLATE 7

a. *Amsinckia intermedia*.
Fiddleneck.

b. *Lophophora williamsii*.
Peyote.

c. *Halogeton glomeratus*.
Halogeton.

d. *Sarcobatus vermiculatus*.
Black Greasewood. Branch with
erect spikes of male flowers.

PLATE 8

a. *Cannabis sativa*.
Marijuana.

b. *Datisca glomerata*.
Durango Root.

c. *Rhododendron occidentale*.
Western Azalea.

d. *Euphorbia milii*.
Crown-of-thorns.

PLATE 9

a. *Abrus precatorius*.
Rosary Pea. Seeds used as beads.

b. *Ricinus communis*.
Castor Bean. Seeds polished
and used as beads.

c. *Astragalus asymmetricus*.
Horse Loco.

d. *Ricinus communis*.
Castor Bean.

f. *Robinia pseudoacacia*.
Black Locust.

e. *Lupinus latifolius*.
Broadleaf Lupine.

PLATE 10

a. *Sesbania punicea*.
Scarlet Wisteria Tree.

b. *Hypericum perforatum*.
Klamathweed. Plant.

d. *Hypericum perforatum*.
Klamathweed. Leaf with black
and translucent glands.

c. *Ligustrum japonicum*.
Japanese Privet.

PLATE 11

a. *Phytolacca americana*.
Pokeweed.

b. *Prunus virginiana* var.
demissa.
Western Chokecherry.

c. *Argemone munita* subsp.
rotundata.
Prickly Poppy.

d. *Papaver somniferum*.
Opium Poppy.

PLATE 12

a. *Aconitum columbianum*.
Western Monkshood.

b. *Delphinium variegatum*.
Royal Larkspur.

c. *Digitalis purpurea*.
Common Foxglove.

d. *Atropa belladonna*.
Deadly Nightshade.

PLATE 13

a. *Brugmansia sanguinea*.
Red Angel's Trumpet.

b. *Cestrum aurantiacum*.
Orange-flowered Cestrum.

d. *Nicotiana glauca*.
Tree Tobacco.

c. *Datura stramonium*.
Jimsonweed.

PLATE 14

a. *Datura meteloides*.
Tolguacha. Flower.

b. *Datura meteloides*.
Tolguacha. Fruits.

c. *Solanum laciniatum*.
Kangaroo Apple.

d. *Solanum pseudocapsicum*.
Jerusalem Cherry.

PLATE 15

a. *Solanum sarrachoides*.
Hairy Nightshade. Fruits.

b. *Daphne odora* 'Marginata'.
Winter Daphne.

c. *Dieffenbachia maculata*.
Spotted Dumb Cane.

d. *Philodendron bipinnatifidum*.

PLATE 16

a. *Veratrum californicum*.
Corn Lily.

b. *Zigadenus venenosus*.
Meadow Deathcamas.

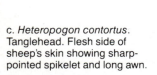

c. *Heteropogon contortus*.
Tanglehead. Flesh side of
sheep's skin showing sharp-
pointed spikelet and long awn.

d. Cattle affected by phalaris
staggers showing weight loss on
the feedlot.

FIG. 31. *Actaea rubra* subsp. *arguta*. Red Baneberry. Lower
left: Flower (enlarged). Lower right: Fruiting stem with an enlarged berry and seed.

ture in northwestern Lassen County; also naturalized near
Klamath Falls, Oregon. Annual plants to $1\frac{1}{2}$ ft tall, unbranched
except toward the top of the plant; leaves alternate, dissected
into linear segments; flowers $1\frac{1}{2}$ in. across, five to eight petals,
bright crimson with a purplish black base, many stamens,
blackened pistils on a raised receptacle; fruits numerous hardened angular beaked achenes borne in an elongated head about
1 in. tall. Entire plant contains a cardiac glycoside known as
adonidin that produces nausea, vomiting, abdominal pain, diarrhea, and blurred vision. Established as a weed since 1930;
poisonings from Summer Pheasant's-eye are not known in California, even though this weed may be quite abundant in some
years. It has caused livestock deaths in other countries, how-

ever, particularly from the dried plants in hay. Species of *Adonis* have been used medicinally in Europe.

Anemone. **Anemone, Windflower, Pasque Flower.**

Six species are native in California; in wooded areas, mostly in the mountains; several species are cultivated as ornamentals. All species of *Anemone* are considered poisonous.

Anemone blanda. Poppy-flowered Anemone.

From southeastern Europe and adjacent Asia Minor.

Anemone × hybrida. Japanese Anemone.

Garden hybrid of forms from Japan and Nepal. Propagated commercially in quantities along the coast of California; very popular garden plants.

Ingestion of plants of species of *Anemone* produces severe gastroenteritis and other symptoms of protoanemonin poisoning. Plants of native species of *Anemone* do not occur in sufficient numbers to cause distress to livestock.

Caltha palustris. Marsh Marigold.

Widespread native of Europe and North America but not in California; cultivated in California gardens. Contains protoanemonin, a highly irritant oil (see p. 212).

Clematis. **Clematis, Traveler's Joy, Old Man's Beard.**

Several species are native to California; others from Europe and Asia are commonly cultivated in California. Climbing vines sometimes woody; leaves opposite, usually pinnately divided into three or five leaflets, rarely undivided; flowers showy and often brightly colored. The leaves of any species of *Clematis* must be regarded as toxic. The native species are not eaten by animals. See discussion of protoanemonin (p. 212).

Consolida ambigua [*Delphinium ajacis*]. Rocket Larkspur, Annual Larkspur.

Native of southeastern Europe and Asia Minor; commonly cultivated. Annual plants to 5 ft tall; leaves dissected three times into linear segments; flowers blue, pink, or white, five petal-like sepals, the upper one with a long spur; fruit a follicle, pubescent, to 1 in. long.

Toxic part: Entire plant, especially the seeds.

Toxin: Ajacine and ajaconine, toxic diterpene alkaloids.

Symptoms: In livestock, initially difficult and painful walking, lowered temperature, and slowness of pulse and respiration. In severe cases, muscular spasms and death from respiratory failure.

The most poisonous parts of the plant are the seeds, which contain a fixed oil; used since ancient times to rid the body of external parasites.

Delphinium. **Larkspur, Cow Poison, Poison Weed.**

About thirty species are native to California; several are cultivated. All species should be regarded as poisonous. In California larkspurs probably cause more loss of cattle than any other group of poisonous plants. Annual herbs or perennials from rhizomes or tubers, 1 to 6 ft tall; leaves alternate, on long petioles, leaf blades palmately divided or lobed; flowers showy in tall clusters above the leaves; five petal-like sepals, one prolonged into a conspicuous spur, variously colored blue, purple, white, or red, two or four small petals, many stamens, one to five pistils, when in fruit becoming erect pods that split down one side.

The larkspurs are similar in appearance to the monkshoods, *Aconitum,* and the two might be confused. Both genera have flowers with small petals and five petal-like sepals. The flower of larkspurs has one sepal prolonged into a conspicuous spur while the Monkshood flower has one sepal enlarged and hood-like. Moreover, the larkspur stem is hollow whereas the stem of Monkshood is solid.

In California some native species of *Delphinium* are widespread and common. The tall larkspurs are found in wet habitats, extending to higher elevations in the mountains, plants 3 to 6 ft tall; the low larkspurs are common in grassy fields in the foothills and valleys and even extending to the desert, plants usually 1 to 2 ft tall.

Plants of any larkspur species that are locally abundant can be regarded as a source of livestock poisoning. The following tall larkspurs are sufficiently abundant to be of concern:

Delphinium californicum, Coast Larkspur, found in the South Coast Ranges from Contra Costa and San Francisco counties south to San Luis Obispo County

Delphinium glaucum, Mountain Larkspur, in the San Gabriel and San Bernardino mountains, growing up to 10,000 ft elevation in the Sierra Nevada from Madera County and northward to Siskiyou County, extending to the Rocky Mountains and Alaska

Delphinium trolliifolium, Poison Larkspur or Cow Poison, at low elevations of Humboldt County and northward in adjacent Oregon

The low larkspurs associated with livestock poisoning are:

Delphinium andersonii, Anderson's Larkspur, east side of the Sierra Nevada from Mono County to Siskiyou and Modoc counties, extending to the north and east of California

Delphinium gypsophilum, Gypsum Larkspur, from Kern County to San Joaquin County, particularly along the west side of the San Joaquin Valley

Delphinium hesperium, Western Larkspur, low elevations of the Central Valley and southward into the mountains of San Diego County

Delphinium menziesii, Menzies' Larkspur (fig. 32), along the coast from Mendocino County northward

Delphinium parryi, Parry's Larkspur, along the inner South Coast Ranges from Santa Clara County south to Santa Barbara County and along the coast from San Luis Obispo County to Baja California

Delphinium patens, Smooth Larkspur, at middle and lower elevations of the Central Valley and south in the mountains of southern California to Baja California

Delphinium recurvatum, Alkali Larkspur, bottomlands of the Central Valley

Delphinium variegatum, Royal Larkspur (pl. 12*b*), grasslands and woodlands of the Great Valley and the Coast Ranges

FIG. 32. *Delphinium menziesii*. Menzies' Larkspur.
A. Mature fruit of three separate follicles.

The following larkspurs are grown in gardens in California:

Delphinium × *belladonna,* a garden hybrid

Delphinium elatum, Candle Larkspur, from eastern Europe to Siberia

Delphinium grandiflorum, Chinese Larkspur, from Siberia and China

Delphinium semibarbatum, from Iran, with yellow flowers

Some of the native species of larkspur may also be cultivated in California gardens.

Toxic part: Entire plant, particularly young plants and seeds.

Toxin: Very toxic diterpene alkaloids: delphinine, ajacine, ajaconine, which are closely related to aconitine in the species of *Aconitum.*

Symptoms: In cattle, the alkaloids act on the nervous system producing weakness. The animal develops excitability, a stiff gait, and a stance with the hind legs spread apart; then the animal suddenly drops, sometimes with muscular spasms. Nausea may result in vomiting, and the ingesta will cause asphyxiation. Instead of vomiting, bloat may develop. Constipation is a consistent feature of larkspur poisoning, and animals treated for it may be expected to recover. In some cases poisoned animals may recover fully in 24 hours.

Cattle are attracted to larkspur plants in the spring and are lethally poisoned by eating these plants at 0.7% of their body weight in an hour. This toxicity is rapidly lost as the plants flower and go to seed. The seeds are toxic, but after they are shed the dried plants are about one-sixteenth as toxic as the young ones. In experiments sheep and horses develop the same symptoms as cattle but rarely do so, if ever, under range conditions. Horses do not graze these plants in sufficient amounts to be poisoned. Sheep are very resistant to larkspur toxicity. Per pound of body weight, about six times as much of the young plants are needed to kill a sheep compared to the fatal dose for a cow. For some unknown reason, however, the toxicity of the plants to sheep does not drop markedly as the season advances.

Good management of grazing may be used to reduce losses from larkspur poisoning. Cattle should not be permitted to eat the very young plants. Grazing of infested areas should be deferred until the pods have shattered and the plants are in the least toxic stage. Using older cows accustomed to the toxic plants and not the young inquisitive stock will help to reduce losses; using sheep in heavily infested areas may also be successful because of their resistance to the larkspur alkaloids. Hungry animals should not be driven through larkspur areas because they will not graze selectively.

Chemical control of larkspurs may be indicated if there is sufficient cause for concern. Enough oil or some other distasteful material should be added to the herbicidal spray so that the spray itself does not attract animals to the poisonous plants. The addition of oil, however, may reduce the herbicide's effectiveness in killing the larkspur plants.

Helleborus.   **Hellebore.**

Several species, native of Europe and Asia Minor, are cultivated in California gardens:

Helleborus foetidus, Stinking Hellebore; plant ill-smelling
and flowers greenish with purple margins

Helleborus lividus subsp. *corsicus,* Corsican Hellebore;
flowers pale yellowish green

Helleborus niger, Christmas Rose; flowers greenish white
becoming pink or purple with age

Helleborus orientalis, Lenten Rose; flowers creamy white
turning brownish

Perennial herbs with stout rootstocks; leaves mostly basal,
palmately divided; flowers 1 to 2 in. across in clusters; five
sepals, green or colored, petal-like; petals reduced to nectar-
producing structures; many stamens; several pistils, becoming
pods that split open along one side.

Toxic part: Entire plant.

Toxin: Ranunculin, the protoanemonin glycoside; also cardiac glycosides, one of which, hellebrin, occurs in the roots of
some species. Both fresh and dried plants are toxic.

Symptoms: See protoanemonin (p. 212). In humans the cardiac glycosides have a digitalis-like effect: burning sensation
in the mouth, nausea, vomiting, acute diarrhea, delirium, and
convulsions. Can be fatal. Fresh juice of the plant can cause
dermatitis. In sheep and cattle acute manifestations are diarrhea with blood and mucus and frequent, excessive urination.
Neither human nor livestock poisonings from hellebores are reported for California.

Ranunculus.   **Buttercup, Crowfoot.**

Of the thirty-one species of buttercups listed as occurring in
California, most are native; several are naturalized introductions from the Old World.

The best-known cultivated buttercup is *Ranunculus asi-
aticus,* Persian Ranunculus or Turban Ranunculus, with tuberous roots and large flowers, 3 to 5 in. across, in shades of
red, pink, yellow, or white. The dried tuberous-rooted crowns
are produced commercially in large quantities along the south-

ern coast of California. The Creeping Buttercup, *Ranunculus repens,* has invasive tendencies and becomes weedy in cool shady gardens; it is naturalized in California mostly in the central and northern parts of the state along the coast.

Presumably all buttercup species contain the same toxic substance and produce the same symptoms. Poisoning by buttercups is rare among grazing animals because the acrid sap and burning taste of the plants make them highly undesirable.

Toxic part: Juice of the entire plant.

Toxin: Ranunculin, a glycoside, forms protoanemonin.

Symptoms: The sap is known to be highly irritating to the skin and mucous membranes, causing blisters and ulcers. It has been reported that beggars in Europe used the juice of buttercups to produce ulceration of their feet in order to arouse sympathy. If eaten the plant causes gastrointestinal irritation similar to that of *Actaea, Anemone, Caltha,* and *Clematis.*

Ranunculus sceleratus. Celeryleaf Crowfoot, Cursed Crowfoot (fig. 33).

Native to Europe; rare weed in California but known for many years in San Joaquin County. A native form, var. *multifidus,* grows in Siskiyou and Modoc counties, extending northward to Alaska and eastward as well. Annual herb 1 to 2 ft tall with stems clustered at the base of the plant; leaves kidney-shaped and divided into three lobes; flowers less than $\frac{1}{4}$ in. across with light yellow petals; fruit with a short straight beak, not the typically conspicuous beak of most buttercups.

In April 1959, in the Manteca district of San Joaquin County, a veterinarian found that Celeryleaf Crowfoot was responsible for the distress of five head of a herd of twenty-two cows. The animals displayed blindness and inability to focus, constipation, lameness, and excessive salivation with grinding of the teeth.

Ranunculus testiculatus [*Ceratocephalus testiculatus*]. Bur Buttercup.

Native of the Old World; rapidly spreading weed in northeastern California. Prostrate annual to 5 in. tall; conspicuous elongated spike of achenes when in fruit. Toxicity not known

FIG. 33. *Ranunculus sceleratus*. Celeryleaf Crowfoot. Lower left: Flower (enlarged). Lower right: Achene
(enlarged).

in California, but reported in 1983 in Utah as causing the death
of 150 sheep out of 800 held in corrals in which the plant cover
was about 50% Bur Buttercup.

Rhamnaceae, Buckthorn Family

Several species of the genus *Rhamnus,* the buckthorns, are
used medicinally as cathartics. Plants of Coyotillo, *Karwinskia
humboldtiana,* are toxic to humans and livestock. It is native
from Texas south into Mexico including southern Baja California; listed in error as native to southern California.

Rhamnus. Buckthorn.

Five native California species are cultivated along with several species from other countries. Mostly deciduous shrubs;
minute greenish flowers; fruit fleshy with several stones or
seeds in each.

Rhamnus cathartica. Common Buckthorn, Purging Buckthorn, Hartsthorn.

Native of Europe and Asia; escaped from cultivation in northeastern United States; cultivated occasionally in California. Shrub to 12 ft tall, some branches tipped with spines; leaves opposite, ovate, three to five main veins from the base; fruits each with three or four seeds or stones.

Rhamnus frangula [*Frangula alnus*]. Alder Buckthorn, Black Alder. [Not to be confused with the native *Rhamnus alnifolia,* Alder-leaved Buckthorn.]

Native from Europe, North Africa, and Central Asia; occasionally cultivated in California. Shrub or small tree to 18 ft tall; branches not spiny; leaves obovate with eight or nine pairs of veins; fruits with two or three seeds or stones in each.

Rhamnus purshiana. Cascara Sagrada [Spanish for "sacred bark"], Bearberry.

Native in northern California and extending to the Pacific Northwest and Montana; sometimes cultivated in California gardens. Tree to 25 ft or more, without spines; leaves oblong, 2 to 6 in. long with ten to twelve pairs of veins; fruits mostly three-seeded.

Toxic part: Fruits and bark.

Toxin: Emodin anthraquinone, the purgative aglycone formed by hydrolysis of the glycosides.

Symptoms: Nausea, dizziness, vomiting, abdominal pain, and watery bloody diarrhea if this drastic cathartic is used in excess.

Although the fresh fruits are particularly harmful to children, it is necessary to eat about twenty to produce a marked effect. Chewing the fresh bark will produce similar symptoms.

In Europe the buckthorns have been used as laxatives since time immemorial. The medical history of *Rhamnus cathartica* dates back in England before the Norman conquest. During the sixteenth century European writers on medicinal plants mention it. In 1650 it was first listed in the London Pharmacopoeia. Because of its drastic action its use in England was gradually replaced during the middle of the last century by the Alder Buckthorn, *Rhamnus frangula.*

Dried bark known as *cascara sagrada* is obtained from native trees of *Rhamnus purshiana* growing in the wild in the Pacific Northwest and is a source of several cathartic drugs available commercially.

Rosaceae, Rose Family

A number of plants in the Rose Family have cyanogenic glycosides and have been implicated in poisonings in various situations. These plants include:

Cercocarpus, Mountain Mahogany
Cotoneaster, Cotoneaster
Eriobotrya japonica, Loquat
Heteromeles arbutifolia, Toyon, Christmas Berry
Malus domestica, Apple
Prunus, Apricot, Peach, Nectarine, Almond, Bitter Almond, Sweet Cherry, Sour Cherry, Plums, Cherry Plum, Japanese Plum, Flowering Cherries, Cherry Laurel, Portugal Laurel, Holly-leaved Cherry, Chokecherry (pl. 11*b*)
Pyracantha, Firethorn
Rhodotypos scandens, Jetbead

Toxic part: Seed or kernel inside the usually hard pit of the fruit and the leaves, particularly when damaged by frost or wilting.

Toxin: Amygdalin and prunasin, cyanogenic glycosides that break down to release hydrocyanic acid, also known by its older name prussic acid. Release of hydrocyanic acid from the cyanogenic glycoside requires an enzyme that may be formed by damaged plant cells or may be supplied by the bacteria in the alimentary canal.

Symptoms: Large amounts of toxic material eaten rapidly may produce death without symptoms in livestock. Smaller amounts are usually encountered, however, and produce symptoms within the first hour of ingesting the toxic plant parts. Breathlessness or deep respiration appear suddenly. Both cattle and sheep die of internal suffocation after a brief period of convulsions. The animal goes down and may turn the head over the flank with a bloody discharge from the nose and mouth. The blood is bright red in color, the result of oxygen being retained in the blood instead of being released to the cells of the body. If this toxic condition continues, the blood will become dark red, the result of a lowered rate of respiration. Animals may be bloated in the final stages of poisoning by hydrocyanic acid. Prompt treatment by a veterinarian will result in almost immediate recovery.

The use of amygdalin, commercially known as Laetrile, as an alternative treatment for cancer presents possibilities of human poisonings. Such amygdalin preparations are usually made from the seed or kernel inside the apricot pits. A report of January 1982 describes the prompt medical treatment that saved a woman in Sacramento who had ingested a lethal amount of amygdalin by grinding up bitter almonds and drinking them in a glass of water. She experienced intense abdominal pain and collapsed 15 minutes after ingesting the material.

Experimentally, injected amygdalin preparations are not as toxic as ingested material, since the bacteria of the intestinal tract supply the enzymes that release hydrocyanic acid from the amygdalin.

Near Modesto, Stanislaus County, a number of house finches died from eating apricot seeds. A gravel driveway had been filled with apricot pits that were crushed by the automobiles, giving the birds access to the seeds.

Many cultivars of species and hybrids of Firethorn, *Pyracantha,* are grown in California. In early winter, flocks of robins behave erratically after feeding on Firethorn fruits, resulting in the birds flying into windows or being killed by passing motor vehicles. Alcohol in the fruits damaged by frost has been postulated as the cause of this behavior, but a test made on such fruits in Sacramento proved negative for alcohol. The seeds of *Pyracantha* are known to produce small amounts of hydrogen cyanide.

Rubiaceae, Madder Family

Cephalanthus occidentalis.   Buttonbush, Buttonwillow.

Native in California, usually along streams and in wet places. Large shrub or small tree; leaves opposite; flowers small, white, in tight heads to 1 in. across, with projecting styles giving the flower head a pincushion effect. Reported for California, but without documentation, as being poisonous to animals when eaten in quantity, producing spasms and paralysis. According to early European sources it contains a bitter glycosidic principle. Buttonbush is listed as a fairly important source of honey; the flowers are worked by bees.

Rutaceae, Rue or Citrus Family

This family is of minor importance as a source of toxicity. The following plants, either native or cultivated in California, may cause photodermatitis in humans:

Citrus aurantifolia, Lime; probably native to Southeast Asia

Citrus aurantium, Sour Orange; probably native to Southeast Asia

Citrus aurantium 'Bouquet', Bergamot Orange

Citrus × *limonia,* Rangpur Lime; hybrid origin

Citrus × *paradisi,* Grapefruit; hybrid origin

Dictamnus albus, Dittany, Gas Plant; native of Europe and Asia

Ptelea crenulata, Hop Tree; native to the Central Valley of California

Ptelea trifoliata, Wafer Ash, Hop Tree; native in eastern United States

Ruta chalepensis, Rue; native of the Mediterranean region

Ruta graveolens, Rue; native of southern Europe.

Toxic part: Entire plant.
Toxin: Psoralens, furocoumarins derived from coumarin.
Symptoms: Localized sunburn.

Plants of these species contain furocoumarins that sensitize the skin to long-wave ultraviolet light. Furocoumarins penetrate the skin more readily if it is wet. Six to 24 hours later, after exposure to sunlight, the affected areas develop a localized sunburn. Furocoumarins bind pyrimidine bases and nucleic acids in the cells to damage the skin. The injury may vary in severity from a mild redness to blisters, but with little itching. The affected area becomes pigmented, persisting in some cases long after the redness disappears.

Trifoliate Orange, *Poincirus trifoliata,* native to China, is cultivated in California as a rootstock on which other citrus may be grafted. Its fruit contains an unidentified toxin, perhaps a saponin. Because of its bitter acrid taste it is not likely to be eaten in any quantity, but presumably it could cause gastro-enteritis.

Sapindaceae, Soapberry Family

Various species of this family have cyanolipids in the seed oils. Cyanolipids decompose to form hydrocyanic acid and one or more groups of fatty acids. *Cardiospermum grandiflorum* forma *hirsutum, Koelreuteria paniculata, Paullinia tomentosa, Sapindus drummondii, Sapindus utilis,* and *Ungnadia speciosa* all contain cyanolipids. Other species, such as *Litchi chinensis* and *Dodonaea viscosa,* do not have cyanogenic lipids in the seeds.

Alectryon excelsus.

Native of New Zealand where it is called by the Maori name, *Titoki;* rarely cultivated in California. Evergreen tree; leaves pinnately compound with four to six pairs of leaflets; flowers small, inconspicuous, in short clusters; fruit a hard capsule, showy when open, revealing the shiny black seed partly enclosed in a red fleshy covering. The plant produces hydrocyanic acid.

Cardiospermum grandiflorum. Balloon Vine.

Africa and tropical America; rarely cultivated in California for ornament. Perennial vine climbing by tendrils; leaves hairy, divided into threes and each part divided further into three leaflets; flowers white, four-parted; capsule much inflated and balloon-like, bearing up to three black seeds. Plants of forma *hirsutum* have longer hairs on the stems and leaves.

Eating seeds of the forma *hirsutum* caused vomiting in children in Riverside, Riverside County, in April 1966. Presumably the cyanogenic lipid could have caused this reaction. No test for hydrocyanic acid was made on the material producing the distress in the children because the occurrence of the cyanolipid was not known in these plants until 1974.

Dodonaea viscosa. Hopbush, Clammy Hopseed, Red Hopseed, Hopseed Bush.

Widely distributed from southwestern United States to South America and scattered areas of the Old World; planted in

California for ornament, particularly a purple-leaved form. Evergreen shrub or small tree; leaves simple, sticky, narrowly lanceolate or oblong to 4 in. long; flowers small, inconspicuous, in a loose cluster; fruits dry, three-winged, cream or pink, showy.

Toxic part: Leaves.

Toxin: An acid resin is the active principle; a saponin, an alkaloid, and a glycoside are also present.

Symptoms: Not described. The plant generally is not associated with poisonings but was suspected in the deaths of calves in Queensland, Australia.

The plant has been used in native medicine for numerous ailments. The leaf is said to have anesthetic properties. In some areas the plant has been used as a fish poison.

Sapindus drummondii.   Western Soapberry, Drummond Soapberry.

Native of southwestern United States and adjacent Mexico; rarely cultivated as an ornamental in California. Deciduous tree; leaves pinnately compound with eight to eighteen lanceolate, pointed leaflets; small flowers in clusters at the ends of branches; fruit a berry with a single seed. The entire plant is poisonous, especially the berries, which contain a toxic saponin and a cyanolipid in the seeds. The crushed fruits have been used to poison fish. The fleshy layer of the berry rubbed between the fingers in water forms a lather, used as a substitute for soap. Plants can cause dermatitis.

Saxifragaceae, Saxifrage Family

This family is of minor importance as a source of poisonous plants.

Hydrangea macrophylla.   French Hydrangea, Garden Hydrangea, Bigleaf Hydrangea.

Native of Japan; widely cultivated as a florists' potted plant and out of doors in California. Shrub, nearly evergreen in parts of California; leaves opposite; numerous small fertile flowers arranged in flattened clusters surrounded by a ring of larger, showy sterile flowers; sometimes nearly all flowers are sterile and showy and arranged in a bun-shaped cluster.

Toxic part: Flower buds and leaves, particularly the young leaves.

Toxin: Hydrangin, a cyanogenic glycoside that on hydrolysis yields hydrocyanic acid. This substance also occurs in the species of *Hydrangea* native to the eastern United States.

Symptoms: Gastroenteritis, nausea, vomiting, diarrhea, abdominal pain, lethargy, and coma.

A case of poisoning is on record from eating buds of a *Hydrangea* put into a salad. *Hydrangea* leaves have been used as a tea substitute. The glycoside is present in greater amounts in young leaves and may be lacking in older leaves of the fall season.

The dried leaf and stem of *Hydrangea macrophylla* contain an alkaloid that is similar to the Chinese antimalarial drug obtained from the related *Dichroa febrifuga*. Experimentally it does have an effect against malaria, but it readily causes vomiting. This alkaloid is an important starting point for investigating related compounds, some of which have proved more medicinally effective than the natural alkaloid.

Scrophulariaceae, Figwort Family

Digitalis purpurea. Common Foxglove (pl. 12*c*).

Native of western Europe; grown as an ornamental throughout California, naturalized along the coast, more commonly the north coast areas but as far south as Santa Barbara County. Biennial, rarely a short-lived perennial, forming a large rosette of basal leaves the first year from seed; the flowering stem to 6 ft tall appears the second year; leaves rough and hairy, reduced in size toward the upper flowering portion of the stem; flowers hanging down on one side of the upper 1 or 2 ft of the stem, corolla inflated tubular, purple, pink, or white, spotted inside the tube; fruit a many-seeded capsule; commonly produces volunteer seedlings in the garden.

Toxic part: Entire plant; the dried leaves are the drug digitalis.

Toxin: A number of steroid glycosides and saponins.

Symptoms: Stimulation of the heart and, indirectly, an increase in the flow of urine. Foxglove has furnished an important cardiac medication since the 1700s. The contraction of the heart is both slowed and strengthened. Continued use or over-dosage will cause toxic symptoms initially of anoxia (lack of oxygen) and vomiting. Disruptions of the heart rhythm and pain below the sternum develop. There may be salivation, abdominal pain, and diarrhea. Color vision commonly becomes disturbed; white appears tan or everything appears bright whitish yellow or green. Headache, delirium, and convulsions may also occur. The presence of saponins may cause vomiting. Can be fatal.

Livestock will not eat this plant because of its disagreeable taste. There are California reports of horses being poisoned before 1900 from both green plants and those dried in hay. Cattle, pigs, and turkeys also have been poisoned. Toxic honey from Foxglove is recorded in Europe.

In May 1977, an elderly couple in the state of Washington died from drinking an herbal tea made with Foxglove. The wife had mistaken Foxglove leaves for those of Russian Comfrey.

Another case of mistaking Foxglove for Comfrey occurred in November 1982 in Marin County, California. Two people ate a soup containing fifteen leaves of Foxglove. One person was nauseated, vomited, and had blurred vision. The other experienced dull pain, irritability, and nausea but did not vomit. Both recovered.

Simaroubaceae, Quassia Family

Ailanthus altissima. Tree-of-heaven, Chinese Sumac.

Native of China; originally cultivated, now widely naturalized and weedy throughout the temperate parts of the world. Rapidly growing deciduous tree; leaves alternate, large, to 3 ft long, eleven to twenty-five leaflets per leaf, leaflets 3 to 5 in. long with one to four teeth at the base of each leaflet, each tooth bearing a conspicuous gland; flowers inconspicuous; fruit a two-winged samara, reddish, up to 2 in. long, borne in huge showy clusters that almost cover the tree during the summer.

Toxic part: Leaves, bark, flowers.

Toxin: Ailanthin, a very bitter substance; also a glycoside and a saponin.

Symptoms: In humans, weakness, vertigo, and nausea from chewing a leaf or piece of bark. In sheep, eating the bark resulted in rapid respiration and paralysis of the hindquarters. Drinking water, when contaminated with flowers, has an unpleasant odor and has caused gastritis and dermatitis.

The foliage has a disagreeable odor; livestock avoid the plants completely.

Solanaceae, Nightshade Family

This family is well known for the large number of alkaloids, some toxic, derived from many of its plants. Several alkaloids have a long history of medicinal use. Hyoscyamine is the principal solanaceous tropane alkaloid. The plants in which it is abundantly found are *Atropa belladonna,* Deadly Nightshade or Belladonna, and *Hyoscyamus niger,* Black Henbane. In addition, scopolamine occurs in species of *Brugmansia,* angel's trumpets, *Datura,* jimsonweeds, and also in *Mandragora officinarum,* Mandrake. Atropine occurs only in trace amounts in these plants, but it is readily produced commercially by a slight rearrangement of the atoms of hyoscyamine.

Nicotine, another toxic solanaceous alkaloid, is found in various species of *Nicotiana,* the tobaccos. Cultivated tobaccos, used for smoking, chewing, and snuff, are some of the world's best-known harmful plants. Because of their alkaloids some plants of the Solanaceae are extremely toxic to humans and domestic animals. These alkaloids remain toxic in the dried plants.

Atropa belladonna.   Deadly Nightshade, Belladonna (pl. 12*d*).

Other plants are called nightshades, especially in the genus *Solanum.*   Native to the Mediterranean region and western Asia; rarely cultivated in California gardens. Perennial with erect branched stems to 5 ft tall; leaves with petioles, alternate or sometimes opposite, ovate or somewhat oblong, to 6 in. long; flowers one or two at bases of upper leaves; corolla bell-shaped to 1 in. long, purplish brown, dull red, or greenish yellow; fruit a glossy black berry ½ in. across, subtended by the enlarged calyx.

Deadly Nightshade has been used for centuries in witchcraft, medicine, and homicidal poisonings. When ingested, the plant material is a strong hallucinogen that can have very un-pleasant side effects.

Toxic part: Entire plant; the shiny black berries are a poten-tial hazard particularly to children.

Toxin: Hyoscyamine, a tropane alkaloid, the tropic acid es-ter of tropanol.

Symptoms: On ingestion, dry mouth and difficulty in swal-lowing and speaking, flushed dry skin, rapid heartbeat, dilated pupils and blurred vision, and neurological disturbances in-cluding excitement, giddiness, delirium, headache, confusion, and hallucinations; more common in children. Long continued use of atropine as a drug may result in greater manifestations of these symptoms, including glaucoma. Continued medica-tion may also result in dependency. If the drug is suddenly stopped, withdrawal symptoms may occur.

Because of its ability to dilate the pupils of the eyes, atro-pine has been used in ophthalmology. It is routinely used as a preoperative medication to prevent salivation and bronchial se-cretions. Another use has been as an antidote for poisoning from parathion, an extremely toxic insecticide. In 1967 atro-pine was used in Tijuana, Mexico, to treat people who had eaten bread that had accidentally been contaminated with parathion.

Atropa belladonna can be fatal; three berries may be enough to cause the death of a child. Humans are more likely to be poisoned by this plant than domestic animals; livestock avoid it even where it is an abundant weed.

Brugmansia. Angel's Trumpet, Tree Datura.

A small genus of five species of shrubs or small trees mostly from the Andes of South America; three species are cultivated as ornamentals in California. Stems to 15 ft tall, usually un-branched at the base, branched above; leaves alternate, 6 to 12 in. long; flowers large to 12 in. long, funnel or trumpet-shaped, nodding or pendulous, white or sometimes yellow, red, salmon, or pinkish; fruit a berry.

The species of *Brugmansia* are sometimes included with the genus *Datura,* which differs in having a herbaceous (non-woody) habit and dehiscent capsular fruits. Plants of both genera contain toxic tropane belladonna alkaloids that cause poisonings in humans more frequently than in livestock.

Brugmansia × *candida* [*Datura* × *candida*].   White Angel's Trumpet.

Native of Peru. Calyx split down one side only and therefore spathe-like; corolla white, trumpet-shaped, 10 to 12 in. long; flared at the open end, the corolla lobes ending in recurved pointed teeth.

Brugmansia sanguinea.   Red Angel's Trumpet (pl. 13*a*).

Native of Peru. Calyx five-angled or smooth, somewhat inflated, ending in one to four teeth; corolla 8 to 10 in. long, narrowly trumpet-shaped with short irregular recurved lobes, red at the mouth, greenish yellow toward the base.

Brugmansia suaveolens [*Datura suaveolens*].   White Angel's Trumpet.

Native of Brazil. Calyx inflated, angled, shortly five-toothed; corolla white or pink, 8 to 12 in. long, funnel-shaped, the lobes ending in spreading but not recurved teeth.

The following paragraphs apply to both *Brugmansia* and *Datura*.

Toxic part: Entire plant but particularly the seeds.

Toxin: Tropane alkaloids, chiefly scopolamine (hyoscine), hyoscyamine, norhyoscyamine, meteloidine, and atropine.

Symptoms: Onset of symptoms varies with the amount and kind of material eaten. In humans, symptoms may appear within a few minutes after ingestion of a decoction made from the plant or not for several hours after eating or chewing plant parts such as leaves, flowers, or seeds. Symptoms include thirst, dryness of the mouth and skin, flushing of the face, visual disturbances, nausea, rapid pulse, fever, delirium, incoherence, and stupor. Sometimes intoxicated patients may pick at imaginary objects in the air or on their bodies. There is usually a loss of memory for this period of time.

Poisonings have occurred in children who have sucked nectar from the base of the tube of the large showy flowers. Fruits and seeds of *Datura* are a potential hazard to children. In November 1980, the parents of a child found the child's eyes to be dilated after playing with a plant of *Brugmansia sanguinea.*

These plants also have been used deliberately by those seeking hallucinogenic effects. In the spring of 1977, three teenage boys in San Francisco became seriously ill after deliberately eating the flowers and perhaps other parts of the Red Angel's Trumpet, *Brugmansia sanguinea.* Jimsonweed, *Datura stramonium,* used in this manner is extremely dangerous, especially to small children.

In South America the tree daturas or angel's trumpets, species of *Brugmansia,* have a long history of being used as ritualistic hallucinogens and as medications by aboriginal peoples. *Brugmansia sanguinea* and several other species native in the Andes from Colombia to Chile are used as such. Throughout the area Indians have used these plants in different ways. In one area in Colombia they ceremonially used the narcotic called tonga, obtained from *Brugmansia sanguinea,* in their Temple of the Sun. Peruvian Indians still believe that tonga enables them to communicate with their ancestors. The Valley of the Sibundoy in Andean Colombia has been designated the "most narcotic-conscious area of the world." Tree daturas are cultivated for their unusually large amounts of scopolamine. Medicine men in the area use them as drugs in disease diagnosis, divination, prophecy, or witchcraft.

Brunfelsia pauciflora.   Yesterday-today-tomorrow, Morning-noon-and-night, Yesterday-and-today.

Native in Brazil; a variable species with several forms cultivated as ornamentals in California. Evergreen shrubs to 10 ft tall; leaves alternate, 3 to 4 in. long, oval, entire; flowers very showy, lavender, purple, or white, sometimes covering the shrub; calyx tubular, enclosing the tube of the corolla; five spreading corolla lobes saucer-shaped, 2 in. across. The common names are derived from the flowers, at first lavender, then purple, quickly fading to yellow or white, and lasting only one day. This species is not usually reported as being poisonous, but plants of *Brunfelsia* species that occur in Brazil and in Andean South America have been used by Indians of Colombia, Peru, and Ecuador as hallucinogens, either alone or added to other hallucinogens.

Capsicum annuum. Chili Peppers.

The original wild ancestor was probably native to Central and South America. To this species belong the various forms of peppers: chili, red, cherry, and others, cultivated for their hot and spicy flavors.

Plants grown as annual herbs; flowers dingy white, ½ in. across; fruits of the various kinds of peppers rounded to elongated, mostly yellow, red, or purplish.

Peppers contain a pungent principle, capsaicin, that can escape as a gas at higher temperatures and be extremely irritating. The smoke of chili peppers irritates the mucous membranes and was once used as a form of torture. Eating an excess of chili peppers can cause gastritis that may lead to enteritis and diarrhea. Chili peppers have an irritant effect on the human skin and have caused severe localized irritations on workers' hands.

Cestrum. **Jessamine, Cestrum** [sometimes also called Jasmine, but this name is more properly used for plants of the genus *Jasminum* in the Olive Family, Oleaceae].

Native in tropical America; cultivated as ornamentals. Shrubs or small trees, erect or somewhat weak and climbing on other shrubs; leaves alternate, occasionally with a strong odor when crushed; flowers in showy clusters at the ends of branches or at the bases of upper leaves, flowers sometimes fragrant, corollas of various colors, white, yellow, red, or purple, tubular, about 1 in. long, with five small lobes, five stamens included in the tubular corolla; fruit a berry. The following species are cultivated in California gardens:

Cestrum aurantiacum, Orange-flowered Cestrum (pl. 13*b*); plant lacking hairs, flowers orange yellow, berries white
Cestrum diurnum, Day Cestrum, Day-blooming Cestrum; plant hairy, flowers creamy white, berries black, shining
Cestrum elegans [*C. purpureum*], Red Cestrum; plant hairy, flowers ruby red and lacking hairs, berries reddish
Cestrum fasciculatum, Red Cestrum; plant hairy, flowers purplish red and hairy, berries reddish
Cestrum nocturnum, Night-blooming Jessamine; plant lack-

ing hairs, flowers greenish white, fragrant at night, berries white

Cestrum parqui, Willow-leaved Jessamine; plant lacking hairs, flowers greenish to brownish yellow, berries black; plants, especially the crushed leaves, having unpleasant odor

Toxic part: Probably the entire plant but especially the fruits, which are particularly noticed by children.

Toxin: Saponins; also traces of nicotine and other alkaloids.

Symptoms: Nausea, dizziness, increased salivation, delirium, muscular spasms, anxiety, fever, and paralysis of the respiratory system.

We have no records of toxicity to humans or livestock in California, but prunings from these shrubs should not be discarded where they might be accessible to children or livestock. *Cestrum fasciculatum,* a generally larger plant than *C. elegans,* is the only species recorded as naturalized in California; two small populations are known in central and southern Humboldt County.

Cestrum diurnum, which has escaped in southern Florida and sometimes is cultivated in California, is the cause of calcinosis in cattle and horses. Glycosides derived from excessive amounts of vitamin D_3 cause the deposition of calcium in the soft elastic tissues and at the ends of the bones resulting in progressive lameness and loss of weight in the animals. Calcinosis has not been observed in California.

Cestrum parqui is regarded as highly toxic to livestock, but reports are conflicting, perhaps because the plants are not always correctly identified. The foul odor of the plant is from a glycoside. The stems, leaves, flowers, and unripe berries contain saponins derived from glycosides.

In East Africa, *Cestrum aurantiacum* has caused the loss of cattle. The symptoms include extreme sensitivity to external noises and objects; often the animal charges anything moving. The animal bellows frequently and just before dying develops paralysis that begins in the hindquarters. These signs appear to be much the same as in bovine rabies. The unripe berries were fatal to sheep fed experimentally, but the leaves, although toxic, were not fatal.

Datura.   **Thornapple.**

A small genus of eight species of annuals or perennials. Native of warm parts of southern North America and Asia; several introduced species grow as weeds in California.

Leaves alternate, entire to lobed, strong smelling; flowers erect or finally nodding, white or strongly tinged with purple, often fragrant, some nocturnal and lasting only a day; calyx splits at a line around the base leaving the lower part as a collar, the upper part falling with the corolla, which is trumpet-shaped; fruit a capsule that splits open regularly lengthwise or bursts irregularly.

This genus is closely related to the genus *Brugmansia,* from which it differs chiefly in its herbaceous habit and its capsular fruits. The species of *Datura* contain the same toxic tropane belladonna alkaloids as *Brugmansia;* see pp. 234–235 for the toxic properties and symptoms.

Datura discolor.   Desert Thornapple.

Native to the Colorado Desert of California and to the east and south. Irregularly branching winter annual; leaves broadly ovate, lobed, 3 to 10 in. long; corolla trumpet-shaped, 7 to 8 in. long, white with light purple throat; capsule 2 in. across, covered with reflexed spines 1 in. long, splitting open regularly lengthwise. The tan-colored dried plants are often found in low places in the desert.

Datura ferox.   Chinese Thornapple.

Native of Asia; a weed of disturbed low places in the Central Valley of California. Erect branched summer annual; leaves to 10 in. long, broad-ovate, lobed; flowers erect, 4 in. long, white; capsule with a few coarse spines 1½ in. long from a broad base, splitting open lengthwise when mature.

Datura meteloides [*D. inoxia, D. wrightii*].   Downy Thornapple, Angel's Trumpet, Sacred Datura, Tolguacha (pl. 14*a, b*).

Native to southwestern United States and Mexico; perhaps introduced into California. Branching perennial to 3 ft tall and

6 ft across, grayish green from the hairs on the plant; leaves ovate, 8 to 10 in. long, entire to sinuately toothed; calyx tubular, corolla white, purplish tinged, trumpet-shaped, to 8 in. long; capsule nodding or reflexed, hairy, with hairy spines ⅜ in. long, splitting open irregularly when mature.

Datura stramonium [*D. tatula*].   Jimsonweed, Jamestown Weed, Thornapple, Common Thornapple, Stramonium, Mad Apple, Stinkweed, Stinkwort (pl. 13*c*).

Presumed native to eastern United States; cultivated and widely naturalized around the world, including California. Strong-smelling summer annual, generally lacking hairs; leaves ovate, coarsely and irregularly toothed; flowers to 5 in. long, smaller than those of the preceding species; capsule erect, sharply spiny (rarely lacking spines), splitting regularly into four parts.

British soldiers sent to Jamestown, Virginia, in 1676 became ill from eating young plants cooked as greens—hence the name Jamestown Weed or Jimsonweed. A number of cases of human poisonings from different species of *Datura* have been reported for California.

In January 1959, a 3-year-old boy opened a spiny capsule of *Datura* and after eating some of the seeds showed typical symptoms of hyoscyamine poisoning. His physician stated that the youngster, upon recovery, behaved like a person recuperating from the effects of marijuana poisoning. A similar case was reported in August 1962 from Compton, Los Angeles County. In October 1981, five boys with severe symptoms of Jimsonweed poisoning were brought for treatment to the Kern Medical Center, Bakersfield. Another poisoning in the Bakersfield area had been reported about 4 years previously. At that time several teenagers were said to be using the plant for its intoxicating effect.

Livestock poisonings from species of *Datura* are rare because the plants are extremely distasteful. In September 1959, four animals died and two were ill in a flock of nineteen lambs near Modesto, Stanislaus County. The only toxic plants that could be implicated were those of a *Datura* present in a 1-acre pasture where the animals had been held for a month.

Hyoscyamus niger.   Black Henbane, Stinking Nightshade.

Native of temperate Europe and Asia; known in California from a single naturalized population in Modoc County and as waif plants adjacent to California in the Reno area, Washoe County, Nevada. Annual or biennial plants to 3 ft tall, strongly scented from glandular hairs; leaves alternate, sessile, to 5 in. long, pinnately lobed; flowers sessile or with short stalks, at the base of upper leaves and more numerous at the tops of the stems, corolla openly bell-shaped, to $\frac{1}{2}$ in. long, tubular with pointed lobes, greenish yellow with dark purple veins and purplish black center; fruit a capsule, circumscissile at the top, covered by the old corolla.

The entire plant is toxic, containing tropane alkaloids (mostly hyoscyamine but also scopolamine and atropine). Symptoms of poisoning are the same as for *Atropa belladonna:* dilation of the pupils, nausea, muscular weakness and inco-ordination, excitability, rarely coma, death from heart failure and respiratory paralysis.

Poisonings from Black Henbane are rare but have been recorded for humans, particularly children. Livestock poisonings occur despite the bad odor and unpleasant taste. The biennial form of this species is more commonly found both in cultivation and naturalized. It can become quite abundant on roadsides and waste ground in dry soil. Black Henbane is cultivated as a source of alkaloidal drugs; it has been used medicinally since the beginning of recorded history.

Lycium barbarum [*L. halimifolium*].   Matrimony Vine.

Native of southeastern Europe and western Asia; rarely cultivated in California, readily persisting around old home-sites. Spreading shrub to 9 ft tall with arching branches, often spiny; thickish leaves lanceolate to oblong-lanceolate tapering to a slender petiole, to $2\frac{1}{2}$ in. long, grayish green; lilac purple flowers, corolla funnel-form with five spreading lobes; fruit a rounded red berry, $\frac{1}{2}$ in. long.

Toxic part: Entire plant.

Toxin: Not known but presumably a steroid glycoalkaloid similar to the solanine complex found in the Nightshade Family.

Symptoms: Excitability and convulsions followed by death.

In the eastern United States where Matrimony Vine is more commonly naturalized, calves and sheep have been poisoned from eating large amounts of the plants growing around abandoned houses.

Lycium ferocissimum.   African Box.

Native of South Africa; cultivated rarely in California and escaped in the Ballona Marshes of Los Angeles County. This spiny shrub to 10 ft tall with repeatedly arched branches has fleshy fruits similar to those of *L. barbarum*. The fruit, listed as edible, has been suspected in Australia of poisoning humans and pigs; the toxic effect, at least in humans, produces narcotic symptoms. The spines are also mildly poisonous.

Lycopersicon lycopersicum [*L. esculentum*].   Tomato.

Native of Andean South America; cultivated widely as a commercial vegetable crop for its edible fruit; sometimes re-seeds itself and persists in waste places, particularly along the warmer parts of the California coast. Grown as an annual, erect to somewhat weak and climbing; leaves divided into leaflets with coarse teeth; corolla yellow with a short tube and widely spreading lobes; fruit an edible, juicy, red or yellow, many-seeded, round or pear-shaped berry.

Toxic part: Stem and leaves.

Toxin: Tomatine (similar to solanine) and its aglycone, tomatidine.

Symptoms: Reported as toxic to livestock eating the plants and green fruit. Children became ill from drinking a "tea" made of the stems and leaves.

Nicotiana. Tobacco.

Six species are known in California, four native and two introduced from South America and now naturalized. *Nicotiana tabacum*, the main commercial tobacco species, and *Nicotiana rusticana*, the industrial tobacco used in insecticides, are not grown as commercial crops in California. Several herbaceous species are cultivated in California as ornamentals: *Nicotiana alata*, Jasmine Tobacco or Flowering Tobacco, from South

America; *Nicotiana* × *sanderae*, of hybrid origin; and *Nicotiana sylvestris*, from Argentina. Annual or perennial herbaceous plants or even small trees, many viscidly glandular and strongly scented; leaves alternate, simple, usually without lobes or teeth on the margins; flowers commonly in leafy clusters at the ends of branches, in colors of white, yellow, rose, red, or purple, corollas funnel-shaped or tubular with five lobes at the open end; fruit a capsule.

Toxic part: Entire plant, especially the leaves.

Toxin: A number of alkaloids but principally nicotine or anabasine, depending on the species.

Symptoms: Smoking tobacco by intolerant individuals results in mild cases of nicotine poisoning with salivation, nausea, vomiting, and diarrhea accompanied by a clammy cold skin, palpitation, and even slight prostration. In severe poisonings from swallowing nicotine there can also be labored respiration, weak irregular pulse, faintness, and collapse. A severely poisoned patient may die within a few minutes.

Tree Tobacco, *Nicotiana glauca* (fig. 34; pl. 13*d*), contains the alkaloid anabasine, an isomer of nicotine, and has caused poisonings and several deaths of humans in California. Seedlings of this species appear very succulent and apparently edible to some people. One case of human poisoning occurred in San Joaquin County, where the seedlings were mistaken for edible greens. The patient recovered but experienced typical symptoms of nicotine poisoning including respiratory paralysis that required use of an iron lung.

Species of *Nicotiana* are unpalatable to livestock, yet poisonings have been recorded for California from plants of *Nicotiana glauca* and two native species: *N. attenuata*, Coyote Tobacco, and *N. trigonophylla*, Desert Tobacco.

The alkaloid anabasine in commerical tobacco plants has been suspected of causing deformed offspring of sows in Kentucky, similar to the effects of *Lupinus* in the Fabaceae. (See the discussion of crooked calf disease on p. 163.)

Physalis. **Groundcherry.**

Only a few species of *Physalis* have caused toxic problems in California. Annual or perennial herbs, rarely somewhat woody at the base; leaves alternate, sometimes in pairs, simple,

FIG. 34. *Nicotiana glauca*. Tree Tobacco.
A. Flowers. B. Fruits. C. Seed (enlarged).

coarsely toothed or angled; flowers at the base of the leaf on slender stems; calyx inflated and persistent, enclosing the fruit, a berry.

Toxic part: Unripe berry.

Toxin: Solanine, a bitter glycoalkaloid.

Symptoms: Mostly gastroenteritis similar to that caused by a bacterial infection. There may be diarrhea and fever with a scratchy feeling in the throat some hours after eating the unripe berries. Children are more likely to be poisoned than adults.

The toxin is destroyed by heat. Ripe fruits are not toxic and often are made into jams or preserves. The following cultivated species of *Physalis* may cause trouble.

Physalis alkekengi [*P. franchetii*].   Chinese Lantern Plant, Wintercherry, Japanese Lantern, Strawberry Tomato.

Native of southwestern Europe and Asia; cultivated in California for its edible fruit and its ornamental bright red calyx resembling a miniature Chinese lantern, to 2 in. long, surrounding the berry.

Physalis philadelphica [*P. ixocarpa*]. Mexican Husk
Tomato, Tomatillo, Jamberry.

Native of Mexico; occasionally cultivated for its edible
fruit; naturalized as a waste-ground weed in warmer areas of
California. Annual; flower bright yellow with a dark center,
1 in. across, much more showy than flowers of other species of
Physalis; fruit a yellow to purple berry, 2 in. across, enclosed
by the papery inflated calyx.

Physalis peruviana. Cape Gooseberry, Barbados
Gooseberry, Poha, Groundcherry, Wintercherry, Cherry
Tomato, Strawberry Tomato, Gooseberry Tomato.

Native of tropical South America; cultivated in California
for its edible fruit. Perennial to 3 ft tall; flower light yellow
with a purple center and dark veins; fruit a yellow berry $\frac{3}{4}$ in.
across enclosed in the enlarged calyx.

Of the native or introduced species in California, only the
following groundcherry has been reported as troublesome to
livestock.

Physalis angulata var. *lanceifolia* [*P. lanceifolia*]. Lanceleaf
Groundcherry (fig. 35).

Native of California; an abundant agricultural weed of the
Central Valley and extending south to San Diego and Imperial
counties, widespread eastward. Summer annual to 2 ft tall,
branched at the base, yellowish green in color, lacking hairs on
the stems and leaves; leaves usually shallowly toothed, 1 to 3
in. long; yellow flowers $\frac{1}{4}$ in. across on slender stems that elongate in fruit to 2 in. long; fruit a green berry enclosed in the
husk of the inflated calyx.

In September 1958, dairy cattle exhibited toxicity from
plants of Lanceleaf Groundcherry that were abundant in hay.
The poisoning occurred in the Loma Rica district of Yuba
County, but the hay came from north of the Sutter Buttes. General signs were loss of appetite, swollen udders, and watery
milk. The animals recovered when this hay was withdrawn.

Near Chico, Butte County, in September 1958, a band of
100 sheep was placed in a 57-acre pasture of trefoil and ryegrass. *Physalis angulata* var. *lanceifolia* was the dominant

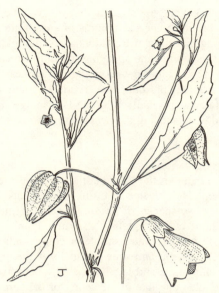

FIG. 35. *Physalis angulata* var. *lanceifolia.* Lanceleaf
Groundcherry. Center left: Fruit enclosed by inflated calyx.
Lower right: Flower.

weed of this pasture and showed evidence of being eaten. Nine
aged ewes died in 7 days; two more succumbed during the next
2 days after being removed from this pasture. There were no
further deaths in the band. Hemorrhagic lesions of the lungs
and tracheal mucous membranes were noted. Liver damage
and cyanosis were also observed.

Solandra maxima [*S. hartwegii;* sometimes incorrectly called
S. guttata].   Copa-de-oro, Cup-of-gold Vine, Chalice Vine,
Gold Cup, Trumpet Plant.
   Native of Mexico; cultivated in the warmer parts of California as an ornamental. Evergreen climbing shrub 10 to 40 ft tall,
requiring some support; leaves alternate, leathery, ovate, to 6
in. long; flowers large, corolla yellow with five purple lines, to
9 in. long, 6 to 8 in. across, chalice-like, with a suede-like
thickish texture.

Entire plant has the toxic alkaloids hyoscyamine, norhyoscyamine, and noratropine. Symptoms produced are similar to those following ingestion of Belladonna, including dry mouth, difficulty in swallowing and speaking, rapid heartbeat, hot dry flushed skin, fever, blurred vision and sometimes dilation of the pupils, excitement, delirium, headache, and hallucinations, particularly in small children. Poisonings in children have resulted from sucking the nectar from the corolla tube of the flower.

Solanum. **Nightshade.**

This genus includes the White Potato and the Eggplant and many ornamentals cultivated in California gardens. There are at least twenty-two species of *Solanum,* native or naturalized, growing without cultivation in California. Plants are sometimes spiny; leaves alternate, simple or compound; flowers in clusters, often colorful and showy, usually white or blue, corolla with a short tube and five spreading lobes, five stamens with anthers held cone-like in the center of the corolla; fruit a black, yellow, orange, or red berry with many seeds.

Toxic part: Entire plant, especially the green berries.

Toxin: A complex of glycoalkaloids and their aglycones: solanine and its aglycone solanidine; solasonine and its aglycone solasodine; or similar compounds of slightly different structures, varying with different species of *Solanum.*

Symptoms: Solanine and solasonine cause the same symptoms: a scratchy feeling in the throat, gastric irritation with nausea, vomiting, diarrhea, fever, headache, dizziness, and rapid heart action followed by slowing of the heart; may result in coma and death. Symptoms appear after a latent period of several hours following ingestion. Solanine is less toxic to adults than to children. Poisoning from this compound may be confused with bacterial gastroenteritis.

The species of *Solanum* most commonly seen in California are the following:

Solanum americanum [*S. nodiflorum*]. American Black Nightshade.

Native of tropical America; now naturalized as a weed around the world, the common weedy black-fruited nightshade

in California. Summer annual or a perennial with inconspicuous whitish flowers; fruit a black berry $\frac{1}{4}$ in. across with small
reflexed calyx lobes about $\frac{1}{16}$ in. long.

Solanum douglasii. Douglas' Nightshade.

Native of central and southern California; frequently found
in disturbed ground as well as in undisturbed areas of native
vegetation. Perennial to 6 ft tall, stems rough-angled, fruit a
black berry with adherent calyx lobes clasping the base of the
berry.

Solanum dulcamara. Bittersweet, Climbing Nightshade.

Native of Europe and Asia; weedy and naturalized in northern California. Climbing shrub; leaves ovate or with two opposite basal lobes; flowers several in a cluster, pale to dark
purple blue; berry red, nearly $\frac{1}{2}$ in. long.

Solanum elaeagnifolium. White Horsenettle, Silverleaf
Nightshade.

Native of central United States and southward into Mexico;
weed in the warmer parts of California. Perennial, $1\frac{1}{2}$ ft tall,
creeping with underground runners; plants covered with orange
spines among silvery stellate hairs; flowers a bright violet blue,
showy, to 1 in. across; fruit a yellow berry turning brown with
age.

Solanum jasminoides. Potato Vine, Jasmine Nightshade.

Native of Brazil; cultivated in California as an ornamental.
Climbing shrub; leaves ovate, pinnately divided, 1 to 3 in.
long; flowers in clusters, bluish white, about 1 in. across.

Solanum laciniatum. Kangaroo Apple (pl. 14c).

Native of southeastern Australia; cultivated as an ornamental in California and occasionally spontaneous in gardens.
Evergreen shrub to 10 ft tall; leaves lanceolate, up to 12 in.
long, unevenly pinnately lobed, petioles extending as wings
down the stem; corolla shallowly five-lobed, blue, $1\frac{1}{2}$ in.
across; fruit an ovate berry 1 in. long with concretions of stone
cells among the seeds, prominently warty in the dried fruit.

The leaves and green berries of the Kangaroo Apple are

toxic, containing solasonine, also called purapurine. The ripe berries are considered edible and have been made into jam. Livestock have been poisoned by Kangaroo Apple plants, but the deaths could not be duplicated experimentally. The related *Solanum aviculare,* also called Kangaroo Apple and native to New Guinea, New Zealand, and eastern Australia, is not definitely known to be cultivated in California.

Solanum mauritianum [*S. auriculatum*].   Woolly Nightshade.

Native of Argentina and southern Brazil; cultivated in southern coastal California, rarely spontaneous in cultivated fields or waste places. Large shrub or small tree to 20 ft tall, soft-wooded, all parts densely covered with stellate hairs; leaves ovate, 2 to 10 in. long, hairier and paler green below than above, usually with one or sometimes two smaller leaflets at the base of the petiole; flowers nearly 1 in. across, violet blue, numerous, in branched clusters; berries round, about $\frac{1}{2}$ in. in diameter, soft and yellow when ripe.

Toxic part: Entire plant, especially the berries.

Toxin: Glycoalkaloids of the solanine group (solauricine, solasonine, solasodamine).

Symptoms: Intense allergic reaction from handling the plant has been reported from South Africa. At first, a dull stinging occurs on the skin of the arms. The following day the skin becomes red and itchy, covered with blisters. Itching and new spots appear during the next several days, and finally the eyes become swollen and stiff but not closed.

In South Africa the fruit is said to have caused human fatalities and has been suspected of poisoning cattle. Evidence of these toxicities, however, is not convincing. In eastern Australia where the plant is widespread and abundant, no such cases have been reported. There have been a few cases of fatal poisonings of pigs and illness of cows in Queensland. Additional suspected cases of livestock poisoning were reported but without reliable evidence. No toxicities have been reported for California.

Solanum melongena.   Eggplant, Aubergine.

Probably native to southern Asia; commonly grown in California for the edible fruits. Somewhat woody perennial grown as an annual, to 3 ft tall, much branched at the base and spreading to 5 or 6 ft across; thickish leaves 6 to 16 in. long, gray hairy, oblong or oval, unequal at the base and angled or lobed; flowers violet, 2 in. across, nodding, in the leaf axils; fruit a shining berry 2 to 12 in. long; round, oval, or elongated; purple, white, yellow, or striped. The bitter principle, abundant in the overripe fruit, has been variously identified as solanine, solanine-m, or solasodine. Experimentally fruits and leaves, either dried or fresh, have produced a marked drop in the blood cholesterol level, as low as 50% of normal. The breakdown of cholesterol is considered the result of the magnesium and potassium salts in the plant material. In one case, however, there was no effect.

Solanum nigrum. European Black Nightshade.

Native of Europe; a rare weed in California. Summer annual; leaves ovate, entire or lobed, 4 to 7 in. long; flowers ¼ in. across, white, several in a cluster; fruit a black berry with the calyx lobes clasping the base of the berry. A polyploid complex of a number of extremely variable species may be called the *Solanum nigrum* group. Often reported as poisonous, but plants in some localities are used as potherbs and the berries made into pies. Heat apparently destroys the glyco-alkaloids.

Solanum pseudocapsicum. Jerusalem Cherry (pl. 14d).

Native of the Old World; grown as a pot plant or outside as a summer ornamental, persisting in warmer situations. Evergreen shrub 2 to 4 ft tall, lacking hairs; dark green leaves oblong-lanceolate to 4 in. long; flowers white, ½ in. across; fruits round, yellow or red, 1 in. across, attractive, poisonous if eaten. A steroid alkaloid, solanocapsine, is present in the entire plant. Although there are differences of opinion regarding the toxicity of this species, children eating three or four berries have been fatally poisoned. Dried alcohol extracts of the leaf and fruit have proved experimentally to be quite toxic to animals, resulting in a wide disorganization of the heart's action.

Solanum pseudocapsicum, an escape from a neglected garden, covered about 160 sq ft of the corner of a Bermudagrass pasture near Pomona, Los Angeles County. In August 1958, five sheep died suddenly in this pasture. The Jerusalem Cherry plants were eaten by the sheep, but none of the plant was found in the one dead animal examined. Only a slight skin hemorrhage was noticed in these previously healthy sheep.

Solanum sarrachoides.   Hairy Nightshade (pl. 15*a*).

Native of Brazil; abundant agricultural weed of the Central Valley of California. Glandular hairy annual herb, branching from the base; leaves ovate and usually toothed, 1 to 2 in. long; flowers white, $\frac{1}{8}$ in. in diameter; fruit a green berry almost half-covered by the clasping calyx. The mature fruit remains green and contains the toxic glycoalkaloids that disappear in ripe fruits of other *Solanum* species. Hairy Nightshade grows in disturbed waste ground and in cultivated crops; it may occur abundantly in first-year alfalfa hay. On one occasion fruits of Hairy Nightshade appeared in processed green peas; both are about the same size and color.

Solanum seaforthianum.

Native to tropical America; cultivated in California as an ornamental. Plants grow as vines to 20 ft long; leaves usually pinnate with three to nine leaflets; many flowers, in clusters, white to pale blue; fruit scarlet but rarely produced in California.

Solanum tuberosum.   White Potato, Irish Potato.

Native to Andean South America; widely cultivated as one of the world's most important food crops. Cultivated as a weak-stemmed annual with edible underground tubers; leaves pinnately compound; small white or bluish flowers about 1 in. across; fruit a rounded yellow berry, seldom produced. Leaves, stems, sprouts on tubers, and green skin on old potatoes contain the toxic solanine or solanidine. Solanine has a bitter taste and is irritating to the throat. It is not lost or destroyed when green potatoes are cooked. Under certain conditions, such as continuous exposure to light, sprouts from the tubers may accumulate a high alkaloid content. Sprouts on tubers should be removed, and green or sunburned potatoes should be peeled before cooking. Spoiled potatoes should be discarded.

Eating green or sprouting potato tubers has caused headache, anxiety, weakness, and circulatory and respiratory disturbances. The liver may be enlarged; hemorrhages may develop; the patient may collapse. Smoking potato leaves over a period of several days has resulted in hemorrhages in the eye, headache, nausea, and vomiting. Solanine from stored potatoes infected with late-blight fungus was suspected, but not actually determined, to be the cause of malformed human embryos.

Livestock may develop "potato eruption" if fed large amounts of raw or cooked potato tubers, leaves, or other commercial potato waste. Blistering and inflammation of the skin occurs, particularly on the lower limbs, as well as a stiff gait and rise in body temperature. Severe cases may result in weakness and death.

Solanum umbelliferum [*S. californicum*].   Blue Witch.

Native to California; one of the conspicuous shrubby species of *Solanum* in the chaparral and other plant communities; sometimes cultivated as an ornamental. Shrub or subshrub; leaves elliptic-ovate, entire or irregularly lobed, hairy; flowers blue, nearly 1 in. across; fruit a rounded berry ½ in. across, white with a greenish base.

Solanum wendlandii.   Potato Vine, Giant Potato Creeper.

Native of Costa Rica; cultivated in California gardens. Vigorous, sparsely prickly vine; leaves usually pinnate or sometimes undivided; flowers blue, to 2 in. across, many, in branched clusters.

Sterculiaceae, Sterculia Family

Cola acuminata and *C. nitida* are evergreen trees grown in West Africa and tropical America as commercial crops. Cola nuts have an alkaloid content of 1% to 2.35% caffeine and have been used to prepare cola soft drinks. Theobromine is present in the nuts as well.

Brachychiton populneus.   Kurrajong, Bottle Tree.

Native in eastern Australia; cultivated in California as an ornamental. Large evergreen tree to 60 ft tall; leaves alternate, ovate, undivided or deeply three to five-lobed, to 3 in. long, on long petioles; flowers cream-colored, often dark-spotted inside, in showy clusters; fruit a woody brown pod on a stalk 1 to 2 in. long, splitting open along one side; many seeds, embedded in the sharp, stiff, irritating hairs lining the inside of the pod.

Toxic part: Seeds.

Toxin: Unknown.

Symptoms: Lameness, tremors, collapse, and death in cattle and sheep.

Poisoning of livestock from Kurrajong is not known in California, but care should be taken to keep animals from these trees or their prunings that have pods and seeds attached. Experimentally the seeds have proved to be toxic. In Australia this species is regarded as one of the best fodder trees for cattle and sheep. Animals have been poisoned only when their diet consisted almost totally of this tree.

Sterculia foetida.   Indian Almond.

Native of the tropics of the Old World; rarely cultivated in the warmest parts of California. Tree to 60 ft tall; leaves palmately compound with five to seven leaflets; ill-smelling flowers, unisexual or bisexual in panicles below the leaves, calyx purplish red, five-lobed; fruit of five dark red follicles. Entire plant contains hydrocyanic acid. The raw seeds are purgative but edible when heated.

Thymelaeaceae, Mezereum Family

Many species in this family are known to be toxic; several in the genera *Daphne, Dirca, Gnidia,* and *Pimelea* are either native or in cultivation in California. In Europe where they are native, *Daphne cneorum* and *D. mezereum* have been recorded as poisonous since the time of Dioscorides in the first century.

The following species of *Daphne,* and several others less frequently seen, are grown in California for their usually fragrant flowers and showy fruits:

Daphne cneorum, Garland Daphne, Garland Flower; native to the mountains of Europe; low evergreen shrub to 1 ft tall

Daphne mezereum, Mezereum, Spurge Laurel, Spurge Olive, Flax Olive, Dwarf Bay, February Daphne; native of Europe and western Asia; deciduous shrub 4 to 5 ft tall; flowers appear before the leaves

Daphne odora, Winter Daphne; from China and Japan; evergreen shrubs to 4 ft tall; *D. odora* 'Marginata' (pl. 15*b*), with leaves bordered with a yellow band, is commonly grown

Toxic part: Entire plant, fresh or dried, but especially the fruits.

Toxin: An irritant resin called daphnetoxin (a diterpene derivative similar to phorbol found in the Euphorbiaceae) and in the bark a coumarin glycoside that has the aglycone dihydroxycoumarin.

Symptoms: In humans, eating the berries or other plant parts causes blistering with a burning sensation and swelling of the lips, mouth, and tongue. This reaction is followed by thirst, nausea, abdominal pain, persistent bloody diarrhea, and a sensation of pricking or tingling of the skin. There may be kidney damage.

In New Zealand, where daphnes are cultivated and one species is also naturalized, animals have been poisoned in rare instances. Additional signs are bloodstained purgation; clinical findings are intense gastritis with white patches and a burned appearance on the lining of the stomach.

Ingestion of one or two bitter berries, which are brightly colored and attractive, may cause severe poisoning in a child. About twelve berries may be fatal to an adult.

Dirca occidentalis. Western Leatherwood.

Native, but uncommon, on wet slopes around San Francisco Bay. Deciduous shrub to 6 ft tall with small yellow flowers in midwinter. There are no reports of its toxicity in California, but the closely related *Dirca palustris,* Leatherwood, of eastern North America has an unidentified toxin and is reported as mildly poisonous, causing gastrointestinal irritation and severe dermatitis.

Gnidia polystachya.

South African deciduous shrub with small yellow flowers in showy dense clusters; occasionally cultivated in California. Most species of *Gnidia* contain an irritant resin, daphnetoxin, and an alkaloid, daphnine.

Pimelea prostrata.

Native of New Zealand; grown in California as an ornamental. Low prostrate evergreen shrub with small crowded leaves about ¼ in. long and small waxy white flowers in dense heads. All parts of the plant contain prostratin, a highly toxic diterpene acetate. There are no records of poisonings in California. In New Zealand horses and cattle have been poisoned, even fatally, but sheep are not particularly affected.

Tropaeolaceae, Nasturtium Family

Tropaeolum majus. Garden Nasturtium, Tall Nasturtium, Indian Cress.

Native of the Andes of South America; commonly cultivated in California, naturalized along the coast. Somewhat succulent annual plants, stems weak, climbing to 6 ft by coiling petioles; leaf blades peltate, on long petioles; flowers to 2½ in. across, spurred and irregular, in shades of maroon, red, orange, yellow, and white; fruit three-lobed, separating at maturity. Young leaves are used in salads; flower buds and seeds are used for their peppery taste. These parts contain a benzyl mustard oil, benzylisothiocyanate, that is different from the isothiocyanates discussed under the Brassicaceae. Benzylisothiocyanate is formed in the plant as a glycoside, glucotropaeolin, and has an antibiotic activity against several kinds of bacteria.

Urticaceae, Nettle Family

Hesperocnide tenella. Western Nettle.

Native in California on shaded canyon slopes in the Coast Ranges from Napa County south and in the Sierra Nevada from Tulare County south into San Diego County. Weak-stemmed annual to 18 in. tall, nettle-like in appearance; resembles the introduced *Urtica urens*. It has stinging hairs on the stems and leaves.

Urtica. **Stinging Nettles.**

Four perennial species of nettles are native to California; another species, an annual, native of Europe, is naturalized. The native species are:

> *Urtica californica,* California Nettle; native to low areas near the immediate coast, San Mateo County to Sonoma County
>
> *Urtica holosericea,* Creek Nettle, Hoary Nettle; the common stinging nettle, native throughout California west of the mountains below 9,000 ft elevation, in low places and along stream banks, rarely found on the edges of the deserts
>
> *Urtica lyallii,* Lyall's Nettle; native to the immediate coast from Mendocino County to Del Norte County and northward
>
> *Urtica serra* [*U. breweri*], Serra Nettle, Brewer's Nettle; native in the Sierra Nevada, 5,000 to 10,000 ft elevation and to the east and south in adjacent states

These four species are perennial, stems to 8 ft tall, stout, bristly with stinging hairs and sometimes also softly hairy; leaves opposite, narrow-ovate, 2 to 6 in. long, margins toothed, covered with scattered stinging hairs; lower surface of leaf grayish and softly hairy, upper surface greener; tiny flowers borne in clusters at the base of each upper leaf.

Urtica urens. Burning Nettle, Small Nettle.

Annual from Europe; naturalized as a weed of gardens and orchards in California, particularly in the warmer coastal areas. Annual plants usually branching from the base, from a few inches to 24 in. tall, with only stinging hairs; coarsely toothed leaves to 1 in. long; flower clusters less than $\frac{1}{2}$ in. long.

Toxic part: Fluid from the stinging hair cells. Tip of the hardened hair breaks off easily, leaving a minute slanted point that can easily penetrate the skin.

Toxin: Acetylcholine, histamine, and 5-hydroxytryptamine.

Formic acid is present but causes only a very small part of the stinging sensation.

Symptoms: In humans, penetration of the skin results in reddening and marked itching of the spot, followed by swelling and an intense burning sensation, usually of short duration. Pollen of *Urtica* is listed as a cause of hay fever.

In December 1963, near Sanger, Fresno County, six beagle pups ran for about half an hour through nettles in a river bottomland area. The pups developed tremors, convulsions, thick ropy excess salivation, and vomiting. They improved with treatment by the veterinarian. The same signs are seen in hunting dogs, especially in December through March, after they have been run for an hour or two. Hunters should work their dogs carefully in low areas along waterways, avoiding places where nettles are abundant.

The Roman Nettle, *Urtica pilulifera,* native of southern Europe and sometimes grown as an ornamental, has more virulent stinging hairs than the foregoing species. The irritating effect from these nettles is mild compared to the painful reaction produced by the Stinging Tree, *Laportea,* of New Guinea and northern Australia.

Verbenaceae, Vervain Family

Aloysia triphylla [*Lippia citriodora*].   Lemon Verbena.

Grown in California gardens and used for a hot tea or in iced drinks, often in herbal mixtures. The volatile oil in the plant is high in citral, an aldehyde, and has been used experimentally to kill mites and aphids. Native to Argentina and Chile.

Duranta repens [*D. plumieri*].   Golden Dewdrop, Skyflower, Pigeonberry.

Native of southernmost Florida, West Indies, Mexico to Brazil; cultivated in California as an ornamental. Much-branched shrub or small tree 10 to 18 ft tall, stems sometimes spiny; leaves ovate, to 2 in. long; flowers blue or lilac, $\frac{1}{2}$ in. across; fruit a rounded yellow drupe about $\frac{1}{4}$ in. across, enclosed in the enlarged calyx with a curved beak at the apex.

Toxic part: Fleshy fruits, leaves.

Toxin: Uncertain; saponins have been reported.

Symptoms: Drowsiness, fever, rapid pulse, edema of the lips and eyelids, dilation of the pupils, irritation of the intestinal mucous membrane, abdominal pain, vomiting, and convulsions. *Duranta repens* is not known to have poisoned humans or animals in California, but it is said to have caused illness and death of children in Queensland, Australia. The plant is very bitter and usually is not eaten by livestock, but it has been suspected in livestock poisoning in Australia.

Lantana camara. Red Sage, Yellow Sage, West Indian Lantana (fig. 36).

Native of tropical America; cultivated widely in California as an ornamental. Shrub to 6 ft tall; leaves ovate, 1 to 6 in. long, dark green, rather thick and rough; small orange or yel-

FIG. 36. *Lantana camara.* Red Sage.

low to red flowers in a dense flat-topped head to 2 in. wide; fruit a small round fleshy blue-black drupe about ¼ in. across.

Toxic part: Entire plant, especially immature fruits.

Toxin: Two pentacyclic triterpenes, lantadene A and lantadene B.

Symptoms: In humans, gastrointestinal irritation, muscular weakness, lethargy, and circulatory collapse. In livestock, photosensitization from liver damage, inflammation and discharge from the eyes and nose, peeling of the nose, which is brick red in color, distension of the gallbladder, necrosis of the kidney tubules, weakness with a staggering gait, usually constipation with a dark ill-smelling manure from decomposed blood; animal may or may not recover. Experimentally, lantana poisoning causes increases in oxidative enzymes and decreases in enzymes associated with drug metabolism. The red blood cells of poisoned animals are more vulnerable to osmotic shock.

In Australia investigations have found toxic and nontoxic forms of *Lantana camara;* of the toxic forms, some are more poisonous than others. Lantana plants have a very bitter taste, and although they are rarely eaten, livestock poisonings have been recorded. In Queensland, Australia, many cases of jaundice seen at slaughter are considered the result of *Lantana camara* ingestion while the animals were held in corrals. Experimentally, ¾ to 1 pound of dried leaves has produced this damage.

Other species, such as *Lantana montevidensis,* Creeping Lantana, with small purple flowers in heads, have been suspected of causing poisonings. In one Australian experiment, however, *Lantana montevidensis* proved nontoxic to sheep.

Violaceae, Violet Family

Viola odorata. Sweet Violet, Garden Violet, English Violet, Florists' Violet.

Native of Eurasia and Africa; cultivated throughout California. Rhizomatous perennial, spreading by stolons; leaves in a basal rosette, with petioles to 6 in. tall, blades heart-shaped and toothed on the margins; flowers borne singly on stems to 6 in. tall, purple, rose, or white; fruit a dehiscent capsule formed from cleistogamous flowers. Large amounts of leaves, rhizomes, and seeds have been used in household medicines as an emetic, purgative, and diuretic. The plant contains a glycoside that breaks down to methyl salicylic acid, salicyclic acid, and a volatile oil containing methyl salicylate.

Viola tricolor. Johnny-jump-up, European Wild Pansy.

Native of Europe; common in cultivation. Annual or perennial, to 1 ft tall, resembling a small pansy. Has the same chemical constituents as *V. odorata* and has had the same uses as an emetic, diuretic, and purgative.

Viscaceae, Mistletoe Family

Also called Loranthaceae, the Mistletoe Family has three genera in California. *Phoradendron* is the largest; plants occur as partial parasites on both angiosperms and gymnosperms. Various species of *Arceuthobium,* the dwarf mistletoes, are found on cone-bearing trees. *Viscum album,* native of Europe and Asia, is established locally in Sonoma County.

In California sprigs of plants of *Phoradendron* with leathery green leaves and pale translucent berries are used as Christmas decorations; plants of *Viscum album* have a limited use in the San Francisco Bay region. The sticky seeds of *Phoradendron* and *Viscum* plants are spread largely from host to host by birds eating the berries. The seeds of *Arceuthobium* are spread by the explosion of the ripe berries.

Phoradendron tomentosum subsp. *macrophyllum*
[*P. flavescens*]. Common Mistletoe.

Native of California, east to western Texas. Principal hosts are cottonwoods, poplars, willows, sycamores, and ash trees; the partially parasitic plants often occur on the host trees in large heavy clumps.

Phoradendron villosum subsp. *villosum.* Hairy Mistletoe.

From California north to Oregon and south to northern Baja California. Principal hosts are oak trees.

These two California mistletoes have simple, oblong, leath-

ery leaves; tiny flowers embedded in the flower stalks are followed by showy whitish berries in spike-like clusters.

When mistletoe branches are used as Christmas decorations, the berries should be kept out of children's reach.

Toxic part: Entire plant.

Toxin: Probably two toxic amines, ß-phenylethylamine and
tyramine, and phoratoxin, a toxic lectin.

Symptoms: Ingestion of a few berries usually is associated
with only moderate abdominal pain and diarrhea. A large
quantity of berries or a tea prepared from the leaves, however,
can produce severe intestinal disturbances and has been fatal.
Plant material of *Phoradendron* has slowed the heartbeat similarly to digitalis.

The European Mistletoe, *Viscum album,* was introduced
into Sonoma County, California, early in this century and still
persists in apple orchards and adjacent native vegetation, having spread only about 4 miles in the 70 years since its introduction. Amines in the leaves of *Viscum album* produce effects
less toxic but similar to those caused by ricin from Castor
Bean, *Ricinus communis.* Ingestion of the berries of this mistletoe produces only moderate abdominal pain and diarrhea.

Cattle usually avoid mistletoe, which is unpalatable to
them, but there are reports of animals poisoned from browsing
on trees with mistletoe. In 1952 in Los Angeles County, thirteen cows with newborn calves died after actively feeding on
plants of *Phoradendron villosum* growing on oak trees. The
losses occurred in a herd of fifty cows over a period of 3
months; the animals died suddenly within 12 hours. All deaths
occurred in animals with 1 or 2-week-old calves. The range
had many small oak trees with mistletoe growing on them. The
berries on the mistletoe plants had been eaten as high as the
animals could reach.

Vitaceae, Grape Family

Parthenocissus quinquefolia. Virginia Creeper, Woodbine,
American Ivy, Five-leaved Ivy (fig. 37).

Native of the eastern coast of the United States to Texas and
Mexico; cultivated in California. Vigorous high-climbing deciduous vine with branched tendrils clinging by disk-like ad-

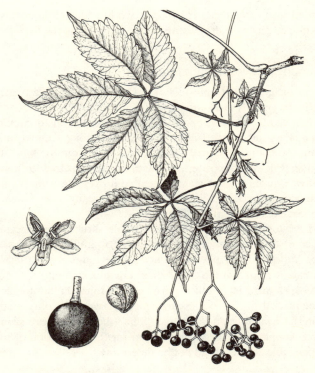

FIG. 37. *Parthenocissus quinquefolia.* Virginia Creeper.
Lower left: Flower, berry, and seed.

hesive tips; leaves divided into five leaflets, leaflets to 6 in.
long, coarsely toothed; flowers minute; fruit a bluish black
grape-like berry with one to three seeds.

Parthenocissus tricuspidata.   Japanese Ivy, Boston Ivy.

Native of Japan; cultivated in California. Differs from *P.
quinquefolia* in that the leaves are 8 in. across and three-lobed
or divided into three leaflets.

The grape-like, inedible berries on plants of these two species contain small amounts of calcium oxalate, a gastroenteric
irritant, but the amount of calcium oxalate is small and its irritant effect usually is not significant. A few cases have been re-

corded for *Parthenocissus quinquefolia* in which children have developed gastroenteric distress from chewing on the leaves or eating the berries.

Zygophyllaceae, Caltrop Family

Larrea tridentata [*L. mexicana*].   Creosote Bush. [Perhaps the same species as *L. divaricata* of South America.]

Native of California and to the east and south; the dominant abundant shrub of the California deserts. Strong-scented, resinous, spreading evergreen shrub, branched from the base, to 10 ft tall; leaves opposite, each with two leaflets to $\frac{3}{8}$ in. long; yellow flowers to $\frac{3}{4}$ in. across, scattered over the entire bush; rounded, hairy, dry fruit $\frac{1}{4}$ in. across. The resinous coating on the leaves gives them a polished look.

Livestock will not eat this unpalatable shrub. A rare occurrence of toxicity was recorded from the foothills of the Antelope Valley in Los Angeles County. At the end of August 1951, six calves from a group of ten died; obvious amounts of Creosote Bush material, including fruits, were found in the stomach contents. At this time of year the shrubs were the only green material available on the range. Since the calves were being given supplemental feed, the reason for their ingestion of the unpalatable shrubs could not be determined.

Peganum harmala.   African Rue, Syrian Rue, Common Harmel.

Shrub native to the Mediterranean region and eastward to northern India, Mongolia, and Manchuria, where it is sold as a commercial spice and used in folk medicine. The seeds contain several ß-carboline or indole alkaloids: harmine, harmaline, and tetrahydroharmine. Reports regarding its use as a hallucinogen are vague, and there is some question whether it actually induces visions. *Peganum harmala* is a serious agricultural weed and very difficult to control; it is illegal to grow it in California.

Tribulus terrestris.   Puncturevine, Goat Head (fig. 38).

Native of the Old World tropics; naturalized almost throughout California, a serious weed of crops and roadsides. Prostrate, silvery hairy annual, stems usually reddish; leaves opposite,

FIG. 38. *Tribulus terrestris*. Puncturevine.
A. Flower. B. Fruit. C. Single nutlet.

pinnately compound with four to seven pairs of leaflets; flowers $\frac{3}{8}$ in. across, borne singly at the base of a leaf, five separate yellow petals; fruit five-parted, separating on drying, each part or nutlet with two bony sharp spines resembling the horns of a goat. One of the large spines of the triangular-shaped nutlet is always upright and can puncture an inflated rubber tire or an animal's foot.

Toxic part: Entire plant.

Toxin: Three sapogenins: diosgenin, ruscogenin, gitogenin; also toxic levels of nitrates, particularly potassium nitrate.

Symptoms: Causes photosensitization or "bighead" in sheep, resulting in important losses in South Africa and Australia. It is less troublesome in the United States and of minor importance in California. Presumably this toxicity occurs mostly from eating vigorously growing Puncturevine plants that are in a vege-tative stage before flowering and have wilted from drought or some other cause. Such poisonings of sheep are difficult to reproduce experimentally. Puncturevine has caused nitrate poisoning in both sheep and cattle.

The hard spiny nutlets cause mechanical injury to livestock, not only externally but also internally when they are included in hay. A homicidal use of the spiny fruits has been reported from South Africa. The spines were generously smeared with the poisonous juice from Bushman's Poison, *Acokanthera oppositifolia*. The poisoned nutlets were then strewn along a path used by the intended, obviously barefoot, victim.

Monocotyledons

Agavaceae, Agave Family

This small family is of minor importance as a source of poisonous plants.

Agave. **Agave, Century Plant.**

Three species are native to the deserts of southeastern California; other species are cultivated throughout the state. Plants succulent, leaves in basal rosettes, often spiny on the margins; yellow flowers on a usually tall, branching, flowering stalk in the center of the rosette of leaves, in some species as much as 20 to 30 ft high.

Agave species are popularly known as century plants because it was erroneously thought that they flowered only after reaching 100 years of age. Most plants flower after 6 to 15 years, although a few in cultivation rarely flower. Each plant flowers only once and then dies, but during its lifetime it may produce suckers that form new plants.

Agave americana. Maguey, Century Plant.

Native of Mexico; cultivated in California, widely naturalized. Leaves up to 6 ft long, with hooked spines along the margins; flowering stalk after 10 or more years, erect, 15 to 40 ft tall. Some plants have leaves striped with white or yellow.

Agave attenuata.

Native of Mexico; cultivated in California. Leaves blue green, 2 to 3 ft long, without marginal spines, numerous flowers densely arranged on an arching stalk, 5 to 10 ft long.

Toxic part: Sap of the leaves or inflorescence.

Toxin: A hemolytic sapogenin, an acrid volatile oil, oxalic acid, and oxalates.

Symptoms: Fresh juice of the plant is cathartic and diuretic. The juice of the leaf and particularly the outer layers of the leaf are highly irritant to the skin.

Agave americana has been known to cause a burning rash and itching welts. Hemorrhagic dermatitis appeared several hours after the juice from an agave splattered on those who cut the plant with a power saw. Plants of this species have been suspected of poisoning stock in the field. Experimentally the plant has been lethal to a sheep and a rabbit.

Mescal de pulque, or pulque, is fermented from the sap collected from the slashed or topped flower stalks of *Agave atrovirens.* This is a beer-like drink with a ropy consistency; it may be consumed as such or it may be distilled to produce mescal. Mescal and tequila are usually distilled from the roasted and fermented bases of the plants of *Agave tequilana,* cultivated for the distillation of liquors particularly along the western coast of Mexico from Sinaloa to Jalisco. Mescal should not be confused with mescaline, the hallucinogenic drug from the Peyote, *Lophophora.*

The saponin in the root of *Agave americana* is used as a soap in Mexico. The plant has been used as a fish poison in Cuba and Venezuela.

Furcraea.

This genus is closely related to *Agave.* From tropical America and rarely cultivated in California, it has several species grown for their fibers and used especially in the manufacture of coffee bags. The fibers are irritant and cause dermatitis in those who handle the plant or the coffee bags.

Amaryllidaceae, Amaryllis Family

The Amaryllis Family is closely related to the Lily Family; the two are sometimes combined into one family, the Liliaceae. Members of this family have a single flowering stalk with a cluster of flowers in an umbel at the top subtended by one or more bracts.

Agapanthus.

Two species native to South Africa are widely cultivated in California gardens and along roadways. Both are perennials with thick rhizomes and showy lily-like flowers that are blue, sometimes white, in an umbel borne on a stem to 3 ft tall.

Agapanthus africanus [*A. umbellatus*], African Lily, Blue African Lily, Lily-of-the-Nile

Agapanthus orientalis [*A. umbellatus*], common names as above

Toxic part: Rhizome, leaves.

Toxin: Yuccagenin, a hemolytic saponin; agapanthogenin, a steroidal saponin.

Symptoms: Ulceration of the mouth from the sticky acrid sap of leaves.

In South Africa, plants of these species are used in native medicine and are suspected of causing human poisoning.

Allium cepa. Onion.

The onion of commerce, known only in cultivation. Bulb formed from the swollen leaf bases; green leaf blades hollow; numerous small greenish white flowers in a dense umbel with one or two, sometimes more, papery bracts at the top of a hollow stem. All parts give off a strong onion odor.

Toxic part: Bulb.

Toxin: *N*-propyl-disulfide. Although an alkaloid was considered the cause of cattle deaths in one case, it was not verified.

Symptoms: Blood in the urine, anemia, and jaundice.

Cattle and horses have been poisoned from eating dumped onions or those left in the field. Fresh, frosted, or rotting onions are equally poisonous. Sheep are more tolerant than cattle or horses and have been maintained for weeks or months on a winter diet including onions.

Over a period of time large amounts of onions will cause anemia, jaundice, and digestive disturbances in humans. Eaten in moderation, onions do not result in serious problems.

Allium canadense, Wild Onion, of central and eastern

North America is known to cause gastroenteritis. In horses it has caused hemolytic anemia.

Allium validum, Swamp Onion, is native in California from Lake County and north in the Coast Ranges and in alpine meadows from Tulare County north in the Sierra Nevada, extending into the Pacific Northwest. Dried flowering plants of this species were fed to sheep experimentally at the rate of 150 to 300 grams per day along with normal rations. Within 2 to 3 days hemoglobin from the breakdown of the red blood cells appeared in the urine. After 3 to 4 weeks the hemoglobinuria ceased, and there was an increase in numbers of red blood cells.

Anyone in California wishing to eat native wild onions should do so with caution.

FIG. 39. *Amaryllis belladonna.* Naked Lady. Flower stalk that appears before the leaves. Lower left: Capsule and seed. Lower right: Bulb and enlarged seed.

Amaryllis belladonna. Belladonna Lily, Cape Belladonna, Naked Lady (fig. 39).

Native of South Africa; frequently cultivated in California, often persisting from cultivation without additional summer water. Plants from a bulb; leaves appear in the spring and die down by the time the flowering stalk appears in summer; umbel of six to ten showy flowers, pink, rose pink, to almost white.

Toxic part: Bulb.

Toxin: Lycorine and smaller amounts of related alkaloids.

Symptoms: See *Narcissus*.

Hippeastrum puniceum [*H. equestre*]. Barbados Lily.

Native of tropical America; cultivated in California. Strap-shaped leaves appear after large flowers that are bright red, two to four in the umbel.

Toxic part: Bulb.

Toxin: Lycorine and several other alkaloids.

Symptoms: See *Narcissus*.

Narcissus pseudonarcissus. Daffodil, Trumpet Narcissus (fig. 40).

Native of southern Europe; widely cultivated in California and often persisting. Bulbous perennial, leaves basal, flowering stalk with a single large showy yellow flower.

Toxic part: Bulb.

Toxin: Lycorine and related alkaloids. Lycorine, a heat-stable alkaloid, is found in *Amaryllis belladonna* and *Hippeastrum puniceum* as well as other genera in the Amaryllis Family cultivated in California: *Clivia, Crinum, Galanthus, Haemanthus, Hymenocallis, Leucojum, Nerine*.

Symptoms: Following ingestion of fairly large amounts, symptoms include nausea and persistent vomiting; diarrhea is usually slight or absent. Plants of these genera also cause both allergic and irritant dermatitis, particularly in florists and growers who are repeatedly exposed to these plants.

Human poisonings from bulbs of these plants are rare. One recorded case resulted from mistaking a daffodil bulb for an onion. Poisonings of livestock occurred in Holland during World War II when bulbs of *Galanthus, Narcissus,* and *Hyacinthus* (of the Liliaceae) were used for emergency food.

FIG. 40. *Narcissus pseudonarcissus.* Daffodil. Lower left: Capsule and enlarged seed. Lower right: Bulb.

Araceae, Arum Family

Many plants in this family are cultivated for ornament; some are especially useful as indoor foliage plants, others can be grown outdoors in protected places in California gardens. The inflorescence, the most characteristic feature of the family, consists of a densely flowered columnar spadix surrounded by a specialized bract, the spathe, often colored and showy, and is mistakenly considered to be the "flower." The flowers are small, arranged along the length of the spadix, mostly unisexual, lacking petals, with the female flowers aggregated in the lower part of the spadix and the male flowers above; in plants of some genera the two are intermixed. The fruit is usually a berry, sometimes brightly colored. Members of the Araceae have the same toxic potential.

Toxic part: Juice of leaves and stems.

Toxin: Calcium oxalate crystals with or without unidentified toxic proteins.

Symptoms: Painful irritation of the lips and the mucous membranes of the mouth and throat. In extreme cases swelling of the throat is sufficient to cause choking and inability to swallow. Usually after 4 days the swelling begins to lessen, eventually disappearing after 12 days. The pain may continue for about 8 days. In addition, contact dermatitis commonly occurs.

The calcium oxalate crystals are insoluble in water and are grouped together in bundles called raphides. When a part of the plant is bitten or chewed, the minute sharp crystals quickly penetrate the lining of the mouth. Because of the intense pain, plant material is rarely swallowed, but if it is it will cause inflammation and irritation of the gastrointestinal tract. Toxic proteins and other harmful materials may also be present in the plant tissue. Fatalities have been recorded from these toxins, but not in California.

The following plants are likely to be encountered in cultivation in California:

Aglaonema commutatum
Aglaonema costatum
Aglaonema crispum [*A. roebelinii*]
Aglaonema modestum, Chinese Evergreen
Alocasia macrorrhiza, Giant Elephant's Ear
Anthurium andreanum, Florists' Anthurium, Flamingo Lily, Oilcloth Flower
Arum italicum, Italian Arum
Arum maculatum, Cuckoo Plant, Wake Robin, Lords-and-ladies
Caladium bicolor, Fancy-leaved Caladium, Heart of Jesus
Colocasia esculenta [*C. antiquorum*], Taro, Elephant's Ear, Kalo, Dasheen (fig. 41)
Dieffenbachia maculata [*D. picta*], Dumb Cane, Spotted Dumb Cane (pl. 15*c*)
Dieffenbachia seguine, Dumb Cane, Mother-in-law Plant
Dracunculus vulgaris
Epipremnum aureum [*Pothos aureus, Raphidophora aurea, Scindapsis aureus*], Pothos

FIG. 41. *Colocasia esculenta*. Elephant's Ear. Lower left: Plant (much reduced). Lower right: Tuber.

Monstera deliciosa [*Philodendron pertusum*], Ceriman, Swiss-cheese Plant, Breadfruit Vine, Hurricane Plant, Fruit-salad Plant, Window Plant, Split-leaf Philodendron, Cut-leaf Philodendron

Philodendron bipennifolium [sometimes called *P. panduriforme* in error], Fiddle-leaf Philodendron, Horsehead Philodendron, Panda Plant

Philodendron bipinnatifidum (pl. 15*d*); frequently seen outdoors in California

Philodendron domesticum [sometimes erroneously called *P. hastatum*], Spade-leaf Philodendron, Elephant's Ear Philodendron

Philodendron erubescens, Red-leaf Philodendron, Blushing Philodendron

Philodendron scandens, Heart-leaf Philodendron (fig. 42)

FIG. 42. *Philodendron scandens.* Heart-leaf Philodendron.

Philodendron selloum, Selloum; frequently outdoors in
    California
Philodendron squamiferum
Philodendron verrucosum
Spathiphyllum hybrids and cultivars
Xanthosma sagittifolium, Blue Elephant's Ear
Xanthosma violaceum, Blue Taro, Purple Elephant's Ear
Zantedeschia aethiopica, Calla Lily, Arum Lily, Florists'
    Calla, Garden Calla, Pig Lily, Trumpet Lily; commonly
    cultivated and occasionally naturalized in California

Plants of these species produce similar symptoms: almost
immediate painful irritation of the lips, mouth, tongue, and

throat; contact dermatitis is common. The descriptions for the following two species of *Dieffenbachia* emphasize the toxic problems associated with plants of the Arum Family:

> *Dieffenbachia maculata* [*D. picta*], Dumb Cane, Spotted Dumb Cane
> *Dieffenbachia seguine,* Dumb Cane, Mother-in-law Plant

Closely related species, natives of the American tropics. Widely used in California and elsewhere as indoor container-grown foliage plants, often seen in public buildings; can be moved to sheltered patios in the summer or rarely grown outdoors in frost-free areas.

Stems erect, 2 to 4 ft tall or taller in *D. seguine,* lower stem showing conspicuous horizontal leaf scars; leaves with clasping or sheathing petiole, leaf blade entire to 15 in. long or more, midrib and ten to twenty secondary veins prominent, blades often variously marked with irregular white spots or blotches; occasionally mature plants form calla-like flowers. Both species have a number of cultivars differing in the varied markings on the leaves.

Oxalate crystals have been found in several kinds of cells in the stems and leaves. Certain spindle-shaped cells, called biforines, have pores at their tips. By some undetermined mechanism, these cells eject needle-like crystals one after another at regular intervals. If bitten or chewed, the juice of the stems and leaves produces a painful burning sensation with copious salivation, severe irritation, and edema of the lips, mouth, and tongue; these reactions may interfere with swallowing and breathing. In Brazil it was reported that with large doses respiratory difficulties may lead to fatal suffocation. The swelling of the tongue may be so great that in severe cases the person cannot speak—hence the common name, Dumb Cane. This condition lasts for several days or longer. In past times it was recorded that in the West Indies slaves were made to chew this plant to render them speechless.

In January 1958, a child was treated at Orange County General Hospital following illness from chewing on a leaf of *Dieffenbachia maculata.*

Experimentally the juice of the stem has proved to be more

toxic than that of the leaves. The toxic substance is unknown, but it is heat-stable and associated with the bundles of insoluble oxalate crystals. Also in the juice is a proteolytic enzyme, similar to trypsin, which increases the hydrolysis of proteins. In rats and mice, indomethacin has proved effective in reducing the intensity of the swellings caused by the juice of *Dieffenbachia* stems. Antihistamine therapy has been successful only at the beginning of the inflammation and does not affect the length or intensity of the swellings.

The juice of the plant may also cause injury to the eyes. A case was reported of accidental injection of the juice into the eye of an 8-year-old boy. This accident caused keratoconjunctivitis, and the needle-like crystals were seen on the cornea. After treatment the crystals disappeared gradually during the following 2 months.

Cyperaceae, Sedge Family

Cyperus brevifolius [*Kyllinga brevifolia*].   Kyllinga.

Native to tropical regions of the New World; naturalized at scattered localities in coastal southern California from Los Angeles to San Diego and in the Sacramento Valley from Red Bluff to the vicinity of Sacramento. Plants are perennial with creeping rhizomes; flowering stems from a few inches tall in mowed lawns to $2\frac{1}{2}$ ft, each stem bearing single-flowered spikelets in a single globose head $\frac{1}{4}$ in. across, subtended by three bracts, one erect and two lateral ones directed downward. Toxicity is unknown in California, but plants are reputed to cause diarrhea in young livestock in Australia. Kyllinga plants are weedy in wet lawns and pastures, effectively competing with Bermudagrass. They should be kept out of pastures where they might cause distress to young stock.

Iridaceae, Iris Family

Only a few members of this family are poisonous.

Homeria.   Cape Tulip, Tulp.

Native of South Africa; two species, *Homeria breyniana* [*H. collina*] and *Homeria ochroleuca,* are occasionally cultivated in California. Underground corms; sword-shaped leaves;

flowering stems shorter than the leaves, 2 to 3 ft tall, orange, yellow, or salmon pink cup-shaped flowers about 1 in. across. Entire plant is toxic; contains a cardiac glycoside with a bufa-dienolide aglycone and five other toxic compounds. Poisoning of cattle results in stiffness of hindquarters, loss of appetite, marked thirst, diarrhea, and death.

There are no records of human or livestock poisoning in California. Humans were poisoned in South Africa and New Zealand from the corms, however. Plants, fresh or dried, are reported as extremely poisonous to cattle in South Africa and also Australia, where it is naturalized. Heavy losses have also occurred in horses, sheep, and goats.

Iris. **Iris, Flag, Fleur-de-lis.**

Large genus of about 200 species mostly in the north temperate zone. The genus is divided into sections based on underground parts and flower structure. California native species and many others from Europe, Asia, and eastern North America are grown in California as ornamentals.

Iris foetidissima. Scarlet-seeded Iris, Stinking Iris, Gladwin, Stinking Gladwin.

Native of Europe and North Africa; cultivated in California gardens. Rhizomatous perennial; leaves when crushed have a fetid smell; flowers dull blue-purple, outer perianth segments not bearded; seeds scarlet, more colorful than the flowers, remaining attached to the open capsule in late summer and fall.

Iris germanica. Flag, Fleur-de-lis.

Nativity unknown; may be of hybrid origin, perhaps originally from the Mediterranean region; widely cultivated. Rhizomatous perennial; flowers in many bright colors, outer broad perianth segments reflexed and bearded.

Iris pseudacorus. Yellow Iris, Yellow Fleur-de-lis, Water Iris.

Native to western Europe and North Africa, growing in wet areas; cultivated and also naturalized in California at a number of wet localities. Perennial from a rhizome; flowering stems branched, to 5 ft tall; flowers bright yellow, reflexed sepals not bearded.

Toxic part: Underground rhizome and leaves.

Toxin: An irritant resin.

Symptoms: Gastroenteric pain, nausea, and pronounced diarrhea; also contact dermatitis.

Iris plants are not usually ingested by either humans or livestock because of the very bitter taste. *Iris missouriensis,* Wild Iris, is a widespread native species found in California in meadows on the eastern side of the Sierra Nevada. Under grazing conditions the iris plants become abundant in meadows because the animals will not eat the bitter leaves. In an experimental area fenced from grazing and used for testing herbicides, the iris plants disappeared from both the sprayed and unsprayed parts in 3 years.

Moraea polystachya. Cape Blue Tulip, Blue Tulp.

Native of South Africa; rarely cultivated in California; naturalized on a roadside and rarely at other places in Santa Barbara. Plants similar to *Iris,* perennial from a corm; narrow leaves to 1 ft tall; branching cymose inflorescence among the leaves, extending somewhat taller, flowers lilac-blue of six perianth segments, the outer three segments somewhat reflexed and having a basal yellow spot. Entire plant is poisonous with alkaloids that make up about 0.8% of the weight of the plant. In livestock, the toxins cause acute gastroenteritis, profuse diarrhea, nervous prostration, and death.

Toxicity of this species of *Moraea* is not known in California, but it causes extensive losses of stock in South Africa. Four and one-half pounds of the plant caused the death of a 4-year-old ox in 20 hours. In a sheep, 250 grams of the fresh leaves from flowering plants proved lethal in 10 hours.

Plants related to *Moraea,* with clumps of narrow iris-like leaves and with flowers like a small Japanese Iris, are cultivated in California. They are species of *Dietes,* Butterfly Iris or Natal Lily, but are not recorded as being toxic to humans or to animals. *Dietes bicolor* has light yellow flowers to 2 in. across with six spreading segments to the perianth; each of the outer three segments has a basal dark brown blotch. *Dietes vegeta* has white flowers to 3 in. across; each of the outer three segments of the perianth has a brown or yellow basal spot.

Romulea longifolia.   Oniongrass, Guildfordgrass.

Native of South Africa; known to be naturalized in California only at Bolinas and Pt. Reyes in Marin County, established in Australia and a common contaminant of Subclover seed planted in California. Narrow grass-like leaves to 12 in. long arising from a corm about 3 in. underground; flowers about $\frac{3}{4}$ in. across formed just above the surface of the ground among the leaves; three sepals and three petals similar, rounded, bright pinkish purple; fruit a capsule with many seeds.

Plants of *Romulea* are suspected of causing infertility, abortion, and paralysis in sheep in Australia. Infection by a leaf-spot fungus, *Helminthosporium biseptatum,* may be responsible; cultures of this fungus have reduced fertility in experimental animals. There is no record of this toxicity in California, as plants of this species are established only in a very restricted area of Marin County, typically found in and along the edges of beaten paths.

Juncaceae, Rush Family

This family is of almost no importance as a source of poisonous plants.

Juncus phaeocephalus.   Brown-headed Rush.

Native of California, mostly along the coast; perennial with creeping rhizomes; stems flattened, two-edged, 1 to 3 ft tall; leaves shorter than flowering stems; many flowers, brownish in color, in one to several spherical heads at the ends of the stem.

A single case of hydrocyanic acid poisoning from this common rush has occurred in California. The only other known toxicity of this kind was caused by a different species of rush in Australia. In both species the plants are widely distributed, yet only a single case is known from each of the widely separated regions.

In December 1958, two dairy heifers died on a pasture near Petaluma, Sonoma County. From the symptoms observed, the veterinarian made a qualitative test for hydrocyanic acid on the rush plants and received a positive indication that this toxin was present. Tests made on the plants at the Chemistry Laboratory of the California Department of Food and Agriculture showed 30 ppm of hydrocyanic acid. Because of the volatile nature of this chemical, the content of hydrocyanic acid at the time of eating the plants could have been much higher.

In this particular case, the animals had been confined to a small pasture where all the forage had been eaten. Only the new growth of the rush plants that appeared above the water level of a drainage ditch were eaten. There were local frosts at this time, and the pasture was in a low-lying area between surrounding hills. The animals may have died after eating the frost-damaged rush plants in the morning before they were fed.

Juncaginaceae, Arrowgrass Family

Triglochin. **Arrowgrass.**

Four species, distinguished by technical characteristics in the flowers, are native in California:

Triglochin concinnum, Common Arrowgrass; var. *concinnum* is found in coastal salt marshes of California and extending to British Columbia and Baja California; var. *debile* [*Triglochin debile*] is found in wet saline soils of interior California from the Mojave Desert to Modoc County and extending eastward

Triglochin maritimum, Seaside Arrowgrass (fig. 43); found in wet alkaline or saline soils along the coast from San Francisco Bay and north and in wet areas below 7,500 ft in the San Bernardino Mountains and the Sierra Nevada; widespread to the Atlantic Coast and Eurasia

Triglochin palustre, Marsh Arrowgrass; occurs in California only in the Sierra Nevada above 7,500 ft in Inyo and Tulare counties but has a worldwide distribution

Triglochin striatum, Three-ribbed Arrowgrass; found in the coastal salt marshes of California from Mendocino County to Ventura County; widespread in the New World

Perennials; grass-like leaves thick and fleshy in basal tufts; leaf blades linear, flattened on one surface, rounded on the other, sheathing at the base, shorter to longer than the flower-

FIG. 43. *Triglochin maritimum.* Seaside Arrowgrass. Left:
Plant. Center: Inflorescence with fruits. Right: Fruit.

ing stems; flowers inconspicuous, greenish, in terminal racemes. Plants of these species are well known for producing
hydrocyanic acid.

Toxic part: Entire plant.

Toxin: Cyanogenic glycosides; plants growing under drought
conditions have the highest concentration.

Symptoms: Typical of hydrocyanic acid poisoning in livestock: rapid breathing, a staggering stiff walk, spasms, bright
red blood from the nostrils, eventual convulsions, and death
from respiratory failure.

In the saline valleys of northeastern California where *Triglochin concinnum* var. *debile* can be abundant or even the only plant in an area, Arrowgrass plants are routinely cut for hay; on drying, the hydrocyanic acid disappears. In 1959, in Surprise Valley near Cedarville, Modoc County, numerous jackrabbits died from eating freshly cut plants of *Triglochin*. Dead animals were found only within a distance of ¼ mile around the mown hayfield.

Liliaceae, Lily Family

This family is closely related to the Amaryllidaceae; superficially the flowers of the two families look similar. Some taxonomists place the two families together in the Liliaceae. The Liliaceae includes many types of inflorescences but not an umbel subtended by bracts, and the flowers of many genera have a superior ovary, rarely an inferior one. Several poisonous plants are found in the Lily Family, some of which are of medicinal use: bitter aloes, *Aloë;* squill, *Urginea;* and hellebore powder, *Veratrum.*

Aloë barbadensis [*A. vera, A. perfoliata* var. *vera*].
Barbados Aloe, Curaçao Aloe, Medicinal Aloe, Unguentine Cactus. [The last two vowels of the botanical name *Aloë* are pronounced separately; the common name, however, is pronounced as two syllables.]

Native of the Mediterranean region; widely cultivated as an ornamental, commercially cultivated for the drug aloe mainly in the Netherlands Antilles; naturalized in Mexico. Rosettes of fleshy succulent leaves to 2 ft long, toothed on the edges, otherwise very smooth, waxy green in color, rosettes at ground level or on stems to 1½ ft tall, spreading by suckers to form dense clumps of plants; flowering stems grow above the leaves, usually unbranched, 3 to 4 ft tall with a dense spike of yellow tubular flowers, 1 in. long.

Toxic part: Juice of the fleshy leaves that may be dried as a yellowish crystalline powder.

Toxin: Aloins (glycosides that break down to anthraquinones); the principal aloin is barbaloin.

Symptoms: Acts as a cathartic in humans. Excessive use as a cathartic or an illegitimate use to induce abortion results in painful bleeding, damage to the gastrointenstinal tract, irritation of the kidneys with pain in the lumbar region, and an alkaline urine colored red.

Aloe has long been used in folk medicine as a cure for many conditions including burns, insect bites, and inflammations of various kinds. Aloin, combined with other cathartics or with an antispasmodic agent, is used in the manufacture of several nonprescription drugs.

Asparagus officinalis.　Garden Asparagus.

In Los Angeles County in 1937, dairy cattle were poisoned when accidentally feeding on fields of nearly mature Asparagus plants. The cause of the poisoning is not known.

Colchicum autumnale.　Autumn Crocus, Fall Crocus, Meadow Saffron, Mysteria, Wonder Bulb (fig. 44). [Not to be confused with the genus *Crocus* in the Iris Family.]

Native of Europe and North Africa; cultivated in California gardens, sometimes grown in containers. Perennial from a corm; several lanceolate leaves to 1 ft long in the spring, which die down before one to four flowers appear in the fall; flowers white to purple with six perianth parts joined at the base into a long tube.

Toxic part: Entire plant, especially the underground corm.

Toxin: Colchicine, an alkaloid.

Symptoms: After a latent period of 2 to 6 hours symptoms begin with burning in the throat, difficulty in swallowing, intense thirst, nausea, abdominal pain, profuse vomiting and diarrhea, and kidney damage. Can be fatal.

Colchicine is a mitotic poison that interferes with cell division. It is used in medicine for treatment of gout. The drug is slowly excreted from the body and thus may accumulate from even small doses. Children have been poisoned from eating the flowers, and the corm has been mistaken for an onion.

Colchicine has been used in plants to stop nuclear division, thereby producing polyploids useful in plant breeding.

FIG. 44. *Colchicum autumnale.* Autumn Crocus. Left:
Flowers from a bulb in the fall. Right: Leaves from a bulb in
the spring.

Convallaria majalis.   Lily-of-the-valley, Mayflower, Conval
Lily (fig. 45).

Native of Europe; grown in California gardens and a popular florists' potted plant. Herbaceous perennial 8 to 10 in. tall;
spreads by rhizomes to form dense clumps; leaves broad ovate;
white cup-shaped fragrant flowers, nodding, in racemes; fruit a
red berry with many seeds.

Toxic part: Entire plant, as well as the water in which the cut
flowers have been kept.

Toxin: Cardiac glycosides (convallotoxin, convallarin, and
convallamarin); irritant saponins.

Symptoms: In humans, irritation to the mucous membranes
of the mouth followed by vomiting and abdominal pain caused
by the saponins. The digitalis-like glycosides have toxic effects
on the heart.

Any part of the plant may be toxic, but this species seems to be less of a hazard because of the foul taste. Although leaves appear to be mildly irritant, cases of dermatitis are rare even among florists who handle them frequently.

Dianella tasmanica.   Blue Flax Lily.

Native of Tasmania; cultivated as an ornamental in California for its attractive blue berry. Herbaceous perennial; stiff, sword-shaped leaves 3 to 4 ft long arising from a rhizome; flowers pale blue about ¾ in. across in a large branched panicle much taller than the leaves; fruit a blue berry. Toxicity is not recorded for California. In Australia and New Zealand this and other species of *Dianella* have been suspected of poisoning

FIG. 45. *Convallaria majalis.* Lily-of-the-valley. Flowering plant with a creeping rhizome. Right: Enlarged flower.

livestock and humans, but reports are circumstantial and inconclusive.

Endymion non-scriptus [*Scilla non-scripta, S. nutans*].
English Bluebell, Harebell.

Native of Europe; cultivated in California. Bulbous perennial; leaves lanceolate to 18 in. long; flowering stem about the same length; flowers bell-shaped, several along one side of upper stem, nodding, blue, about ½ in. long. Entire plant is poisonous; the toxins are unknown, probably digitalis-like glycosides. Ingestion results in abdominal pain, diarrhea, and slow pulse.

Fritillaria meleagris.   Snake's Head, Checkered Lily,
Guinea-hen Tulip.

Native of Europe; cultivated for its showy flowers. Herbaceous perennial; stem to 15 in. tall; leaves linear to lanceolate; usually one or two flowers, bell-shaped, 1½ in. long, nodding, in shades of red and yellow on outside, more highly colored with lines and spots on inside, looking like a snake's head. Entire plant is toxic; contains imperaline and other alkaloids related to those of *Veratrum*, which depress cardiac action. No cases of human or animal poisoning in Europe have been recorded in recent years.

Gloriosa rothschildiana.   Gloriosa Lily, Glory Lily,
Climbing Lily (fig. 46).

Native of tropical Africa; cultivated as an ornamental climber for its showy red flowers, outdoors in warmer areas or as a summer potted plant. Climbing perennial herb from a tuber; stem weak; leaf lanceolate to ovate-lanceolate with the tip extended into a coiled tendril by which the plant climbs; striking red and yellow flowers with recurved perianth segments margined yellow; six spreading stamens and a long green style. The plant dies back after flowering. A closely related species, *Gloriosa superba*, sometimes cultivated, is also extremely toxic.

Toxic part: Entire plant.
Toxin: Colchicine and similar alkaloids.

FIG. 46. *Gloriosa rothschildiana.* Gloriosa Lily. Lower right: Tuber.

Symptoms: After a latent period of 2 to 6 hours, symptoms in humans begin with burning in the mouth and throat, difficulty in swallowing, intense thirst, nausea, abdominal pain, profuse vomiting and diarrhea, and kidney damage. Can be fatal. Poisonings have occurred when the tubers were mistaken for sweet potatoes.

Hyacinthus orientalis.   Hyacinth, Dutch Hyacinth, Garden Hyacinth.

Native to southeastern Europe and North Africa; one of the best-known bulbous plants, grown commercially by the bulb industry. Perennial herb with a bulb; strap-shaped basal leaves; numerous flowers in various shades of blue, pink, yellow, or white, in a dense raceme on a hollow stem about 1 ft tall.

Toxic part: Entire plant, especially the bulb.

Toxin: Reported to be alkaloids.

Symptoms: Severe gastroenteritis with pain, nausea, vomiting, and diarrhea.

When planted in gardens, the bulbs may be mistaken for onions. In Holland during World War II, Hyacinth bulbs as well as those of Daffodils and Snowdrops were used as emergency food and caused livestock deaths. Needle-shaped calcium oxalate crystals about ⅛ in. long in the scales of the bulbs have caused irritant dermatitis. Oil of hyacinth used in perfumes also has produced dermatitis.

Ornithogalum umbellatum. Star-of-Bethlehem, Summer Snowdrop, Nap-at-noon.

Native of Europe and North Africa; cultivated and readily persisting in the garden; may become weedy. Bulbous perennial; basal narrow grass-like leaves; cluster of greenish white flowers 1 in. across, on a stem 12 in. tall, petals green with white margins.

Toxic part: Bulbs and flowers.

Toxin: Digitalis-like glycoside, similar to that in *Convallaria majalis*.

Symptoms: Nausea, vomiting, and intestinal disturbances.

Bulbs of related species have similar toxic properties:

Ornithogalum arabicum, Star-of-Bethlehem; native of the Mediterranean region; cultivated in California; about 2 ft tall, flowers white, each with a black pistil

Ornithogalum caudatum, Sea Onion, German Onion, False Sea Onion, Pregnant Onion; native of South Africa; cultivated as a novelty for its green bulbs, 3 to 4 in. across, partially on top of the ground; in East Africa, tribal groups recognized the toxicity of this plant

Ornithogalum thyrsoides, Chincherinchee, Wonder Flower, African Wonder Flower; native of South Africa, occasionally cultivated in California; numerous creamy white flowers, frequently used as a cut flower by florists because of its long-lasting qualities; the plant contains a highly toxic resin

Scilla peruviana.   Cuban Lily, Peruvian Jacinth, Hyacinth-of-Peru, Star Hyacinth, Squill.

Native of the Mediterranean region (not Peru); cultivated for its attractive flowers. Bulbous perennial to 18 in. tall; leaves strap-shaped; flowers blue, purple, or white, with spreading segments in a dense cone-shaped cluster of 50 to 100 flowers at the top of the stalk.

Toxic part: Entire plant.
Toxin: Glycoside with a digitalis-like action.
Symptoms: Same as those caused by digitalis.

Tulipa species, hybrids, and cultivars.   **Tulip.**

About 100 species are native to southeastern Europe and western Asia. Many species and cultivars, some of hybrid origin, are cultivated in California gardens. A considerable commercial industry markets tulip bulbs. Bulbous perennials; leaves basal or on the stem; flowers in a vast array of colors; perianth segments held together to form bell-shaped flowers or spreading.

Ingestion of the bulbs may cause gastrointestinal irritation. Tulps or Cape Tulips, *Homeria,* in the Iris Family, should not be confused with Garden Tulips. Handling Tulip bulbs causes dermatitis, particularly among those collecting, sorting, and packing them.

Urginea maritima [*Scilla maritima*].   Sea Onion, Squill, Red Squill.

Native from the Mediterranean region east to Syria; cultivated in California as a source of red squill, used in medicine and as a rodent poison; also sometimes grown as an ornamental. A number of toxic species of *Urginea* are native in South Africa.

Bulbous plant with basal strap-shaped leaves; leaves with a waxy coating and somewhat fleshy; flowering stem 4 to 5 ft tall; white flowers $\frac{1}{2}$ in. long in a dense mass at the top 18 in. of the stem.

Toxic part: Bulb; the drug is made from the fleshy inner scales of the bulb.
Toxin: Digitalis-like glycosides.

Symptoms: Burning of the pharynx, gastroenteritis and vomiting, diarrhea with colic, pain in the legs, and convulsions. Larger doses may produce digitalis-like actions on the heart and result in a heart block. Fatal cases have been reported.

Rats cannot vomit red squill. Humans and domestic animals find it distasteful, vomit almost immediately, and thus are not likely to be harmed by it. Therefore red squill is regarded as a "safe" rodent poison.

Veratrum californicum.   Corn Lily, Skunk Cabbage, California False Hellebore, Swamp Hellebore, Cow Cabbage, Wild Indian Corn, Indian Poke (pl. 16*a*)

Native in California, mountain meadows in the North Coast Ranges and Sierra Nevada south through mountains of southern California; extends into Baja California, Washington, the Rocky Mountains. Perennial from a rhizome, leafy stems to 6 ft tall; leaves broad, clasping the stem, strongly parallel veined and pleated; numerous greenish white flowers ½ in. long in showy branched clusters; fruit a three-sided capsule.

Three other species of *Veratrum* occur in northern California in restricted areas:

> *Veratrum fimbriatum,* Fringed False Hellebore; in Sonoma and Mendocino counties
> *Veratrum insolitum,* Del Norte False Hellebore; in Del Norte County and adjacent Oregon
> *Veratrum viride,* Green False Hellebore; restricted in California to a part of Siskiyou County but extends across the continent to the Atlantic and to Alaska

These species of *Veratrum* in California and North America are called false, swamp, or green hellebores and should not be confused with the hellebores of the Eurasian genus *Helleborus* discussed under the Ranunculaceae.

Toxic part: Entire plant.

Toxin: A number of alkaloids; cyclopamine is regarded as the most important alkaloid since it occurs in much higher concentrations than the others.

Symptoms: In humans, shortly after ingestion a burning sensation and pain in the upper abdominal area, light-headedness, followed by salivation, pronounced nausea and vomiting, dry cough, difficulties in focusing, talking, or writing, weakness, and numbness in the fingers. Slow heartbeat and hypotension might appear to an adult as a heart attack. Most symptoms disappear in 24 hours.

In sheep, deformed lambs are produced when the pregnant ewes ingest this plant on the fourteenth day of pregnancy. Abnormalities may vary from mild to very severe deformities in which the lambs have a single median eye, protruding lower jaw, a shortened upper jaw, and a peculiar nose-like projection below the single eye. Epidemics of malformed lambs occurred in the western states until the cause was found: Pregnant ewes were feeding on the plants of *Veratrum*. They are not generally palatable to livestock, however, and have increased in numbers in the mountain meadows of California as livestock prefer other plants. *Veratrum* plants are the first to appear in the spring and therefore are present in sufficient quantity to be dangerous if sheep are turned into the meadows early in the season before other forage plants have appeared.

Veratrum viride plants contain several alkaloids that have had medicinal uses for their hypotensive, cardiac depressive, and sedative properties. In recent years, however, better drugs have become available. The European White False Hellebore, *Veratrum album,* has been used medicinally in Europe since the time of Dioscorides in the first century A.D. It contains alkaloids similar to those of *V. viride*. The alkaloids in the seeds of both Green and White False Hellebores have been used in insecticides.

Green False Hellebore was used by Indians of eastern North America as a remedy for various ailments. Its medicinal uses were recognized early in the history of the United States, and under the name *Veratrum viride* it was listed in the United States Pharmacopoeia (1820–1942) and in the National Formulary (1942–1960). Rhizomes were used in the preparation of the drug.

Zigadenus. Deathcamas, Zigadene, Squirrel Food, Wild Onion, Poison Sego.

Zigadenus is sometimes incorrectly spelled *Zygadenus*. Called Deathcamas, it should not be confused with plants of the genus *Camassia,* commonly called Camas, which have edible bulbs and attractive blue flowers.

Zigadenus fremontii. Chaparral Deathcamas (fig. 47).

Native in the Coast Ranges of California and extending into Baja California and southern Oregon. Similar to the Meadow Deathcamas with perianth segments $\frac{1}{4}$ to $\frac{1}{2}$ in. long.

Zigadenus venenosus. Meadow Deathcamas (pl. 16*b*).

Native almost throughout California from sea level to above 8,000 ft elevation, extending to Baja California, British Columbia, and Utah. Perennial plants with onion-like bulbs but lacking the onion odor; slender channeled linear leaves; flowering stem to 24 in. tall, flowers in a raceme, perianth of spread-

FIG. 47. *Zigadenus fremontii.* Chaparral Deathcamas. Left: Bulb. Right: Inflorescence with leaves.

ing yellowish white segments to $\frac{3}{16}$ in. long, a green gland at the base of each segment, six stamens longer than the perianth; fruit a three-parted capsule.

Four other species of *Zigadenus* of more restricted distribution also occur in California.

Toxic part: Entire plant, particularly the bulbs.

Toxin: Zygadenine (an alkaloid) and several other steroidal alkaloids related to those of *Veratrum*.

Symptoms: In humans, burning in the mouth, thirst, dizziness and headache, persistent vomiting, slow heart action, low blood pressure, and convulsions. Drowsiness and staggering progress to a coma with slow and irregular respiration.

Human poisonings have occurred when the bulbs have been mistaken for those of wild onions or the edible Camas. Such cases occurred during the early settlement of the western United States. American Indians apparently were aware of the poisonous properties of the zigadenes. Bulbs of a Deathcamas made into flour caused serious illness to members of the Lewis and Clark Expedition.

Signs of Deathcamas poisoning are generally the same for all kinds of livestock. Excessive salivation is typical for the duration of the poisoning. Vomiting, weakness, rapid labored breathing, and collapse may follow. Some animals die within a few hours; others may remain in a coma for several days.

In spite of the name Deathcamas, these native herbs are not usually eaten by animals and therefore are not the threat to livestock that the common name would imply. In California it is usual to find spring pastures eaten down closely to the ground and Deathcamas plants in full bloom untouched by the animals. Because Deathcamas plants are the first to appear in the late winter or early spring, they can be a hazard to animals if adequate forage is not available. Sheep are the most susceptible to this poisoning when forced by hunger to eat them. One to two pounds of Deathcamas plants may cause the death of a 100-pound sheep. A 1,000-pound cow would require a much larger amount as a lethal dose. Dried hay from native meadows can be a source of poisoning as the toxicity of Deathcamas plants is not lost on drying.

In late April 1960, a snowfall of 6 to 18 in. covered the forage in a large rangeland pasture near Manton, Tehama County.

Cattle heavily grazed the abundant growth of Deathcamas plants, *Zigadenus* sp., in a small rocky swale that was in the open and unshaded by trees. Several of the animals died. Cattle usually will not dig through snow for forage as do horses or other animals, but in this open swale the snow had melted early, exposing the toxic plants.

In March 1963, three horses of a group of ten became ill from eating native meadow hay that included plants of *Zigadenus venenosus* with immature seed pods. The horses showed colic, salivation, depression, cramping, and intermittent diarrhea. The hay had been cut from meadows in the Bieber area of northern Lassen County and adjacent Modoc County.

Ninety head of sheep died while being trailed through an area of *Zigadenus paniculatus* at the beginning of April 1979 in southern Modoc County.

Poaceae, Grass Family

Also called Gramineae, members of this cosmopolitan family are of minor importance as sources of plant toxins that affect domestic animals. A few accumulate nitrates, however, and others are cyanogenic, producing hydrocyanic acid. In humans, pollen of many California grasses causes hay fever in sensitive individuals and a few species cause contact dermatitis.

Annual or perennial herbs; only the bamboos have woody stems. Stems or culms hollow except at the nodes from which the leaves arise; leaves consisting of a sheath surrounding the culm and an expanded blade with the ligule as a membrane or line of hairs at the top of the leaf sheath; small flowers, called florets, subtended by two modified bracts, the lemma and palea, usually consisting of three stamens and a pistil, lacking petals and sepals but with two small thin scales below the pistil, the lodicules.

Identification of grasses relies mostly on the technical structure of their floral parts. Because the florets are small and highly specialized, careful examination is best done with the aid of a dissecting microscope, but identifications can be made with a hand lens. Vegetative characteristics also are important; the plant should be removed from the soil to determine whether it is annual or perennial. The many variable characteristics of the perennial rhizomes aid in identification. Leaf structures often are confirming characteristics.

Mechanical injury and irritation to the skin of humans and livestock may be caused by a number of California grasses that have grains, mature florets, or entire spikelets with sharply pointed ends or barbed awns. Livestock are injured when these sharp structures, usually upwardly barbed, penetrate the tissues inside the mouth, nostrils, ears, or eyes. Extreme cases of such injury are discussed under *Heteropogon* and *Setaria*. Grasses causing mechanical injury include various species of the following genera:

Aristida, Threeawn
Bromus, Bromegrass
Heteropogon, Tanglehead
Hordeum, Barley
Setaria, Bristlegrass
Stipa, Needlegrass

Cynodon dactylon. Bermudagrass, Devilgrass.

Native of the Old World; widely naturalized in California, common in fields, orchards, and lawns. Creeping perennial with extensive rhizomes; difficult to eradicate if unwanted although it makes a good lawn; three to five lateral flowering spikes are formed at the tip of erect culms.

Bermudagrass produces contact dermatitis, and its pollen is an important cause of hay fever. In some parts of the United States, cattle grazing on Bermudagrass have on occasion developed three diseases: photosensitization; paralysis known locally as "Bermudagrass poisoning"; and a nervous disease characterized by muscular twitching and eventual inability to stand. None of these diseases has been recorded for California.

Eragrostis cilianensis [*E. megastachya*]. Stinkgrass.

From the Old World; in California a naturalized weed in areas of summer water. Annual; several culms from the base; glands on culms below the nodes, on leaf sheaths and lower surface of the blades, and on the spikelets; each spikelet many-flowered, flattened, to $\frac{1}{2}$ in. long, arranged in panicles; florets without awns. The glands give the plant a peculiar odor. Reported in Kansas to cause sickness in horses when ingested fresh or dry over a long period of time.

Festuca arundinacea.   Tall Fescue, Reed Fescue.

Native of the Old World; widely cultivated and naturalized; grows best in moist soils along roadsides and in ditches. Its strong root system makes it useful in preventing soil erosion. Robust clumped perennial, culms 2 to 6 ft tall, each spikelet with several flowers, spikelets arranged in panicles to 15 in. long; florets not awned. In plants that have not formed seed heads, leaves and stems may contain alkaloids similar to those of ergot. They cause a disease in cattle called "fescue foot," which has been found in the Pacific Northwest to the southeastern United States but has not been reported for California. The condition shown by cattle is typical gangrene with sloughing of the hooves, tip of the tail, and tips of the ears. These chronic symptoms have not been seen in horses. Tall Fescue is resistant to drought, and during such periods cattle may be forced to eat almost a complete diet of this grass. In a mixed pasture, well fertilized, Tall Fescue seems to be harmless.

Festuca rubra var. *commutata.*   Chewings Fescue.

The species is native to the cooler parts of the northern hemisphere including California; this particular variety has been chosen for its useful characteristics because it forms a dense sod and the more erect stems make a better hay crop than other forms of this species. Chewings Fescue is a commonly available crop plant in California, useful in the cooler sections. Clumping perennial with stems to 3 ft tall; four to six florets per spikelet; florets with a small awn less than $\frac{1}{8}$ in. long. Toxic nematode galls are formed in place of the grains in the seed head. The toxin is unknown, but it is not destroyed by heat and is soluble in alcohol. It produces nervousness and death in animals.

Grass-seed nematode poisoning of livestock has not been reported for California, but it is known from central Oregon where there is commercial production of grass seed on a large scale. Cattle and horses have died from eating these grass-seed galls; when tested on chickens and rats, they proved fatal. Livestock have been poisoned from feeding on Chewings Fescue hay with seeds in the immature dough stage and many galls; grass-seed screenings have proved toxic as well. Typically the animal knuckles over on the forelegs, tucking the head between the legs. Convulsions and death follow.

The grass-seed nematode, *Anguina agrostis,* lays eggs in the immature pistil in the floret. Numerous larvae develop in a single ovary that is transformed into a gall in place of the grain. This nematode can infest many different grasses including Orchardgrass, bentgrasses, and Kentucky Bluegrass.

Glyceria striata.　Fowl Mannagrass, Fowl Meadowgrass.

Native to California in meadows and wet places, 5,000 to 8,000 ft elevation in the mountains of the northern part of the state and in the Sierra Nevada; widespread outside California. Perennial, tufted; culms slender, 1 to 4 ft tall; spikelets arranged in spreading panicles 4 to 8 in. long; florets not awned, lemma strongly seven-nerved. Fowl Mannagrass may cause cyanogenic poisoning in livestock and was incriminated in the death of several calves in Maryland.

Heteropogon contortus.　Tanglehead (pl. 16c).

Native in warm areas around the world; in the United States from Arizona to Texas and in Hawaii. At present Tanglehead is not known in California. At one time a small population was established in northeastern Imperial County, but it has been eradicated. Another collection of Tanglehead was made near Dehesa School, San Diego County, in 1938, but it apparently has not persisted. Leafy clumped perennial plants to 3 ft tall. The seed head is a tangled mass of spikelets with long twisted awns—hence the common name Tanglehead. Each spikelet resembles a floret of the native needlegrasses, *Stipa,* but consists of an entire spikelet of two glumes and a single floret, the floret with an awn 5 in. long bent at two places. The upwardly pointed scabrous hairs make the spikelets serious contaminants of wool. If the sharp point of the spikelet penetrates the skin of the sheep, serious economic losses result because the meat must be rejected for human consumption.

In October 1968, more than 2,000 feeder lambs were brought from western Texas to the Imperial Valley to be raised during the winter. Examination of a few animals that died from various causes revealed extensive contamination of the wool, particularly around the neck, by numerous spikelets of Tanglehead. It appeared that the animals had reached past the seed heads to graze on the stems and leaves of the plants. Entire embedded spikelets of glumes and attached awns were seen on the flesh side of the skin. The sharp end of the spikelet often had made a hole in the skin about ½ in. across.

Hilaria rigida.   Big Galleta, Woolly Galleta.

Native in California, on both deserts, and eastward. Coarse perennial, spreading by rhizomes; culms rigid, solid, woolly, 1½ to 3 ft tall; spikelets sessile, forming a spike-like inflorescence, florets not awned.

Although considered to be a valuable forage grass, Big Galleta has been implicated in California in sudden death of cattle. Losses of cattle occurred in January 1956 when a number were moved to desert rangeland on the eastern slope of the Laguna Mountains in San Diego County. The actual cause of death is not known. Examination of the animals showed that their stomach contents were compacted and very dry. Big Galleta had been the main forage. Such losses might be avoided by gradually adapting the animals to the desert range with feed supplements.

Holcus lanatus.   Velvetgrass.

Native of Europe; naturalized throughout much of California, particularly along the coast, growing best on moist soils or soils with subsurface moisture during the growing season. A poor forage grass that dries to a hay of almost no weight. Perennial, usually tufted; leaves velvety with a dense covering of soft hairs; culms 10 to 30 in. tall; spikelets sparsely to densely arranged in panicles often reddish purple to pinkish, pale green, or whitish; florets with a short hooked awn or awnless.

Reported to produce hydrocyanic acid poisoning; there are no cases of this poisoning on record in the United States.

Lolium perenne. Perennial Ryegrass.

Native of Europe; grown in California on a lesser scale than Italian Ryegrass, *Lolium multiflorum,* or the hybrids between the two species. *Lolium perenne* is frequently included in seedings of native areas after fires but rarely persists for any lengthy period. Short-lived perennial, often purplish at the base, to 2 ft tall; sparsely leafy with leaf blades less than ⅛ in. broad; six to ten flowered spikelets, formed on either side of the stem into a flattened terminal spike; florets without awns. The entire plant, especially the leaf sheaths and the grains, is considered toxic. The unknown toxin is the result of infection inside the tissues of the grass by *Phialea temulenta,* an ascomycete fungus. In livestock, the infected plant material produces ryegrass staggers, a neuromuscular disease characterized by an unsteady gait, swaying of the body, and weaving of the head while standing. This disease has not been reported from California but has been found in other regions of the world. Cows, horses, and sheep are all affected by the toxins of this fungus. First reported in the 1940s, this fungus has been the cause of distress to livestock in the Willamette Valley, Oregon, where the grass is grown for seed. The condition appeared after animals grazed on ryegrass stubble and were given seed screenings of Perennial Ryegrass. In New Zealand, signs of ryegrass staggers have been greatly relieved by treatment with Epsom salts, $MgSO_4·7H_2O$, and potassium chloride, KCl.

Lolium temulentum. Darnel, Darnel Ryegrass.

Native of Europe; introduced and naturalized but not an abundant weed in California. Annual plants, culms 2 to 3 ft tall; each spikelet with few flowers, sessile, fitted into notches in the main stem or rachis to form a flattened somewhat zigzag terminal spike; florets awned. The grains, and probably the entire plant as well, are poisonous. The toxin is unknown but is caused by a fungal infection in the plant and transmitted in the seeds. The fungus, *Endoconidium temulentum,* is known only from the sterile stage of growth. Humans eating bread made with wheat contaminated with Darnel have experienced vertigo or dizziness, stiffness of the back, extreme lassitude, abdominal pains, diarrhea, and frequent urination. Although very severe poisoning has been rare, it can cause convulsions and delirium followed by death.

There are no California records of Darnel poisoning of humans or livestock. In June 1961, some grains of Darnel were found in birdseed that had been mixed at Redding, Shasta County, but the origin of the Darnel seed could not be determined.

Pennisetum clandestinum.   Kikuyugrass.

Native of Africa; a pasture plant in tropical and subtropical areas; a serious weed in California in turf and waste places, along the coast from San Diego County to the San Francisco Bay region, but recorded as a turf weed as far north as Redding, Shasta County. Vigorous perennial spreading rapidly by tough rhizomes and stolons; leaf sheaths closely overlapping on the stolons, leaf blades elongated to 18 in. long; flowers at the tips of the stems hidden in the leaf sheaths except for the stamens, which are prominently exerted from the sheaths by long silky filaments bearing the anthers; tips of the pistils are minute, barely extending out of the leaf sheaths.

There are no records of cattle deaths from Kikuyugrass in California, but it is not grown in pastures here; a number of deaths have occurred in New Zealand, however, more than 200 in 1972. The disease is described as a ruminal engorgement; the rumen commonly contains two to three times the normal amount of forage. The cause of this condition is not known but appears to be more prevalent when the grass has been mechanically injured or damaged by insects.

Phalaris aquatica [*P. tuberosa* var. *stenoptera*].   Hardinggrass.

Probably native to the Mediterranean region but first noticed in Australia; cultivated and naturalized in California at lower elevations, a useful forage grass. Densely clumped perennial; culms stout, to 4½ ft tall; numerous spikelets forming a dense, lobed, spike-like inflorescence; florets not awned.

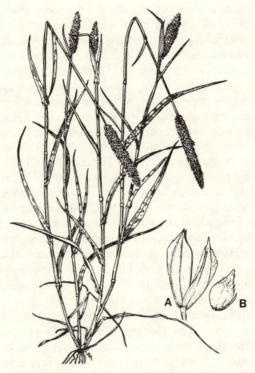

FIG. 48. *Phalaris minor.* Mediterranean Canarygrass.
A. Spikelet. B. Grain.

Phalaris minor. Mediterranean Canarygrass (fig. 48).

From Europe; naturalized as a common weed in California. Annual, culms to 2 ft tall, rarely taller, not as robust as Harding-grass, terminal seed head not lobed.

Both species have caused poisonings of livestock.

Toxic part: Entire tops of the plants, particularly the young parts.

Toxin: Unknown in some cases, tryptamine alkaloids in others.

Symptoms: Phalaris staggers (pl. 16*d*), a neurological disease involving brain damage to cattle in California. When driven, the animals collapse and exhibit hyperexcitability and rapid pulse. This chronic condition is characterized by stiff-legged walking, apparent blindness, and inability to find water and eat feed concentrates. Injured cattle eat only hay, resulting in commercial losses in a feedlot. Death may be caused by de-hydration, starvation, or both. There is a characteristic bluish granular pigmentation of the cells in the brain. Pigmentation also occurs in the medulla of the kidney, less often in the liver.

The toxin causing this complex condition is unknown. It is usually associated with animals eating the first flush of new growth of *Phalaris* plants. Animals can be protected from this disease by placing a slowly dissolving cobalt bullet in the rumen. Experimentally, however, it has been shown repeatedly that this toxicity is not merely a cobalt deficiency in the forage.

In Australia where sheep, and to some extent cattle, develop chronic phalaris staggers, there are also acute cases of poison-ing from Harding-grass forage, either a sudden death syn-drome or only transient disorders of the neurological system. High levels of tryptamine alkaloids are found in plants of *Pha-laris* associated with these acute poisonings.

Although there have been a few unverified reports of sheep and cattle poisoned by Harding-grass in California, only one case of phalaris staggers from this grass has proved valid: in a pasture near Garberville, Humboldt County. Tryptamine alka-loids were found in the Harding-grass in this pasture. The high-est concentrations of alkaloids occurred early in the grazing season, but the greatest losses of sheep occurred in midseason. The actual cause of the toxicity in this pasture is unknown, but several sheep died and others manifested typical phalaris poi-soning. Experimental study of this situation found that those sheep without a cobalt bullet in the rumen were the only ones that died and only when kept on the part of the pasture that had received phosphorus fertilizer. It was postulated that phos-phorus did not become a part of the actual toxins but produced greater amounts of toxic grass.

Phalaris minor is an abundant winter weed of alfalfa and sugar beet fields in the Imperial Valley. Cattle, often from dry rangeland in Mexico, are placed in great numbers in a field to remove the grass weeds quickly and then are moved to another

field. Phalaris staggers has occurred in the Imperial Valley between December and March. In December 1960, one field
caused phalaris staggers in some 200 head of cattle, although
the same field was grazed a few weeks later with no signs of
toxicity. See also "Polioencephalomalacia" (pp. 366–367).

Setaria. **Bristlegrass, Bristly Foxtail.**

From warmer parts of the New World and Old World but
mostly from Africa; introduced into California and now widespread weeds. Annuals or clumped perennials, sometimes
branched from the base; spikelets together with surrounding
bristles in dense terminal spikes, the bristles remaining on the
main stem or rachis after the spikelets fall.

The barbed awns (bristles) on the seed heads of these
grasses cause ulcers in horses' mouths. Such injury was seen
in 1970 near Blythe in Riverside County and in 1977 near
Petaluma in Sonoma County. The horses had eaten hay containing dried Bristlegrass plants. An unusual case occurred in
Kern County in a herd of 400 cows with clinical signs of a viral
infection. They were fed alfalfa hay contaminated with Bristlegrass and developed ulcers in their mouths; a bristle was found
in the center of each ulcer. Normally cattle have tough mouths
and should not have been affected by these minute bristles.

In Stanislaus County during the fall months of 1952 a series
of mouth lesions occurred in cattle and horses. The damage
was caused by hay containing about 2% by weight of dried
plants of *Setaria lutescens* [*S. glauca*], Yellow Bristlegrass or
Yellow Foxtail, grown under drought conditions. Similar lesions were reproduced experimentally in a calf and in a horse
using this hay.

Sorghum bicolor [*S. vulgare*]. Sorghum, Grain Sorghum.

Native of the Old World; in cultivation in warm regions for
centuries; grains used as food for cattle and poultry and, in
some parts of the world, for human food; one form having
a high sugar content is used in the manufacture of syrup. In
California cultivated mostly as a grain for animal feed, sometimes as forage or silage. Robust annual; culms usually stout;
height differs in cultivated varieties, 2 to 12 ft tall; leaf blades

1 to 2 in. wide; inflorescence a compactly branched, terminal panicle.

Sorghum halepense.   Johnsongrass.

From the Old World, probably North Africa; purposefully introduced in California years ago as a crop plant but has become widely naturalized as an aggressive weed in low elevations where there is sufficient moisture. Rhizomatous perennial; culms to 7 ft tall; leaf blades 1 in. or more wide with a white midvein; panicle terminal, spreading, often reddish purple; short racemes break up at the joints; florets with conspicuous awns.

Sorghum sudanense.   Sudangrass.

From Africa; cultivated in California for forage and hay, sometimes as a seed crop; occasionally escapes from planted areas. Annual differing from Johnsongrass in the lack of extensively creeping rhizomes; frequently tall plants, to 12 ft in some varieties; narrow leaf blades, usually less than 1 in. wide.

When heavily fertilized, Sudangrass plants may accumulate toxic quantities of nitrates. Plants with a light green color should be safe; unusually dark green plants should not be used. Plants containing toxic levels of free nitrates retain their toxicity when dried as hay. Johnsongrass and Grain Sorghum may also accumulate toxic amounts of nitrates.

Johnsongrass, Sudangrass, or Sudangrass hybrids can cause poisoning by hydrocyanic acid. This danger can be avoided by using extreme caution in grazing animals on young plants or plants recovering from drought or frost. Well-fed animals can tolerate some hydrocyanic acid in the plants. Hay that is thoroughly dried loses the hydrocyanic acid and is safe to use. New cultivars of Sudangrass are being bred with lesser amounts of hydrocyanic acid.

In the southwestern United States, including California, horses receiving sublethal doses of hydrocyanic acid from sorghums have developed degeneration of the lumbar and sacral segments of the spinal cord. This disease results in incoordination of the hind legs and acute or chronic inflammation of the bladder with large quantities of sediment in the urine. If forced to move, affected horses sway from side to side and dribbling of urine is intensified. Damage is permanent; some horses have shown these signs 3 years after the initial injury. Cattle grazing the same areas or having the same freshly cut hay, however, have not shown any signs of injury. When feeding on fresh plants of Johnsongrass or other sorghums, mares in early stages of pregnancy have had abortions or produced dead foals with permanently flexed joints. Completely dried hay of these plants does not contain sublethal amounts of hydrogen cyanide.

A photodynamic pigment in Sudangrass plants may cause photosensitization. In July 1930, during a spell of hot weather in the Sacramento Valley, 42 of 150 young white-faced rams developed photosensitization after 10 days of grazing a Sudangrass pasture 6 days after it had been irrigated. The animals showed greatly swollen ears and eyelids, abundant nasal discharge, labored breathing, and urticaria with pronounced itching of the affected parts. The animals recovered when they were removed from direct sunlight and were later turned into an adjacent pasture without the Sudangrass.

Pollen of Johnsongrass can cause hay fever. Reports of contact dermatitis caused by Johnsongrass have been proved invalid.

Stipa robusta [*S. vaseyi*].   Sleepygrass.

Native of the southwestern states. When ingested in moderate amounts, Sleepygrass causes a drowsiness or stuporous condition in horses, which is not fatal. Sampson and Malmsten (1942) discussed this toxicity but were in error in stating that Sleepygrass occurs in California.

Zea mays.   Corn, Indian Corn, Maize.

Native of the New World, origin not known; cultivated as a food crop plant for humans and domesticated animals. Tall robust annual with thick stems; male flowers in terminal tassels; female flowers on a lateral thickened elongated axis, the cob, enclosed in a series of modified leaves forming the "ear." Corn may accumulate free nitrates reaching the highest amounts before tasseling. The greatest amount of nitrate, expressed as potassium nitrate, is found in the stems or culms. It may be as high as 8% of the dry weight of the plant. Corn is not usually used as a forage plant, but nitrate poisoning has occurred in the central part of the United States during drought years. When the fields were not making enough growth to produce grain, they were grazed.

In addition to nitrate poisoning, corn of high nitrate content causes nitrosamine poisoning. In the rumen, nitrates are reduced to nitrites that then react with amines to form nitrosamines. When hemoglobin unites with the nitroso fraction (NNO) of the nitrosamines it cannot transport adequate amounts of oxygen. The nitrosohemoglobin lacks the dark brown color of methemoglobin that is found in nitrate poisoning. Young animals are especially susceptible to nitrosamine poisoning and show excitement followed by terminal depression. Extensive internal hemorrhages are typical.

In the first week of corn silage fermentation, nitrogen dioxide and nitrogen tetraoxide gases are produced. These reactions are beneficial because they remove the nitrates in the silage, reducing the danger of possible nitrate poisoning. In a few cases, however, usually involving tightly closed silos, these gases, heavier than air, have been lethal to humans and to farm animals.

Production of cyanogenic glycosides has been suspected in corn because it is related to Sudangrass. Actual findings of hydrocyanic acid in corn plants are quite rare, however, occurring only under unusual climatic and soil conditions.

PLANT TOXINS AND DERIVATIVE DRUGS

The kinds of poisons found in plants are summarized here. More detailed accounts of these toxic substances may be found in the references listed at the end of the book. Only toxins produced or accumulated within the plant body are included. Although humans and animals may also be poisoned by chemicals accumulated on the surface of plants, these substances are beyond the scope of this treatment.

Inorganic Toxins

The following accumulations or deficiencies of inorganic substances occur within plants and have resulted in death or distress to livestock. The simple abnormal amounts of chemical elements in the diets of humans or animals are not included.

Nitrates

Nitrate poisoning is a serious problem in cattle, but sheep and horses are rarely affected. Although hogs are more susceptible than cattle, they are rarely exposed to it. Nitrates are absorbed by plants from the soil and usually are rapidly reduced within the plant to nitrites and then to ammonia. Ammonia plus sulfates and carbohydrates unite to form amino acids; these are united into proteins with the addition of phosphorus in some instances. Nitrates are normal in plant growth. The Division of Chemistry of the California Department of Food and Agriculture has yet to find a plant giving a completely negative test for nitrates.

Any disruption of the normal metabolism of the plant, such as drought, low temperatures, or a deficiency of sulfur or phosphorus, results in the accumulation of excessive amounts of nitrates in plant tissues. Reduction of nitrates does not occur in the dark; therefore low light intensities slow or stop the change of nitrates into ammonia.

Toxic part: The entire plant, but the amount in different parts of the plant may vary. One study has shown that more nitrates are present in the stem of a goosefoot plant, *Chenopodium,* than in the leaves.

Toxin: Nitrates that are converted to nitrites in the rumen of a cow. The nitrites are absorbed by the bloodstream and convert the hemoglobin of the red blood cells into chocolate-brown-colored methemoglobin that cannot carry oxygen.

Symptoms: A cow receiving a fodder with a high amount of nitrates will show depression and increased cardiac and respiratory rates. A drop in milk production, abortion, or stillborn or weak calves may result.

An obvious cause of high nitrate content in forage plants is the excessive use of fertilizer containing nitrogen and a deficiency of sulfur and phosphorus—as, for example, when rangeland is fertilized without sufficient rainfall. The present-day costs of fertilizers do not permit overfertilization as a general practice, however.

Warm soil temperatures in the spring favor the rapid absorption of nitrates by the plants. If this period of warm sunny weather is suddenly followed by a prolonged period of cold air temperatures, as in very cloudy or foggy conditions, nitrates can accumulate rapidly in rangeland plants.

Individual plants of the same species vary greatly in nitrate content from one part of the field to another. In March 1984, four cows died and another aborted with nitrate poisoning. They were pastured in a field where plants of Milk Thistle, *Silybum marianum,* were eaten. One Milk Thistle plant had 12.25% nitrates; another from the same field had 0.26% nitrates. On a dry-weight basis, plants containing 1.5% nitrate, expressed as potassium nitrate, are considered lethal to cattle; those with lesser amounts, down to 0.5%, are suspect. Plants high in nitrates in dried hay are equally toxic.

Plants not killed by 2,4-D herbicides may accumulate toxic levels of nitrates. A level of 4.5% nitrates on a dry-weight basis has been found in Sugar Beets, *Beta vulgaris,* treated with this herbicide.

Nitrate poisoning under range conditions has been observed a number of times. The most severe case occurred in the upper Salinas Valley in San Luis Obispo County at the end of winter and into the early spring of 1952. The weather had been warm and was followed by a prolonged period of foggy cold conditions. An estimated 2,000 to 3,000 range cattle were lost. The abundant range plants with high nitrate contents, expressed on a dry-weight basis, were California Mustard, *Thelypodium lasiophyllum* (3.26%), and two species of Fiddleneck, *Amsinckia douglasiana* (2.02%) and *Amsinckia intermedia* (4.46%).

Nonlethal cases of nitrate poisoning in California have not been reported as commonly as the lethal ones. In Imperial County, a dairy cow that aborted from high-nitrate feed displayed very hostile behavior, quite different from its usual placid nature. A normally well-behaved herd of beef heifers, being driven in western San Luis Obispo County, went through three barbwire fences. The animals had been feeding on range plants that were found to be high in nitrates. There were, however, no deaths in this herd.

Nitrosamines may cause poisoning of livestock without the formation of the methemoglobin of nitrate poisoning. In the warm acid conditions in the stomach of the mammalian body, nitrosamines are formed when atoms of NNO are attached to an alkyl or aryl group of atoms forming such compounds as diphenylnitrosamine, $(C_6H_5)_2NNO$. These strong liver toxins are mentioned under Corn, *Zea mays.* A large number of nitrosamines are known to cause cancer in small laboratory animals. Small children and animals have also been seriously affected by nitrosamines in prepared foods and feeds.

Appendix C lists plants that are capable of accumulating amounts of nitrates sufficient to cause death or distress to livestock, primarily cattle.

Selenium

Selenium concentrations found in California plants are not as high as those found in other western states. When placed under irrigation, desert soils lose the soluble forms of selenium. Distress to domestic livestock from excess selenium in plants has occurred in California only when animals have remained on dry, nonirrigated pasture for a considerable length of time. Where drainage water from irrigated land has accumulated, there can be a signficant buildup of selenium salts. Nevertheless, selenium deficiency in forage plants probably causes more problems than excessive amounts of selenium. In 1981 blood samples from 1,600 cows in twenty-three northern counties in California showed that 40% of the animals were considered deficient in selenium.

Toxic amounts of selenium in California plants have been found mostly in areas not used by livestock. In the Colorado Desert *Astragalus crotalariae,* Rattle-box Milkvetch, has the typical bad odor or garlic smell associated with the accumulation of selenium. After several successive wet years, Rattlebox Milkvetch can become quite abundant.

Toxic amounts of selenium, causing abnormalities in humans and animals, have been found in various parts of the world. Marco Polo described the loss of hooves from animals in Central Asia in an area now known to have excess selenium in the forage. In 1560 and later, Spanish explorers in Central and South America reported a form of paralysis and the loss of hair, fingernails, and teeth in humans; they also observed deformed children and chicks. Investigations in the western states were the first to prove that excess selenium in the diet of animals causes such abnormalities.

Selenium is present in soils in the forms of soluble salts such as selenates. Plants absorb these salts and convert them into proteins. With the death of the plants, soluble selenium compounds accumulate on the surface of the soil, and forage plants growing in these areas will poison livestock. Although wheat and other grains grown on this soil have greater than normal amounts of selenium, no human or animal toxicity has been found from using these grains.

In the plant's metabolic process, selenium replaces sulfur in the formation of amino acids. The resulting polypeptides form enzymes that do not function properly. Thus the metabolism of the cells is disrupted sufficiently to stunt or kill the plant. A few species of plants, however, accumulate large quantities of selenium by converting it into insoluble simple organic compounds. By this mechanism selenium does not interrupt the metabolism of the plant. The highest concentrations of selenium have been found in plants that can shunt selenium into these insoluble compounds.

In most cases excessive amounts of selenium are toxic to humans and animals. Above 0.40 ppm, selenium in a plant may cause some symptoms of toxicity; 5.0 ppm is considered to be toxic to livestock; lethal effects may occur above 20 ppm.

Some species of plants, called indicator or obligate species, require selenium for normal growth. The presence of indicator plants in an area does not necessarily mean that there are toxic levels of selenium in the plants, but merely that there is enough selenium in the soil to permit plants of these species to grow there. In California in the Pea Family, Fabaceae, *Astragalus crotalariae* and *A. preussii* are indicators of selenium. California specimens of *Astragalus crotalariae,* Rattle-box Milkvetch, have been found to have 42 to 2,175 ppm selenium. Other species of *Astragalus* in other western states have accumulated over 14,000 ppm selenium.

In the Brassicaceae, all species of the prince's plumes, *Stanleya,* have been found to be indicators of selenium. Tests of California plants of *Stanleya* have found from 2 to 46 ppm selenium. Some species of woody asters, *Machaeranthera* [*Xylorrhiza*], have been found to be indicator plants in other states, but California plants of this genus commonly are negative for selenium accumulation; only very small amounts have been found in *Machaeranthera orcuttii.*

Other species of plants are secondary or facultative selenium accumulators and grow equally well in soils lacking selenium. Therefore these species are not indicators of selenium in the soil. Some species of *Aster* as well as species of the related woody asters, *Machaeranthera* in the Aster Family, Asteraceae, are such secondary accumulators. Moreover, plants of various species of *Atriplex* in the Goosefoot Family, Chenopodiaceae, have the capability of accumulating selenium and give off the same unpleasant garlic-like odor. This disagreeable smell usually prevents animals from eating these toxic plants so long as other forage is available.

There are three general levels of accumulation of selenium

in different kinds of plants: The indicator or obligate species may have from 100 ppm to as much as 14,000 ppm selenium; the facultative species have intermediate amounts of 25 to 100 ppm selenium; the forage plants, including grasses and grains such as wheat, barley, and corn, have 1 to 25 ppm selenium. These three levels of selenium content give rise to three different expressions of livestock poisoning.

Acute selenium poisoning occurs when an animal consumes large quantities of obligate selenium plants in a very short time. In cattle and horses this toxicity appears within several hours to several days and includes stupor and frequent urination; death is caused by failure of the cardiac and respiratory systems. There is bloody inflammation of the intestine and damage to the liver and kidneys. In sheep there is sudden death with marked congestion and edema of the lungs.

When animals ingest plants with intermediate amounts of selenium, the poisoning occurs a week or more later. The signs are somewhat similar to those of acute poisoning. The animals develop "blind staggers," or wandering and excitability, followed by depression, blindness, and labored breathing. Death is usually from respiratory failure.

"Alkali disease" is the name given to slow chronic poisoning from selenium. The plants causing this condition are those with the lesser amounts of selenium but above 5 ppm. The poisoning develops over a period of weeks or several months, causing lameness and loss of hooves, general emaciation, and death from thirst, starvation, or respiratory failure.

Human poisoning from excessive amounts of selenium in water or food has not been conclusively demonstrated in the United States. Human toxicity from selenium has been known in areas of Colombia, however, since the beginnings of recorded history.

Lead

Lead can accumulate in forage plants, but the amount rarely is sufficient to be toxic to animals. Lead can be a cause of poisonings, however, and exposure over a long period of time produces a chronic toxicity. The primary sources of lead poisoning come from commercial uses of lead. Lead in paint remains a hazard to human and animal health.

There is an interaction between lead and other elements. Large amounts of zinc in the diet, for example, apparently prevent the development of signs of lead poisoning in horses. Also low amounts of phosphorus and calcium have permitted greater accumulations of lead in young growing horses.

Lead poisoning caused the deaths of twenty horses on two ranches between Vallejo and Benicia in Solano County in 1969 through 1971. In late summer and winter these animals had grazed on the dry grasses in the smoke zone of the Selby lead smelter, which is no longer in operation. The old dry grasses contained the highest concentrations of lead (up to 350 ppm). Calculated on the basis of the forage consumed daily, this amount of lead is significantly higher than the cumulative level considered toxic to horses. Horses were poisoned in this same area after the very dry year of 1913. Reports made in 1915 and in 1973 contained a description of "roaring." Following brief exercise the animals had great difficulty breathing and emitted a loud roaring sound. The nostrils became enlarged and trumpet-shaped, remaining so until the animals rested. No poisonings from lead were found in cattle, sheep, and hogs pastured in this same area in 1913.

Small amounts of lead in forage were reported in one study as possibly adding to the complex of lupine poisoning of cows producing "crooked calves." The toxin in lupine plants is now regarded as the only cause of crooked calf disease. See *Lupinus* in the Fabaceae.

Molybdenum

An excess of molybdenum in forage has caused trouble in California along the old floodplains of the Kings and San Joaquin rivers in the San Joaquin Valley. On these poorly drained, heavily alkaline soils excessive amounts of molybdenum accumulate in forage plants in both dry native rangeland and irrigated pastures. Cattle show severe diarrhea or scouring, roughened coats, and poor weight gains. If this condition persists, there is progressive loss of hair color, continued weight loss, and eventually death. The interaction of molybdenum and copper in the metabolism of the animal is somewhat complex as it is also related to the amount of sulfates present. During the grazing period an imbalance between copper and molybdenum can be overcome by allowing livestock to self-feed on copper sulfate in their salt ration or by adding this compound to the water. Copper compounds such as copper gluconate, injected subcutaneously, prevent this condition for 3 to 9 months. If forage plants contain a great excess of molybdenum, three injections of copper compounds may be needed to carry an animal through the entire pasture season.

In 1952, an excess of molybdenum was found in legume plants from an irrigated pasture in the vicinity of Calabasas, Los Angeles County. This area had a previous history of accumulation of molybdenum in plants. The unthrifty cattle were removed from the pasture and responded to treatment with copper sulfate and dry feed.

Bromides

An unusual case of poisoning from bromides accumulated in oat plants grown on fumigated soil occurred in Napa County in 1973. The soil was treated with methyl bromide, CH_3Br, commonly used as a fumigant against insects, plant diseases, plant parasitic nematodes, and weeds. Methyl bromide boils at 38.4°F (3.56°C), is only slightly soluble in water, and usually dissipates quickly when the covering tarps are removed. In this case, however, the methyl bromide was injected into the soil at a greater depth than usual and therefore remained in the soil instead of evaporating.

In Napa County and adjacent Marin County, cattle, horses, and goats died when fed hay made from oat plants that had been grown on this soil. Tests showed 6,800 ppm bromide in the hay plants. Animals eating the hay had 381 ppm bromide in the urine; they developed blind staggers with immobility of the hindquarters and hypersensitivity around the kidneys. Although a number of animals died, others recovered when feeding of this oat hay was discontinued.

Magnesium

Although grass tetany, a disease caused by magnesium deficiency, has been recognized as a metabolic disorder of cattle in California, it was not considered of great importance until the winter of 1963–1964.

Toxic part: The aboveground parts of various grasses and other range plants lack magnesium. Although there is sufficient magnesium in the roots of the plants, it is not translocated to the parts eaten by livestock.

Toxin: A deficiency of magnesium in the blood of the affected animals. Since there is no mechanism in an animal's body for storing magnesium, the animal depends on a continual daily supply in the forage. Although some magnesium is eliminated through the feces and urine, the greatest losses normally take place with milk production. Most deaths from grass tetany occur in cows, aged 5 to 8 years, that are the healthiest and the heaviest milk producers.

Symptoms: The only noticeable signs are excitability, convulsions, prostration, and very rapid death—within 6 to 10 hours after the first signs of toxicity.

Grass tetany is often complicated because calving cows tend to hide on the rangeland and cannot be found in time to be treated. A completely healthy cow in the evening may be dead the next morning. The nervous system of the animal is rapidly affected not only by the low level of magnesium but also by the low calcium content of the blood, which is at its lowest normal limit. This interaction of minerals appears to be related to the use of carbohydrates and the release of energy in the body.

Injections of magnesium salts with calcium gluconate have been effective only in the first few hours after the onset of symptoms. Losses of animals can be prevented by giving the cows magnesium oxide mixed with a ground feed supplement of cottonseed meal, ground barley, and molasses. The intake of the supplement can be controlled by increasing or reducing the salt content. Losses from grass tetany seldom occur on clover pastures, as these plants have a much higher content of magnesium than grasses even during winter and early spring.

In early October 1963, rains started excellent growth of range grasses in the western foothills of the Sierra Nevada, followed by a dry winter and early spring of very cold temperatures with a dense cover of clouds or fog. The cattle preferred to eat the tender young grasses in the drainage ways and avoided the coarser grasses on the dry hill slopes. Losses of cows were reported from as far north as Chico and as far south as Madera. More than 6,000 cows killed by grass tetany were received at tallow works from Marysville in Yuba County to Modesto in Stanislaus County. Dead animals left in the fields were not counted. Most of the deaths occurred below the 500 ft elevation, but losses were recorded as high as the 1,500 ft elevation. The deaths occurred primarily on large range areas where no supplemental feeding was used. On small ranches where the stocking rate was very high and supplements were fed, the animals generally were not affected.

Manganese

Manganese deficiency in forage plants has been listed as a possible cause of deformed calves, but crooked calf disease is now known to be caused by the toxic alkaloid anagyrine occurring in lupine plants. Earlier California literature refers to these calves as "acorn calves" since they were born in the oak belt of the western slopes of the Sierra Nevada and their abnormalities were thought to have resulted from pregnant dams eating acorns. It was found, however, that this condition occurred in other areas not associated with oak trees. Crooked calf disease is discussed under *Lupinus* in the Fabaceae.

Organic Toxins

Numerous organic chemical compounds formed in plants have caused poisoning of humans or animals.

Carbohydrates

Most carbohydrates can be regarded as aldehydes (hydroxyaldehydes) or ketones (hydroxyketones) or related compounds derived from aldehydes or ketones.

Aldehydes

Acetaldehyde is the toxic substance produced in the human body from the incomplete breakdown of alcohol when a person has eaten Inky Cap mushrooms before drinking an alcoholic beverage. See "Coprine Poisoning" (pp. 32–33).

Ketones

These carbohydrate derivatives are rarely toxic. Camphor is a crystalline ketone obtained from the wood of the Camphor

Tree, *Cinnamomum camphora*. Pulegone, a ketone volatile oil, is found in Pennyroyal, *Mentha pulegium*. The toxic compound of *Myoporum laetum* is a furanoid sesquiterpene ketone.

Carboxylic Acids

In plants these acids are derived from carbohydrates, fats, or proteins.

Oxalic Acid

Oxalic acid is a dicarboxylic acid, the only naturally occurring carboxylic acid that is poisonous. It may occur as the acid, as in Rhubarb, *Rheum rhabarbarum,* in the Polygonaceae, or as salts of the acid such as calcium oxalate, potassium oxalate, or sodium oxalate, as in *Halogeton* in the Chenopodiaceae. Plants of some species can accumulate large amounts of oxalates.

Livestock poisoning results from the oxalic acid ion uniting with calcium to form insoluble calcium oxalate. As long as this acid ion is present, the reaction continues to remove calcium from the blood. Crystals of calcium oxalate are then deposited in the kidneys and other organs, causing mechanical damage. The lack of calcium and uremic poisoning caused by dysfunction of the kidneys result in convulsions, coma, and death. Calcium oxalate occurs in certain plants in bundles of sharp, needle-like crystals that puncture the lining of the mouth and throat resulting in swelling and burning sensations. This feature is discussed under *Dieffenbachia* in the Araceae.

Generally plants containing salts of oxalic acid at 10% of the dry weight of the plant are toxic to livestock. Yet plants with as low as 2% oxalates have caused significant losses of calcium in experimental animals.

Bacteria of the rumen can become adapted to the intake of oxalates. Feeding sheep for 4 or 5 days with small amounts of plants containing oxalates changes the bacteria in the rumen so that these animals can digest up to 30% more oxalates without toxic effects. Apparently the bacteria decompose the oxalates in the rumen before the toxin is absorbed into the body. The acid potassium oxalate found in the Oxalidaceae and in *Bassia* is not detoxified by bacteria in the rumen, and therefore its toxicity presents a much more serious danger.

Animals sometimes readily eat plants with high concentrations of oxalates, but cattle, sheep, and horses differ in their responses to them. Cattle are more susceptible than sheep to oxalate poisoning, but cattle range more freely and can thus be more selective in grazing forage plants. Sheep, however, are kept in dense numbers and are more frequently poisoned by inadvertently being driven into an area of toxic plants. Horses feeding on oxalate plants develop a degeneration of the bones from the lack of calcium.

Acute poisoning of cattle and sheep results in a marked deficiency of calcium in the blood, and death may occur in a relatively short time. Chronic poisoning over a longer period results in kidney damage and death from this organ's dysfunction. There is also damage both to the blood vessels of the rumen and to the central nervous system that may result in paralysis.

Experimentally, species of the fungus *Aspergillus* have caused the formation of oxalates in the decay of wheat, oats, or Bermudagrass hay in sufficient amounts to precipitate all the available calcium. There is a consistent relationship between the amount of oxalates produced and the amount of calcium present in the plant material.

Lactones

Lactones are made from the hydroxyl forms of carboxylic acids. Halogenated carboxylic acids are changed into lactones by the removal of the halogen parts such as chlorine or bromine. Some lactones formed in plants are very toxic to animals; other lactones serve a protective function in removing harmful breakdown products of metabolism. The odor of celery is the result of a lactone. A lactone group of atoms forms part of the molecular structure of the aglycone of cardiac glycosides and of coumarin glycosides.

Alcohols

Alcohols are derived from carbohydrates. Ethyl alcohol is the familiar compound formed by the fermentation of carbohydrates in the making of beer and wine. Toxic alcohols are rare in plants. The poisons in the waterhemlocks, *Cicuta,* in the Apiaceae, are highly complex alcohol molecules. Tremetol, a toxic alcohol, occurs in *Haplopappus heterophyllus* and *Eupatorium rugosum* in the Asteraceae. These plants are not in California, but they have caused serious poisonings of humans and livestock in other parts of the United States.

Phenols

Phenols differ from alcohols in that they are acid in character and form salts with bases. Gossypol in cottonseed is a complex phenolic compound. The toxins of Pacific Poison Oak, *Toxicodendron diversilobum,* in the Anacardiaceae, are assumed to be the same phenols as those of other species of *Toxicodendron* that have been studied. Phenol produces the odor found in toxic species of mushrooms of the genus *Agaricus* described under "Gastrointestinal Irritants" (pp. 41–43).

Tannins

Tannins are phenols; the active part is gallic acid, a polyhydroxylphenol. Tannins bind up proteins and therefore stop the actions of enzymes, bringing the metabolism of the living cells to a stop. Although widely distributed in flowering plants, tannins are found in sufficient quantities to cause trouble to livestock only in oaks, *Quercus,* in the Fagaceae and in certain species of *Acacia* in the Fabaceae.

Glycosides

Glycosides are organic compounds formed in plants and may have different functions. Their greatest benefit is probably to prevent plants from being eaten by herbivorous animals and invertebrates. The amount of food used in forming these compounds is considerable, but it has not been substantiated that glycosides may provide some type of food storage mechanism for the plant.

Glycosides break down, or hydrolyze, to form sugar molecules (the glycone part of the glycoside) and various other non-sugar components (the aglycone part of the glycoside). When glucose is the sugar produced by hydrolysis, the parent compounds are called glucosides. Since hydrolysis of many glycosides forms other kinds of sugars having larger molecules, the all-inclusive term glycoside is used.

The decomposition of glycosides occurs inside plant cells only when the tissues are damaged. This process, accomplished by enzymes, may be caused by mechanical injury to the tissues or by damage from frost or drought. Acids and the action of bacteria in the rumen of cattle cause hydrolysis of the glycosides, as well.

Glycosides are widely distributed in different plant families, even more so than alkaloids. Toxicity of the glycosides to humans or to animals is caused by the aglycones or nonsugar components and not by the glycones, the sugars. The toxic glycosides involved in such poisonings include cyanogenic, coumarin, cardiac, saponin, protoanemonin, mustard oil, and anthraquinone glycosides.

In plants there are two types of cyanogens: cyanogenic glycosides and cyanogenic lipids. Both break down to form hydrocyanic acid, also referred to as hydrogen cyanide or its older name, prussic acid.

Cyanogenic Glycosides

These glycosides decompose or hydrolyze into one or more sugars, a carbonyl fraction of one or more aglycones, and hydrocyanic acid (HCN). These glycosides are widespread in distribution in different plant families.

Cyanogenic lipids break down to fatty acids, a carbonyl aglycone, and hydrocyanic acid. They are very restricted in distribution and are found only in the seed oils of some members of the Sapindaceae and in one species in the Boraginaceae.

Cyanogenic glycosides have been known for centuries, and their decomposition to form hydrocyanic acid is known worldwide. They occur in more than 800 species of plants in eighty different plant families. Hydrocyanic acid is found not only in flowering plants but in a few gymnosperms, ferns, and fungi also. The commonest plants having toxic cyanogenic glycosides are in the Rose Family, Rosaceae, the Pea Family, Fabaceae, and the genus *Sorghum* in the Grass Family, Poaceae. Cassava, *Manihot esculenta,* in the Spurge Family, Euphorbiaceae, well known for having hydrocyanic acid, is used as a food plant in tropical regions after removal of the toxin. The occurrence of hydrocyanic acid is rare in the animal kingdom;

it is common in the millipedes and centipedes and is found in three species of butterflies, four species of moths, and three species of beetles.

The intact cyanogenic glycoside and the enzyme for its hydrolysis occur in plant cells. Following damage to the cells from drought or wilting, the enzyme in the plant cells can cause release of hydrocyanic acid. Ruminant animals are more susceptible to hydrocyanic acid poisoning because the inherent acids and enzymes of the bacterial flora in the rumen contribute to rapid hydrolysis of cyanogenic glycosides.

Hydrocyanic acid is rapidly absorbed through the walls of the rumen into the blood. The ultimate effect of the hydrocyanic acid is to stop the action of enzymes in the utilization of oxygen in cellular respiration. The cells thus die from lack of oxygen even though oxygen is plentiful in the blood. The bright red bloody froth from the muzzle of a poisoned animal is significant in determining that poisoning from hydrocyanic acid has occurred.

Amygdalin, the cyanogenic glycoside used in the preparation of Laetrile, occurs in the kernels of apricots, bitter almonds, cherries, and plums, all belonging to the genus *Prunus* of the Rose Family, Rosaceae. The use of Laetrile as an alternative treatment for cancer is highly controversial. A case of hydrocyanic acid poisoning in a woman who ingested a large quantity of bitter almond pits is described under the Rosaceae.

Pseudocyanogenic Glycosides

These substances occur in the seeds of the cycads, small palm-like trees that belong to the Gymnosperms. In structure these pseudocyanogenic compounds are related to glycosides. When treated with a base, they form hydrocyanic acid. When treated with an acid, they form nitrogen, formaldehyde, and methyl alcohol.

Coumarin Glycosides

These glycosides decompose to form aglycones that are forms of coumarin. Coumarins and their precursors are widely distributed in higher plants and also are formed by the metabolism of some species of fungi. They have been isolated from such plants as Sweet Vernalgrass, *Anthoxanthum odoratum;* sweetclovers, species of *Melilotus;* and Red Clover, *Trifolium pratense.* Recent investigations with compounds labeled with radioactive C^{14} have shown that a wide variety of organic compounds are precursors to coumarin. Apparently coumarins are derived from the amino acid phenylalanine through a series of intermediate compounds.

Hydroxylated derivatives of coumarins are more restricted in distribution; they are found in a number of unrelated plant families. Umbelliferone, benzo-alpha-pyrone, occurs in several genera of the Apiaceae and in the hawkweeds, *Hieracium,* in the Asteraceae; aesculin occurs in *Aesculus* in the Hippocastanaceae; daphnin occurs in *Daphne* in the Thymelaeaceae.

Dicoumarol, the first coumarin-containing compound to be isolated, is another coumarin derivative found in improperly cured or spoiled sweetclover hay. Coumarol (the coumarin glycoside) and the enzyme for its hydrolysis are natural components of plants of the sweetclovers, *Melilotus.* The intermediate product formed by hydrolysis of coumarol is changed further by the action of fungi in improperly cured hay to form 4-hydroxycoumarin. This latter compound reacts with formaldehyde in the spoiled hay to produce the toxic dicoumarol.

Dicoumarol prevents blood from clotting and leads to internal bleeding, lethargy, paralysis, and death. Synthetically prepared dicoumarol is used as an anticoagulant drug for certain cardiac problems and has been used to produce the rodent poison Warfarin.

Furocoumarins are formed by the fusion of coumarin with a furan ring; some references call these compounds furanocoumarins. The most common form of furocoumarins is called psoralen. The name psoralen is derived from a species of *Psoralea,* in the Fabaceae, which has been used medicinally in India since ancient times.

Five plant families contain most of the furocoumarins: Apiaceae, Fabaceae, Moraceae, Orchidaceae, and Rutaceae. These compounds are also found in the Asteraceae in Yarrow, *Achillea millefolium;* in the Brassicaceae in Charlock, *Brassica kaber;* and in the Convolvulaceae in Field Bindweed, *Convolvulus arvensis.* Two psoralens with phototoxic properties are produced by the fungus *Sclerotinia sclerotiorum,* which causes the pink rot of celery. These psoralens are not produced by healthy celery plants or by the fungus grown in cultures. This topic is discussed further under *Apium* in the Apiaceae.

Furocoumarins cause photosensitization and poisoning in humans and domestic animals but have been used medicinally as well. For more than 3,000 years Orientals, Hindus, Turks, and Egyptians have applied poultices of plant juices containing furocoumarins to defective whitened areas of the skin called leucoderma or vitiligo. This condition is an inherited tendency occurring in about 2% of the world's population. Synthetic furocoumarins are now used to tan the skin without sunburn by inducing formation of the black melanin pigments in the cells of the skin. Melanin production is related to disruption of the synthesis of deoxyribose nucleic acid (DNA).

Much experimental work in the treatment of psoriasis has been done using furocoumarins, especially psoralen and 8-methoxypsoralen. Psoriasis affects about 2% of the world's population. In this disease the cells of the epidermis divide every 3 or 4 days instead of the normal rate of about every 28 days, an abnormality related to the production of DNA. Other photosensitizing materials have been used in treatment of this disease. Excellent initial results in curing psoriasis have been obtained by using 8-methoxypsoralen orally and exposure to specific wavelengths of ultraviolet radiation.

Numerous other effects of furocoumarins are being studied because of their ability to bind DNA. They have been demonstrated to be effective against gram-positive bacteria. Studies have shown that excessive amounts of furocoumarins are related to the formation of tumors and the production of mutations.

Steroid Glycosides

These glycosides have an aglycone possessing the structure of a steroid. They may be divided into two groups according to their physiological activities: the cardiac glycosides that stimulate the heart and the saponin glycosides that have foam-forming qualities.

Cardiac Glycosides

These glycosides have an aglycone or genin with the structure of a steroid and an additional modified or unsaturated lactone, giving rise to the cardiac activity of the aglycone. Certain cardiac glycosides have been used in medicine for a long time because of their ability to regulate and strengthen the action of heart muscles. The use of the drug digitalis, obtained from the leaves of *Digitalis purpurea*, goes back to the late eighteenth century in England. Digitalis in prescribed doses has had wide use in the treatment of cardiac diseases. Overuse or injudicious use of the drug, however, can result in adverse reactions and poisoning. Symptoms include nausea, vomiting, abdominal pain, diarrhea, blurred and disturbed color vision, and variations from the normal heart rhythm.

Among the approximately 400 cardiac glycosides, some of the best known are those obtained from *Digitalis*, Scrophulariaceae; *Helleborus*, Ranunculaceae; *Scilla, Ornithogalum*, and *Convallaria*, Liliaceae; *Acokanthera, Nerium, Strophanthus, Thevetia*, and other genera, Apocynaceae; and *Asclepias* and *Calotropis*, Asclepiadaceae.

Ouabain, a cardiac glycoside that occurs in *Acokanthera* and *Strophanthus* in the Apocynaceae, has been used by South African tribes for countless centuries as an arrow poison for hunting animals and also for homicide. Effective only if the toxin is injected directly into the bloodstream by the arrow point, it acts quickly and has been reported to kill large animals within minutes.

Saponin Glycosides

These glycosides can be separated into two classes on the basis of their aglycones or genins: the steroidal sapogenins and the triterpenoid sapogenins. Saponins are characterized by their ability to lower the surface tension of aqueous solutions; when shaken they produce a nonalkaline soap-like froth. The amount of foam produced is a rough indication of the quantity of saponins present.

Saponins usually have a bitter, acrid taste and commonly irritate the mucous membranes. These substances alter the permeability of cell membranes and may react with proteins in a nonspecific way causing general toxicity. Cholesterol in the cell membranes apparently has an influence on the toxic reactions of some saponins. Since red blood cells are broken down by hemolysis in saponin solutions, a saponin injected into the bloodstream can be fatal.

Another characteristic of these glycosides is their high toxicity to cold-blooded animals. Indeed, plants containing them have been used as fish poisons around the world. Sap from crushed plants, agitated in a pond or stream, stuns the fish, which float to the surface, making their collection for food simple. The toxin acts on the respiratory organs of fish without affecting their edibility. Saponins are not comparably toxic to warm-blooded animals, however, since the saponin is only absorbed if the gastrointestinal tract has been injured.

Symptoms of saponin poisoning include acrid taste, irritation of the mucous membranes producing a flow of saliva, nausea, vomiting, and diarrhea. Cases of severe poisoning manifest systemic effects such as dizziness and headache, chills and heart disturbances, and eventually convulsions and coma. In past years seeds of the Corn Cockle, *Agrostemma githago* in the Caryophyllaceae, have occurred as contaminants in wheat made into flour. As little as $\frac{1}{10}$ ounce of the seeds of Corn Cockle has produced symptoms of poisoning. The contamination of these weed seeds in wheat has been eliminated by mechanical screening, and modern herbicides have removed such weeds from the wheat fields.

Saponins are widespread in occurrence. They are probably found in most flowering plants and have been identified in more than 500 species in eighty different plant families. The pronounced bitter taste of saponins deters animals from feeding on these plants. In fact, this characteristic of saponins may be part of the chemical mechanism for protecting plants against invertebrate pests and plant diseases. Commerical preparations of saponins are made from:

Sapindus saponaria, Soapberry, Sapindaceae, a small tree native of tropical America; the small ($\frac{3}{4}$ in.) rounded berries are used for the saponin

Quillaja saponaria, Soap-bark Tree, Rosaceae, a small tree that is native to Chile and sometimes grown in California; the bark is used for the saponin

Saponaria officinalis, Bouncing Bet, in the Pink Family, Caryophyllaceae, a perennial herbaceous species spreading by rhizomes to form dense clumps of plants; native of Eurasia and naturalized in waste areas in cooler parts of California; the juice of the plants is used for the saponin

Saponins are discussed further under the Common Corn Cockle, *Agrostemma githago* in the Caryophyllaceae, and Alfalfa, *Medicago sativa* in the Fabaceae.

Protoanemonin Glycoside

This glycoside named ranunculin is found in plants of many species in the Buttercup Family, Ranunculaceae. By enzymatic action it breaks down to form the aglycone protoanemonin, a highly irritant oil. Protoanemonin is unstable and is rapidly converted to the harmless anemonin. The occurrence and effects of this toxic glycoside are discussed under Ranunculaceae.

Mustard Oil Glycosides or Glycosinolates

These compounds occur abundantly in plants of the Mustard Family, Brassicaceae. They are found also in the closely related families Capparaceae, Moringaceae, and Resedaceae and in some other families as well, Caricaceae, Limnanthaceae, and Tropaeolaceae.

Glycosinolates are similar to glycosides; they break down to a sugar fraction and an aglycone of an irritant oil. About eighty different glycosinolates are known. Glycosinolates are hydrolyzed by the enzyme thioglucosidase to form a glucose sugar, a sulfate fraction, and isothiocyanates (mustard oils) or organic nitriles. Isothiocyanates may further break down to release the thiocyanate ion, SCN^-, which is combined to form an organic thiocyanate. Isothiocyanates or mustard oils have a pungent odor and taste and are irritating to the skin and the mucous membranes lining the gastrointestinal tract. Liver and kidney damage may result from these substances as well. Mustard oils are not to be confused with the mustard gas of World War I, a synthetic compound so-named because of its odor.

Isothiocyanates are metabolized to the thiocyanate ion, and therefore certain effects of the isothiocyanates are probably the result of this ion. Isothiocyanates also inhibit the uptake of iodine in the thyroid. Thyroid problems in humans are not caused, however, by the small amounts of these compounds received by consuming large quantities of cabbages and related edible plants. Investigations have shown that milk from cows on diets high in isothiocyanates may have a temporary drop in iodine content. The iodine-deficient milk could possibly cause goiter in young calves.

Rapeseed meal may cause liver damage in chickens. Moreover, chicks may develop goiters as the result of insufficient iodine in the eggs from hens fed rapeseed meal.

Mustard oil glycosides, glycosinolates, have been found in all species of the Brassicaceae that have been examined. The principal crop plant with these compounds is Cabbage, *Brassica oleracea*. *Brassica campestris,* Field Mustard, and *B. napus,* Rape, are important species used as livestock feed. Forage crops are made from Marrow-stem Kale, a cultivar of *Brassica oleracea,* and Dwarf Essex Kale, a cultivar of *B. napus.* Isothiocyanates are also found in weeds belonging to the Brassicaceae and may form part of the forage for livestock.

Condiments from Mustard Seed, *Brassica hirta* and *B. nigra,* and from Horseradish, *Armoracia rusticana,* are so pungent that it is quite improbable that enough could be eaten to cause anything besides tears and irritations of the mouth and upper intestinal tract.

Anthraquinone Glycosides

These glycosides are found in the Fabaceae, Polygonaceae, Rhamnaceae, Liliaceae, and other families of flowering plants as well as in certain fungi, lichens, and bacteria. Anthraquinones are aromatic compounds, a large group of naturally occurring quinones. On hydrolysis the glycosides break down to form a large number of anthraquinone aglycones that are cathartics and act on the large intestine.

The reduced form of anthraquinone, an anthrone or dianthrone, is joined with glucose molecules to form the glycosides found in plants. Glycosides are more effective as drugs since this form is not metabolized as readily and permits the release of a small amount of the active drug in the intestine.

Anthraquinones are found in the bark of plants of *Rhamnus purshiana* and other species in the Rhamnaceae; in the leaves of *Aloë barbadense,* Liliaceae; in the root of *Rheum officinale* and other species but not *R. rhabarbarum* [*R. rhaponticum*], Rhubarb, Polygonaceae; and in the seeds and leaves of *Cassia,* Fabaceae. Various cathartic drugs have been prepared commercially from some of these plants. Cascara bark from *Rhamnus purshiana* is stored for a year before being processed as a drug. Apparently this delay allows for an oxidation of the anthrones and dianthrones to anthraquinones that reduces the griping effects in the large intestine to a more gentle action.

Mild poisonings have occurred, particularly in children, from eating fresh portions of the plants that contain anthraquinone glycosides. These compounds are gastrointestinal irritants; symptoms include nausea, vomiting, bloody diarrhea, dizziness, abdominal pain, and in severe cases kidney damage.

Emodin anthraquinones are slightly modified anthraquinone molecules. These are the anthraquinones found in species of buckthorns, *Rhamnus,* Rhamnaceae, and in the Golden Shower, *Cassia fistulosa,* Fabaceae. Xanthones are derivatives of anthraquinone, formed by splitting of the ring system of the anthraquinone through several intermediate compounds. Xanthones form the purplish black pigments found in the sclerotial bodies of the ergot fungus, *Claviceps purpurea.*

Proteins

Proteins are formed by combining amino acids into long chains. The peptide bond of C, O, N, H forms a link between one amino acid and the next one. When a small number of amino acids are joined together by peptide bonds, the molecule is called a polypeptide or is referred to as a peptide chain. Large protein molecules have a number of amino acids joined together in long chains that may be straight, coiled, or folded. Many common proteins are composed of at least 100 amino acids linked together. Additional side chains may be added to form a particular protein.

Amino Acids

Essentially carboxylic acids, amino acids have an amino group of atoms (NH_2) and a carboxyl group of atoms (COOH). Some amino acids also include atoms of sulfur. Of the many known amino acids, only twenty different ones are used in the makeup of proteins and are referred to as protein amino acids. More than 240 nonprotein amino acids are known; they occur in the free state in living cells and are derived from the protein amino acids.

A few nonprotein amino acids are toxic and occur in a small number of plants. Free amino acids are the toxins in the seeds of some species of *Lathyrus* and *Vicia* in the Fabaceae. Other species in this family are suspected of having toxic amino acids.

Amines

These organic compounds are derived from ammonia, NH_3. Toxic amines are known in the European Mistletoe, *Viscum,* in the Viscaceae and in the vetchlings or wild peas, *Lathyrus,* in the Fabaceae. Moreover, toxic amines accompany the poisonous alkaloids in the ergot bodies of the fungus *Claviceps purpurea.*

Polypeptides

These compounds are rarely toxic. Some of the poisons in the freshwater blue-green algae are polypeptides. The toxins of the mushrooms in the genus *Amanita* are polypeptides arranged in cyclic rings. In flowering plants, two toxic polypeptides are found in the seeds of Akee, *Blighia sapida* in the family Sapindaceae, and in the plants of the European Mistletoe, *Viscum album* in the Viscaceae.

Goitrogens

These substances, producing goiters in humans and in animals, block the formation of the thyroid hormone. In *Brassica,* cabbages and related species, the compounds causing goiter formation are glycosides. In the soybean, the goitrogen is a peptide of two or three amino acids.

Alkaloids

These compounds are formed mostly in plants, rarely in animals, and are derived from amino acids. By definition an alkaloid has a specific pharmacological action. Alkaloids have been used in medicine, but they can also be poisons that are harmful to humans and livestock. Most alkaloids (the name means alkali-like) have a bitter taste.

Alkaloids are classified in different ways. Those derived from a particular plant may receive the name of the plant, such as nicotine from *Nicotiana*. Others may be named for their pharmacological action, such as morphine ("inducing sleep"). They may be arranged into groups and given names such as indole alkaloids or pyrrolizidine alkaloids based on the structure of the cyclic rings of carbon atoms.

Alkaloids are organic nitrogenous compounds with at least one nitrogen atom in a carbon ring. Protoalkaloids have a nitrogen atom in the molecule but not in a heterocyclic ring. Because of their pharmacological action, however, they are usually included in discussions of alkaloids.

Because of the nitrogen atoms, most alkaloids are basic, forming positively charged ions, cations, in aqueous solutions. In most biological cells they are water soluble and occur as salts combined with organic acids. In an alkaline solution they exist as free bases. Moreover, there are the so-called neutral alkaloids that have sufficient carboxyl groups of atoms adjacent to the nitrogen atom in the alkaloid molecule to neutralize the positive effects of the nitrogen atom.

Alkaloids are derived from amino acids or from the acetate metabolism of the cells. Many are formed from two very similar amino acids, phenylalanine and tyrosine. These amino acids form phenylethylamine, $C_6H_5C_2H_5NH_2$, and its derivatives, which are present in many alkaloids and protoalkaloids.

Alkaloids are accumulated in significant amounts in about 15 to 20% of the vascular plants, mostly in the dicotyledons but also in the Liliaceae, Amaryllidaceae, and Poaceae in the monocotyledons; they occur rarely in the gymnosperms and in the club mosses, Lycopodiaceae, and horsetails, Equisetaceae; and there are exceptional cases of alkaloids in the ergot fungus *Claviceps* and in several mushrooms. They occur abundantly in the dicotyledonous families of Apocynaceae, Fabaceae, Papaveraceae, Ranunculaceae, Rubiaceae, and Solanaceae. More than forty other plant families have species that manufacture minute quantities of alkaloids.

The Poppy Family, Papaveraceae, is unique in that plants of all species of the family have been found to produce alkaloids. Most plant families have alkaloids restricted to a certain genus or a group of related genera. Any plant accumulating more than 0.05% dry weight as alkaloids is considered an alkaloid-bearing plant. Annual herbaceous plants have greater amounts of alkaloids than perennials; trees have even lesser amounts and usually of simpler structure.

There is some variation in the amounts of alkaloids present in different parts of an individual plant and at different stages of growth, but not nearly the variation found with other toxic compounds. Nicotine is produced in the root of the tobacco plant and accumulates in the leaves. Using radioactive tracer carbon, however, there was no transfer of steroidal glycoalkaloids from the unlike root to the unlike stem and leaves in reciprocal grafts of potato and tomato.

Alkaloid content varies within a species and changes also through its range. Cultivated varieties (cultivars) differ as to amount, but the parent alkaloids are usually present. The Opium Poppy, *Papaver somniferum,* has been bred for centuries for oil and alkaloids, yet the oilseed varieties still contain a low percentage of morphine in the plants.

Alkaloids are poisonous primarily to animals, causing drastic effects on the nervous system. In general they are not toxic to plant cells, although purified forms cause damage when applied to plant tissues, even on the plant that produced the alkaloid. Moreover, their toxicity in plant cells may be avoided by being accumulated in certain ways as in the vacuoles of the cells. Morphine is accumulated in the Opium Poppy in the latex cells formed into a system that extends throughout the plant. These latex tubes are lacking in the parts of the pods where the seeds are being formed, however, and thus do not affect the developing embryos.

Protoalkaloids or amine alkaloids are different from alkaloids in that the nitrogen atom is attached to a chain of carbon atoms that are not in a ring.

The European Mistletoe, *Viscum album,* Viscaceae, is established as a localized population in California. It is known to have the simple compounds of ß-phenylethylamine, $C_6H_5CH_2CH_2NH_2$, and tyramine, a hydroxy form of phenylethylamine. In the mammalian body these toxic amines cause a pressor action resulting in an increase in blood pressure and in contractions and dilations of the blood vessels.

Oxidation of a dihydroxy derivative of phenylethylamine produces the familiar black color of bananas exposed to the air. The black pigment in hair and skin is a very similar form or polymer. Hordenine, a dimethylhydroxy form of phenylethylamine, is the alkaloid in barley, *Hordeum vulgare.*

Using radioactive labeled precursors the protoalkaloids mescaline, ephedrine, and cathine were found to be derived from amino acids. The hallucinogenic drug mescaline is from Peyote, *Lophophora williamsii* in the Cactaceae. Ephedrine, a drug useful in treating allergic conditions including bronchial asthma and hay fever, is obtained commercially from European and Asiatic species of *Ephedra* in the Ephedraceae, a family of gymnosperms; its use has been known in China for thousands of years. Cathine, from *Catha edulis* of the Celastraceae, is found in chat or khat, used for centuries by Muslims as a central nervous system stimulant; branchlet tips with young leaves from fresh plants are chewed for their stimulating effects.

Alkaloids have a nitrogen atom included with the carbon atoms attached together in a ring. In the following sections, the common structure of the ring in each type of toxic alkaloid is diagrammed. A carbon atom with its four shared bonds occupies the corners of the ring. These atoms are omitted from the diagrams as they are constant, permitting emphasis on the other atoms in the ring structure. Additional atoms or chains of atoms are added to the ring structure, producing the different specific alkaloids. Research has found that shifting the positions and kinds of these side chains or single atoms markedly changes the pharmacological action of an alkaloid.

Pyrrolidine Alkaloids

The pyrrolidine alkaloids have the ring structure diagrammed here. The alkaloid pyrrolidine is a very strong base and occurs in the Wild Carrot, *Daucus carota,* and as a lesser alkaloid in tobacco. Its structure is pictured as:

The carbon atoms and their associated hydrogen atoms can be removed from the diagram to emphasize the peculiarity of a molecular structure. Therefore the basic structure of the carbon ring of the pyrrolidine alkaloids can be diagrammed as:

Forms of methyl pyrrolidine are found in tobacco and the Deadly Nightshade, *Atropa belladonna.*

methyl pyrrolidine

Cuscohygrine occurs in Jimsonweed, *Datura stramonium,* and in Deadly Nightshade, *Atropa belladonna.* The total number of atoms in the molecule can be summarized as $C_{13}H_{24}N_2O$.

cuscohygrine

Pyrrolizidine Alkaloids

For the most part the pyrrolizidine alkaloids are esters of amino alcohols and necic acids, derived from monoterpenes. They cause liver damage in a mammalian body, but the alkaloids may be hydrolyzed before absorption and then are no longer toxic.

Pyrrolizidine alkaloids occur in plants of many families: the Boraginaceae, Asteraceae, Poaceae, Fabaceae, Orchidaceae, and others. In California, these alkaloids are found in species of *Senecio,* Asteraceae, and *Amsinckia, Echium, Heliotropium,* and *Symphytum,* Boraginaceae. These alkaloids are cumulative poisons affecting both humans and livestock. Human toxicity may be evident as slight personality changes that are apparent only to persons who know the patient well. In cattle and horses the animals appear sleepy, walk aimlessly, or may become belligerent.

In livestock, pyrrolizidine alkaloids cause severe damage to the liver, less damage to the kidney and lungs, and frequently cause death. Normally in the breakdown of proteins the liver converts ammonia to urea. Injury to this organ causes an accumulation of ammonia that is toxic to the central nervous system. Ammonia also affects the ability of the brain to utilize oxygen and may result in coma and death.

Livestock poisonings from pyrrolizidine alkaloids are recorded every month of the year in California. The principal causes are the seeds of fiddlenecks, *Amsinckia*, in first-cutting hay or in seed screenings or the dried plants of groundsels, *Senecio*, included in hay. *Amsinckia* plants may be abundant weeds of first-year plantings of alfalfa. If these weeds are not controlled, the first cutting of the new alfalfa is not only useless but dangerous to cattle, horses, and pigs. Since sheep and goats are resistant to these alkaloids they can eat this contaminated hay as part of their rations.

Human poisoning from pyrrolizidine alkaloids is from two sources: accidental contamination of foodstuffs containing the alkaloids and intentional long-term use of plants that contain them. An accidental contamination of epidemic proportions occurred in Afghanistan and India in 1976. The source of the poisonings was wheat harvested from fields infested with vast numbers of plants of two weedy species of *Heliotropium* and *Crotalaria*, both containing pyrrolizidine alkaloids. In these two countries deaths resulted from veno-occlusive disease of the liver.

Intentional consumption of plants containing pyrrolizidine alkaloids is a public health problem in parts of Africa, Central and South America, and Jamaica, West Indies. In these areas several species of plants containing these toxic alkaloids have been used in preparation of teas and medications. This use, based on tradition and ignorance of the toxicity of the preparations, has caused various manifestations of poisonings. In Jamaica a "bush tea" is prepared from leaves mostly from species of *Crotalaria*, Fabaceae, found in the island's scrubland. These preparations are so commonly used in Jamaica that veno-occlusive disease of the liver has almost become epidemic.

Two fatal cases of human poisoning from pyrrolizidine alkaloids, reported in 1977 from Arizona, called attention to the occurrence of the disease in the United States. The cause of poisoning in both cases, involving small children, was from drinking tea prepared from *Senecio longilobus,* a native plant known locally as Gordolobo Yerba. Because of the difficulty in diagnosing this poisoning, other cases may have been overlooked. No cases of human poisoning from pyrrolizidine alkaloids have been reported in California.

Honey produced from the flowers of Tansy Ragwort, *Senecio jacobaea,* an introduced weed established in northwestern California, Oregon, and Washington, contains pyrrolizidine alkaloids. Although in small amounts this honey would not be immediately toxic to humans, the effects of these alkaloids are cumulative and therefore dangerous. Since the honey is off-color and not salable, beekeepers use it for winter feed for their bee colonies.

The greatest concentration of pyrrolizidine alkaloids is in seeds and flowers. In humans, following ingestion of small amounts daily over a prolonged period of time, symptoms are not evident for several weeks or even months. During this period the effects on the liver are cumulative and progressive, but the damage is not evident. Symptoms include enlarged liver, accumulation of fluid in the body cavity, abdominal pain, nausea, vomiting, diarrhea (possibly bloodstained), headache, apathy, and emaciation.

Cattle, horses, and pigs are quite susceptible to this toxicity, but sheep and goats are resistant. Signs of pyrrolizidine poisoning are liver damage resulting in jaundice with yellowish discoloration of the mucous membranes, dullness and loss of appetite, and diarrhea, often bloody, or constipation. Poisoned animals stand with feet apart, chewing on objects, or wander aimlessly pushing against fences. Horses commonly develop colic. Frequently in cattle there is distension of the rectum, and the milk as well as the skin has an unpleasant odor. In some cases the odor seems associated with areas of swelling beneath the skin, and in severe cases serum seeps from below these surfaces. The alkaloid content of the milk has never been proved to be damaging to the offspring of cows or goats feeding on the toxic plants.

Piperidine Alkaloids

Piperidine is the volatile base that is formed by hydrolysis of the pyridine alkaloid piperine, $C_{17}H_{19}NO_3$, which occurs in black pepper, the dried unripe fruit, and in white pepper, the dried unripe fruit with the pericarp removed, of *Piper nigrum,* Piperaceae.

Coniine, $C_8H_{17}N$, is the most important alkaloid of Poison Hemlock, *Conium maculatum,* Apiaceae. It occurs in all parts of the plant and is a colorless, strongly basic oil with a mild pepper or piperidine odor and a burning taste. Moreover, four other piperidine alkaloids are found in Poison Hemlock. Coniine is lethally toxic to all forms of animal life.

Lobeline is the major alkaloid of Indian Tobacco, *Lobelia inflata,* of eastern North America. The alkaloid is similar to nicotine. This plant was used by the Indians as a substitute for tobacco from species of *Nicotiana.* More recently Indian Tobacco has been added to commercial preparations, in tablet or lozenge form, intended to help those with the tobacco habit to break their addiction. It has also been used medicinally in the treatment of bronchitis and asthma. Lobeline stimulates the respiratory center and induces coughing.

Tropane Alkaloids

Tropane alkaloids are condensed derivatives of the pyrrolidine–piperidine group of alkaloids. These are found in the Potato Family, Solanaceae, and include a number of therapeutically important alkaloids. The best known of these are atropine, hyoscyamine, and scopolamine. Plants in California containing these substances are thornapples, *Datura,* angel's trumpets, *Brugmansia,* and Henbane, *Hyoscyamus niger.* The important drug species not in California, or perhaps cultivated only in collections of drug plants, are Belladonna or Deadly Nightshade, *Atropa belladonna,* and European Mandrake, *Mandragora officinarum.* The latter is associated in Europe with many superstitions and folktales.

These alkaloids are extremely poisonous; like many plant toxins, however, they are used in the preparation of a number of useful drugs. The most prominent feature of the tropane alkaloids is their ability to produce hallucinations and delirium.

All the alkaloids found in the Solanaceae are esters of the dicyclic amino alcohol, tropine or 3-tropanol. The commonest alkaloid in the Solanaceae is hyoscyamine, $C_{17}H_{23}NO_3$. Atropine is merely a structural rearrangement of the same atoms of hyoscyamine, which is readily tranformed for drug use into atropine after being extracted from the plant material. Atropine as such probably does not occur in the plant cells.

Scopolamine, also called hyoscine, $C_{17}H_{21}NO_4$, occurs in members of the Solanaceae along with hyoscyamine but is obtained commercially from *Datura metel,* native to China. It is established as a weed in eastern North America but not on the Pacific coast. Scopolamine is a weaker base than hyoscyamine and has been used as a drug for travel sickness.

Another important tropane alkaloid is cocaine, obtained from Coca or Divine Coca, *Erythroxylon coca* of the Erythroxylaceae, a shrub native to and widely cultivated in the Andean highlands of South America. During Inca times the shrubs were considered the property of the royal family and were dispensed according to their wishes. After the Spanish conquest the Coca shrubs were widely distributed among the Indian workers. Today Coca leaves are chewed by thousands of workers throughout the Andes for their stimulating effects. Overindulgence may lead to physical and mental deterioration.

Cocaine, $C_{17}H_{21}NO_4$, was first isolated in 1862, but much was already known about its properties. Cocaine was first used in eye surgery, where a few drops prevented the eye's involuntary movements. Later it was used in mouth and throat surgery and became particularly useful in dentistry. Cocaine is injected along with adrenalin. Because adrenalin contracts the blood vessels, the cocaine remains in the area and acts as a local anesthetic.

Cocaine exerts a strong toxic effect on the central nervous system. Severe poisoning causes convulsions, unconsciousness, and death from respiratory failure. The great problem with this drug lies in its addictive properties and its harmful effects to those addicted. Chronic symptoms, the worst of any drug addiction, cause not only physical deterioration but wild hallucinations and eventual insanity and death.

The Coca shrub is not to be confused with Cacao, *Theobroma cacao*, in the Sterculiaceae or sometimes placed in the family Byttneriaceae. The seeds of *Theobroma* are the commercial source of cocoa and chocolate.

Quinolizidine Alkaloids

These alkaloids are most commonly present in members of the Fabaceae with pea-shaped flowers, but they are found as well in the Berberidaceae, Chenopodiaceae, and Papaveraceae.

Lupinine, $C_{10}H_{19}NO$, the simplest alkaloid of this group, is found in the forage lupines used in Europe but rarely in the United States.

Sparteine, $C_{15}H_{26}N_2$, from the Spanish Broom, *Spartium junceum*, has been used as a diuretic for centuries and was described for such use in ancient Greek literature. It was not realized in those times that sparteine actually increases or strengthens the heartbeat, causing a greater flow of blood to the kidneys, and therefore results in more urine production. The sulfate of sparteine has been used as a treatment of heart disease, but it has been replaced mostly by the digitalis glycosides. Sparteine has proved valuable in treating certain cases of cardiac fibrillation.

Cytisine, $C_{11}H_{14}N_2O$, is known to be toxic but has no hallucinogenic activity. It occurs in the following plants found in California: Scotch Broom, *Cytisus scoparius;* Easter Broom, *Cytisus × racemosus;* Kentucky Coffee Tree, *Gymnocladus dioica;* Golden Chain or Golden Rain, *Laburnum anagyroides.* The latter species of *Laburnum* is considered the second most toxic plant for animals in Great Britain after the yew, *Taxus.* Although these plants are not usually eaten by animals, deaths have been recorded from their ingestion. Symptoms consist of incoordination and excitement, convulsions, and death from respiratory failure.

Anagyrine, $C_{15}H_{20}N_2O$, is very similar to cytisine, and these two quinolizidine alkaloids usually occur together in the same plants. Anagyrine is found in *Lupinus caudatus, L. latifolius, L. sericeus,* and *L. laxiflorus;* plants of the last species have the highest concentrations of this alkaloid. Using purified preparations of the separate lupine alkaloids, anagyrine has been found to be the cause of crooked calf disease (discussed under *Lupinus,* Fabaceae).

Heimia salicifolia of the Lythraceae, cultivated in California, contains an alkaloid called cryogenine or vertine to which psychotropic activity may be attributed. Auditory hallucinations may occur from its use.

Pyridine Alkaloids

Arecoline, $C_8H_{13}NO_2$, is the main alkaloid found in betel nuts, the fruits of a palm tree, *Areca catechu,* Arecaceae, cultivated in the tropics, particularly southeastern Asia. Millions of people chew the ground nuts mixed with lime and wrapped in a leaf of the Betel, *Piper betle,* Piperaceae. This mixture of alkaloids produces a feeling of well-being and contentment but leads to addiction as well. Arecoline is hydro-lyzed to methyl alcohol and a carboxylic acid that is a form of nicotinic acid.

Simple forms of pyridine are found in the Field Horsetail, *Equisetum arvense* (a fern relative), and in some species of the false lupines, *Thermopsis,* in the Fabaceae.

Ricinine, $C_8H_8N_2O_2$, is a pyridine alkaloid occurring in plants of Castor Bean, *Ricinus communis,* the only species found to have a single alkaloid. In all other plants forming alkaloids, they occur as mixtures. The main toxin of Castor Bean is the toxalbumin ricin and not this mildly toxic alkaloid.

Pyridine–Piperidine Alkaloids

nicotine

Nicotine, $C_{10}H_{14}N_2$, has both pyridine and piperidine ring systems in the makeup of its molecule. Nicotine is a very power-ful toxin that is lethal to all forms of animal life. A small amount of this alkaloid stimulates breathing, but absorption of a larger amount depresses the respiratory rate. Death from nicotine poisoning results from respiratory failure. Humans are able to smoke tobacco only because the body is capable of breaking down and eliminating this alkaloid rather than accumulating it. The deleterious effects of smoking are cumulative and cause poor physical conditions, among them lung damage, nervous-ness, and heart and digestive problems. Nicotine is used as an insecticide in agriculture and in home gardens, thereby becom-ing a cause of accidental poisonings.

Anabasine is very similar to nicotine and is the principal alkaloid in the Tree Tobacco, *Nicotiana glauca,* a widespread weed in the warmer parts of California.

Indole Alkaloids

Tryptamine and gramine have the indole structure. Derivatives of tryptamine are the alkaloids found in Harding-grass, *Phalaris aquatica;* its toxicity is discussed under the Poaceae.

Bufotenine is very similar to tryptamine. It is hallucinogenic and occurs in the plants of the genus *Piptadenia* in the Fabaceae, in fungi such as *Amanita citrina* [*A. mappa*], in human urine, and in the secretions of the skin of the toad, *Bufo vulgaris.*

Psilocybin and psilocin are very similar to bufotenine. These indole alkaloids are discussed under "Psilocybin–Psilocin Poisoning" (pp. 37–38).

Ergot alkaloids are much more complex molecules containing not only the indole structure but also other ring systems and side chains; in fact, these alkaloids can be classified as polypeptides. Most of these alkaloids contain thirty to thirty-five carbon atoms, although a few have a smaller number. On alkaline hydrolysis, all ergot alkaloids form lysergic acid as well as a number of other compounds. Lysergic acid diethylamide, LSD, is the most powerful compound known that produces imitations of severe mental disorders. Ergot alkaloids are found in the sclerotial bodies produced by the ergot fungus, *Claviceps purpurea,* and in seeds of the Morning-glory, *Ipomoea tricolor.*

Two indole alkaloids, gelsemine and sempervirine, that occur particularly in the roots in the Carolina Jessamine, *Gelsemium sempervirens* of the Loganiaceae, are depressors of the central nervous system.

From *Catharanthus roseus* [*Vinca rosea*], Rose Periwinkle, of the Apocynaceae, cultivated as an ornamental in California, more than sixty indole alkaloids have been obtained. Two of the purified alkaloids, vinblastine and vincristine, have shown some success experimentally in cancer chemotherapy using a five-drug combination program. They have proved useful in the treatment of Hodgkin's disease, lymphosarcoma, and reticulum cell sarcoma. It is postulated that, by modifications of the molecules, different forms of these alkaloids may be found that will be more effective in cancer treatment and less toxic to the human body cells. In Jamaica a tea is made from the leaves of this species of *Catharanthus* for treating cases of diabetes, but no beneficial effects have been proved.

Quinoline Alkaloids

Angostura bark from *Galipea officinalis* in the Rutaceae has been used as a febrifuge and is a component of the commercial flavoring bearing that name. There are two alkaloids of quinoline derivation in angostura bark. Others in angostura bark have much more complex molecular structures based on the quinoline ring system.

Quinoline–Quinuclidine Alkaloids

quinuclidine

The quinuclidine ring is found combined with the quinoline ring system in the makeup of the molecules of quinine and cinchonine. There are some thirty alkaloids in cinchona bark, but the useful alkaloids are quinine and cinchonine and their stereoisomers quinidine and cinchonidine. The few people who are very sensitive to quinine have been successfully treated with quinidine. Cinchona bark is obtained from trees of *Cinchona* and *Remijia* in the Rubiaceae, native to the wet eastern slopes of the Andes in South America and also in Central America. Cultivars yielding large amounts of alkaloids have

been developed mainly from one species, *Cinchona calisaya;*
these varieties are cultivated particularly in Indonesia. The
genus *Cinchona* was named after the Countess of Chinchon,
wife of the Viceroy of Peru, who was successfully treated for
malaria in 1638. The powdered bark of these trees was used
medicinally until about 1820 when purified alkaloids were obtained; quinine, a mixture of these purified alkaloids, is used in
place of the crude bark preparations.

Isoquinoline Alkaloids

This group is derived from the amino acids phenylalanine
and tyrosine and contains the phenylethylamine structure in the
isoquinoline ring system. Whereas tyrosine is converted to
phenylalanine in the mammalian body, this reaction does not
occur in plant cells.

Isoquinoline alkaloids are found in many widely separated
plant families including the Ranunculaceae, Rubiaceae, Berberidaceae, Fumariaceae, and Papaveraceae. The Papaveraceae
has a large number of these alkaloids in the genus *Papaver,* and
many are found in the genera *Argemone, Chelidonium,* and
Sanguinaria. The largest number of alkaloids obtained from a
single species in the family comes from *Papaver somniferum,*
the Opium Poppy, including those opium alkaloids from which
are derived the narcotic drugs morphine, codeine, and heroin.
The same alkaloids are also found in several California plants:
California Poppy, *Eschscholzia californica;* the prickly pop-
pies, *Argemone;* and *Corydalis* and *Dicentra* in the closely related Fumariaceae.

Benzylisoquinoline Alkaloids

The benzylisoquinoline alkaloids are one of the many forms of the isoquinoline ring structure. With the use of radioactive tracers it has been found that both segments of these alkaloids are derived from tyrosine and not in part from phenylalanine as are the others. Papaverine, occurring in opium, is an example of this group.

Morphinan Alkaloids

Another group of isoquinoline alkaloids comprises the morphinan alkaloids including morphine, codeine, and heroin, which are all derived from opium, the dried latex from the Opium Poppy, *Papaver somniferum,* discussed under the Papaveraceae. Opium has two narcotic properties: It is hypnotic in producing an altered state of consciousness and analgesic in blunting pain. Opium also produces a number of unwanted side effects; the most serious is addiction. There is an initial stimulation followed by a dream-filled sleep that is replaced by depression leading to additional use of the drug. Physical and mental deterioration result from its continued use. Cure of opium addiction is commonly followed by relapses. Acute poisoning from opium results in pinpoint contractions of the pupils of the eyes and in progressively slowed, deep breathing, which is eventually followed by coma and death.

Of the large number of alkaloids in opium, morphine was the first to be purified. It requires only about 15 minutes for a small dose of morphine hydrochloride to induce sleep and dull pain, and this state lasts about 6 hours. Care must be exercised in dosage for a severely injured person since a large amount will induce shock and may be fatal. The illegal use of morphine has even more pronounced addictive effects than opium.

Heroin, a derivative of morphine, has the same results as morphine but the addictive habit is even harder to break.

Codeine is the methyl ether of morphine. As a narcotic it is approximately one-tenth as strong as morphine. Since its effect on the respiratory center of the brain is about one-fourth that of morphine, it has been used in suppression of coughing. Because codeine can lead to addiction, its use is carefully regulated. In the body codeine undergoes demethylation to form morphine.

The isoquinoline alkaloids of the Amaryllis Family, Amaryllidaceae, have been treated as a group; more than a hundred are known. All are derived from a precursor compound norbelladine, which is formed from phenylalanine and tyrosine.

Ipecacuanha alkaloids are a specialized group of isoquinoline alkaloids. Ipecacuanha is the name given in Brazil to a small creeping shrubby plant of the tropical rain forests, *Cephaelis ipecacuanha,* a member of the Madder Family, Rubiaceae. Brazilian ipecac, which is the root bark of this species, has a long history of effectiveness against dysentery. Cartagena ipecac is obtained from *Cephaelis acuminata,* mainly from Central America and Colombia. Five similar isoquinoline alkaloids are found in the root bark of both Brazilian and Cartagena ipecac; emetine, $C_{29}H_{40}N_2O_4$, is the most important. Purified emetine has been shown to be highly toxic to the amoeba *Entamoeba histolytica,* the seriously infective organism that causes amoebic dysentery. Because the effects of emetine are cumulative, emphasis has been placed on the manufacture of related compounds that do not have this cumulative toxicity. Some quinoline derivatives are used to treat the invasive form of the amoeba damaging the liver. Ipecac is also well known as an emetic and may be given to induce vomiting in a person who has ingested a toxic substance.

Steroid Alkaloids

Steroid alkaloids have been found in approximately 250 species of plants in the families Solanaceae and Liliaceae. They occur mostly with sugars as glycosides and have the properties of saponins. Some are very toxic, causing severe gastroenteritis that may be fatal.

Solanine was the name given to the first alkaloidal glycoside found in *Solanum*. Now solanine is known to consist of six different glycosides; each hydrolyzes to the sugar peculiar to its particular glycoside and to the aglycone solanidine, a steroidal alkaloid.

The many alkaloids of false hellebores, *Veratrum*, in the Liliaceae are very similar and can be converted easily to solanidine. False hellebore alkaloids have been used internally for the treatment of high blood pressure, externally for the relief of neuralgia, and also as an insecticide. The effects of these alkaloids, which produce malformed lambs, are discussed under the Liliaceae.

The alkaloids of the various species of deathcamas, *Zigadenus*, have steroidal structures. They cause severe damage to the gastrointestinal tract and may cause death within a few hours after ingestion.

Diterpenoid Alkaloids

These alkaloids, derived from diterpenes, are found in *Aconitum* and *Delphinium* in the Ranunculaceae. From these alkaloids are derived the most toxic drugs of plant origin that are known. Two to five milligrams are a lethal dose to humans.

Purine Alkaloids

Sometimes these compounds are not classed as alkaloids even though they have similar structures. Three of them, caffeine, theobromine, and theophylline, are the purine alkaloids found in coffee, tea, and cocoa. These alkaloids have stimulating effects on the central nervous system.

caffeine

Caffeine is found in coffee beans of species of *Coffea*, Rubiaceae; in leaves and young shoots of Tea, *Camellia sinensis*, Theaceae; in seeds of Cacao, *Theobroma cacao*, Byttneriaceae or Sterculiaceae, which produce the commercial products of cocoa and chocolate; in maté or Paraguay tea made from the leaves of *Ilex paraguariensis*, Aquifoliaceae; and in the seeds or nuts of several species of *Cola*, Sterculiaceae. Acute poisoning from the consumption of large quantities of beverages containing caffeine results in stimulation of the central nervous system causing symptoms of dizziness, ringing in the ears, abdominal pains, and vomiting. In severe cases, additional symptoms may include rapid pulse, lowered blood pressure, and constriction and nonreaction of the pupils of the eye. There is an increased flow of urine that may contain sugar and acetone. Patients recover rapidly from acute caffeine poisoning. Chronic poisoning from ingestion of large amounts of such beverages over a period of time may result in headaches, restlessness, insomnia, tremors, and constipation.

The purified theobromine alkaloid has been used as a diuretic and as a stimulant of the heart muscle.

Toxalbumins or Phytotoxins

These generalized names are given to poisonous proteins. Most of these compounds can be described as enzyme inhibitors or as lectins.

Enzyme Inhibitors

These substances are found in seeds of certain flowering plants. They are proteins that slow down or prevent the action of enzymes that digest the proteins in dormant seeds. Raw or partially cooked seeds, particularly soybeans, can cause distress to livestock but not to humans. These enzyme-inhibiting proteins are destroyed by the heat of cooking.

Lectins

These compounds are proteins usually combined with carbohydrates as glycoproteins. They produce an apparent immune reaction against an antigen, but a lectin is not an antibody as generally defined. Lectins collect at the surfaces of cells where they bind glycoproteins and glycolipids. Lectins or plant hemagglutinins apparently protect the plant cells against pathogenic bacteria and also bind symbiotic bacteria to the cells of root nodules. These compounds are called plant hemagglutinins because they can agglutinate red blood cells outside the animal body. Toxic lectins, which affect both humans and livestock, are highly poisonous on injection, less so on being ingested, but they are absorbed easily in the digestive tract.

Lectins are found throughout the plant kingdom and are probably of universal occurrence. More than 600 species and varieties in the Fabaceae have been found to contain lectins. Although the highest concentrations of lectins are in the seeds, they occur throughout the plant.

In structure, lectins are mostly very large protein molecules composed of four subunits, each capable of binding a sugar. Some lectins do not have this binding capability and are very poisonous.

Among the best known of these toxic proteins are ricin (in the Castor Bean, *Ricinus communis*), robin (in the Black Locust, *Robinia pseudoacacia*), and abrin (in the Rosary Pea,

Abrus precatorius). Black Locust and Castor Bean are introductions frequently seen in cultivation in California and naturalized in local populations. Rosary Pea does not grow in California, but the toxic showy red and black seeds are sometimes brought into California as beads.

These three toxic plant lectins inhibit protein synthesis of the intestinal wall. They are similar to some bacterial toxins in structure, physiological reaction, and action as an antigen. Because these compounds are absorbed slowly from the gastroenteric tract, there is a latent period between the time of ingestion and the appearance of symptoms from the toxin. Abrin and ricin, both more toxic than robin, can cause death of humans or livestock.

The lectin in Pokeweed, *Phytolacca americana,* acts as a strong mitogen affecting the activity of the white blood cells.

Lipids

Lipids, including oils, fats, and waxes, are formed mostly from combinations of fatty acids and alcohols. The simpler fats are combinations of three fatty acids and glycerine; glycerine is actually an alcohol, glycerol. Waxes are the results of combining fatty acids with long-chain alcohols, forming such substances as the waxy coverings of leaves and fruits. Lipids may be many other combinations forming complex organic compounds. Lipids break down to form numerous compounds of different structures. Vitamins are such derived compounds.

Terpenes

Terpenes are probably the most numerous organic compounds in plants, exceeding the total number of all other compounds. Terpenoid compounds are products of acetate metabolism, acetates usually derived from fatty acids. The variations of using isoprene, C_5H_8, have produced the myriad numbers of terpenoids. The following terpenes are involved in toxins made in plants, but commonly these terpenoids are joined with other molecules or parts of molecules to form complex compounds.

Essential Oils

Essential oils are monoterpenes, C_{10}, and sesquiterpenes, C_{15}; they are responsible for the fragrance of flowers. Some oils have been used as vermifuges for the gastrointestinal tract. Citrus oils, and a monoterpene limonene found in them, apparently have a slight capability of producing tumors in mice.

Thujone is an essential oil of widespread occurrence in which the active compounds are monoterpenes. It occurs in some evergreens of the Cupressaceae: *Thuja occidentalis, Chamaecyparis thyoides, Juniperus sabina,* and several species of *Cupressus.* The resins of cone-bearing trees, consisting primarily of monoterpenes, are the main defensive mechanism against the bark beetles that are the major destructive insects of the coniferous forests in North America. Thujone also is found in the Asteraceae in *Achillea millefolium, Artemisia absinthium,* and *Tanacetum vulgare.* Wormwood oil derived from *Artemisia absinthium* was the principal flavoring of absinthe liquor, the cause of much human distress. Taken only once, thujone usually acts as a cathartic. Used repeatedly, however, it may cause personality changes such as inattentiveness, forgetfulness, negligent appearance, and hypersensitivity to criticism. A dose of 30 milligrams per kilogram of body weight can produce convulsions and lesions of the cortex of the brain. The toxicity is cumulative and may even result in death.

Pyrethroids are monoterpenes found in species of *Chrysanthemum* in the Asteraceae. They have been used since ancient times as insecticides; formerly made from the dried flower heads, they are now prepared synthetically.

The leaves of the California Bay, *Umbellularia californica,* contain as much as 4% of their dry weight as irritating oils, the most abundant of which is umbellulone, an essential oil. Crushing the leaves and inhaling the volatile oil can cause a severe headache and even unconsciousness in some people. Headache Tree is another common name for this species of the Lauraceae.

Resins

Resins contain resin acid, derived from diterpenes, C_{20}, as the principal nonvolatile part, but resins also contain monoterpenes or sesquiterpenes that permit the resins to flow. Cicutoxin, the toxin of the waterhemlocks, *Cicuta,* is a highly unsaturated aliphatic alcohol composed of acetylenic fatty acids. Species of *Cicuta* of the Apiaceae are among the most violently poisonous plants native to North America. Cicutoxin may be stored in the plants in relatively large amounts. It acts directly on the central nervous system of a human or animal and usually is fatal.

Oenanthetoxin is the name given to an isomer of cicutoxin found in the genus *Oenanthe,* also in the Apiaceae, responsible for poisonings in Europe. The California species, *Oenanthe sarmentosa,* Water Parsley, is not known to be toxic.

Lipids may have a phenolic ring as part of the molecule as in urushiol, the resinoid that is the principal toxin in the sap of the poison oaks and other members of the Anacardiaceae. Resins or resinoids of diterpene structure are found in the Euphorbiaceae, Ericaceae, and Thymelaeaceae. Phorbol is the most abundant toxic principle of the latex of species of *Euphorbia,* the cause of irritation of the skin. Croton oil, containing phorbol, is made from the seed of *Croton tiglium.* It is a drastic cathartic with many side effects including damage to the gastrointestinal tract, vomiting, diarrhea, and delirium. It causes severe blistering of the skin and acts as a cocarcinogen, even after a long delayed time. Experimentally a cancer-producing hydrocarbon was placed on the skin of mice; 1 year later, when croton oil containing phorbol was applied, skin tumors were produced.

The resins in the Heath Family, Ericaceae, have been termed andromedotoxins. These andromedane diterpenoid derivatives are closely related. In *Rhododendron japonicum,* four toxic diterpenoids called rhodojaponins have been found. In *Leucothoë grayana* the diterpene resins have been named grayanotoxins; seven different ones are known. Andromedotoxins cause cardiovascular disturbances in which there is prickling and tingling of the skin similar to that seen in alkaloid poisonings.

All parts of *Daphne mezereum* of the Thymelaeaceae and other *Daphne* species, grown in California as ornamentals, are highly toxic, but most poisonings are caused by the fruits. The bark and fruits contain a toxic diterpene, daphnetoxin; mezerein is the diterpene causing the toxicity in the seeds. Both of these diterpenoid resins cause dermatitis with blistering and swelling of the skin in humans. Mezerein is also a cocarcinogen similar to the phorbol derivatives in the Euphorbiaceae. Shrubs of the native Western Leatherwood, *Dirca occidentalis,* and the South African *Gnidia polystachya,* both in the Thymelaeaceae, have irritant resins.

In New Zealand and Australia, livestock poisonings from species of *Pimelea* in the Thymelaeaceae have been reported. Although these species are grown in California as ornamentals, no such poisonings have occurred here. The toxin is a tetracyclic diterpene called prostratin.

The rhizomes of several irises, *Iris pseudacorus, I. foetidissima,* and *I. germanica,* also contain resins that act as gastrointestinal irritants and cause contact dermatitis. Another plant containing resins is Marijuana, *Cannabis sativa,* with a series of tetrahydrocannabinols. These substances act primarily on the central nervous system and produce delirium or hallucinations.

The Chinaberry Tree, *Melia azedarach,* in the Meliaceae, is usually listed with resin-containing plants. Its slightly fleshy fruits are toxic. Although not brightly colored these fruits have been eaten by children. Animal and human poisonings have also occurred from the leaves or teas made from them. The resins affect the central nervous system causing convulsions and paralysis.

Saponins

Saponins are triterpenes, C_{30}, that are formed by linking terpene units basally into rings or cyclic structures. Saponins are discussed under "Saponin Glycosides" (pp. 322–324).

Steroids

Steroids too are derived from the triterpenes. Steroids are discussed under "Steroid Glycosides" (p. 321).

Isoflavanoids

Isoflavanoids, including coumestrol, are found almost totally in the Pea Family, Fabaceae. These compounds are derived from terpenes and reduce fertility in sheep. Some isoflavanoids occur combined as glycosides. The related flavanoids are widely distributed in plants and are harmless to humans or animals.

Phytoalexins

Phytoalexins are isoflavanoids derived from terpenes or furocoumarins. Phytoalexins are formed by plant cells when invaded by fungi or bacteria, presumably restricting further growth of the infecting organisms. These induced isoflavanoids produced by the diseased plant tissues can cause distress to humans and animals. A sesquiterpene phytoalexin is formed in infected tuberous roots of Sweet Potato, *Ipomoea batatas,* Convolvulaceae.

Herbal Teas

Herbal teas or hot water infusions have become popular during recent years. Because they are considered to be healthful natural products, it may seem out of place to discuss them in a book on toxic or poisonous plants. Certainly many herbal teas are aromatic and flavorful and have become pleasant substitutes for the caffeine and tannins of coffee and for China tea made from leaves of *Camellia sinensis*. However, a few plant ingredients included in herbal teas contain substances that are potentially or actually harmful. Because these teas are sold in health food stores the public considers them healthful and harmless. Another assumption is that they have been tested and approved for human use. Such is not the case, however, for herbal teas are neither foods nor drugs and do not fall within the scope of state or federal regulatory agencies. Not all herbal teas are obtained from commercial sources; some are gathered by the users from wild or semiwild plants growing in fields, roadsides, or gardens. The purpose in discussing herbal teas here is to point out the problems in identifying their plant ingredients, regardless of their sources, and to call attention to their physiological effects that are not described by their distributors nor understood by their collectors.

Plants used in herbal teas and sold commercially are identified only by common names on their packages. Since the same common name is often used for several different plants and a single plant may have several different common names, these are a great source of confusion. Nevertheless, even though these names are not always reliable, they sometimes provide a clue in the identification of the plant in question. The common

name Eyebright, used for at least three different plants, illus-
trates this point. Plants of the genus *Euphrasia,* Scrophulariaceae, with several species in eastern North America, are called
Eyebright, probably because a European annual species known
as *Euphrasia officinalis* formerly was used for treatment of eye
diseases. No toxic properties are known for this genus of
plants. *Hydrastis canadensis,* Ranunculaceae, also is sometimes called Eyebright, although perhaps it is better known as
Golden Seal. This plant of eastern North America contains
several alkaloids and has had various medicinal uses. Taken
internally it can cause inflammation and ulceration of the
mucous membranes. Another application of the name Eyebright is for *Lobelia inflata,* Lobeliaceae, also known as Indian
Tobacco. It contains the alkaloid lobeline and has been used
medicinally but it is also toxic. The question is: Which Eyebright is used in herbal teas?

The identity of the plant called Gotu Cola, Gota Cola, or
Kotu Cola is also confused. These names have been used for
Hydrocotyle asiatica, Apiaceae, a creeping herbaceous perennial long included in Chinese herbal medicine and eaten raw or
cooked as a food in the Orient. These names, however, have
also been confused with Cola or Kola, *Cola acuminata* and *C.
nitida,* Sterculiaceae. Kola nuts, seeds of these African trees,
contain a considerable amount of caffeine and are chewed by
African people for their stimulating effect. They are used in the
United States for their caffeine content and for flavor in the
manufacture of stimulating, nonalcoholic beverages. The question again is: Which of these plants is used in the herbal tea
called Gotu Kola?

Another problem with herbal teas concerns incorrect identification of plant materials. Such errors may occur both in
herbal teas offered for commercial sale and in plants gathered
for teas from home gardens. Two cases of misidentification of
ingredients in herbal teas occurred in Arizona in 1977 and involved two small Hispanic children, one of whom died. Both
had been given large amounts of a tea prepared from a locally
marketed product called Gordolobo Yerba, a name that is used
for two different plants, a *Gnaphalium* and a *Senecio*. In both
cases, however, the tea given to the infants was made from

leaves of *Senecio douglasii* var. *longilobus* [*S. longilobus*], Asteraceae, which contain pyrrolizidine alkaloids and are hepatotoxic; see "Pyrrolizidine Alkaloids" (pp. 332–334).

Another case of misidentified plant material was reported in the state of Washington in 1977 and involved an elderly couple. They drank a tea prepared by the woman from leaves of a plant she believed to be Comfrey, either *Symphytum officinale* or *S. × uplandicum*, Boraginaceae, but actually they were leaves of the Common Foxglove, *Digitalis purpurea*, Scrophulariaceae. Both she and her husband died from drinking this tea.

The contamination of commercially prepared herbal teas presents yet another problem. A case was reported in 1978 of poisoning from a commercially packaged burdock root tea. The patient who drank the tea exhibited symptoms similar to those caused by an atropine-like alkaloid. Burdock root, obtained from *Arctium lappa*, Asteraceae, is much cultivated in Japan as a food and is found in Oriental food markets in California. Although burdock root does not itself contain such an alkaloid, examination revealed that this herbal tea did contain atropine. There was no way of knowing how the tea became contaminated.

The number of individual plants used in herbal teas, and obtained from commercial or noncommercial sources, is difficult to estimate but perhaps it is 200 or 300. Many companies prepare and market them. Usually each tea package contains a mixture of several kinds of plants, infrequently material of a single plant. There is no indication of the relative amounts of each plant in the packaged tea; nor is there much information regarding any possible effects.

Apart from their use in teas or infusions, herbal preparations are also made for smoking in pipes or cigarettes or made into capsules. Since some of the plant materials may cause nausea when ingested or smoked, it is recommended that they be taken in capsular form. In a few cases smoking these cigarettes has caused poisoning severe enough for hospitalization.

Our major concern here is the physiological reactions to certain plants used in herbal teas. Several plant products act as strong diuretics and cathartics, others have allergenic properties, still others have stimulant and hallucinogenic properties, and some are toxic. Further discussion of other plants used in herbal teas will be found under individual species in their respective families.

Acorus calamus, Sweet Flag or Calamus, Araceae, contains alkaloids chemically similar to those of mescaline and myristicine. Indians in North America made a tea from the rhizomes for medicinal purposes as a tonic and a stimulant.

Aloë barbadensis [*A. vera*], Barbados Aloe, Curaçao Aloe, and one or two other species of *Aloë,* Liliaceae, contain anthraquinone glycosides that are well-known cathartics.

Arctostaphylos uva-ursi, Bearberry or Uva-ursi, Ericaceae, contains the glycoside arbutin, tannins, and other substances. The plant was used in the past for its diuretic and astringent properties.

Argemone mexicana, Prickly Poppy, Papaveraceae, contains several toxic isoquinoline alkaloids that are sources of useful drugs. The plant has been smoked as a euphoriant for its narcotic-analgesic effects.

Artemisia absinthium, Absinthe or Wormwood, Asteraceae, is a perennial from southern Europe. Thujone is the major component of wormwood oil or absinthe, an essential oil, which is obtained from the plant by distillation of the leaves and upper stems. It is the chief ingredient of the French liquor known as absinthe. Because it is habit-forming and can cause delirium, hallucinations, and permanent brain damage, its manufacture was prohibited in France in 1915 and later in the United States. Imitation absinthe is obtainable, usually flavored with anise seeds and containing no wormwood. When smoked or in a tea, wormwood has a narcotic-analgesic effect.

Calea zacatechichi, Asteraceae, is a shrub occurring from Oaxaca, Mexico, to Costa Rica. The Chontal Indians of Oaxaca, who call the plant Zacatechichi, use the leaves as a hallucinogen. Because this use was not reported until 1968, few chemical studies have been made on the plant. Its use in a tea has been reported in California.

Cassia acutifolia, Alexandria Senna, and *Cassia angustifolia,* Tinnevelly Senna, Fabaceae, are Old World shrubs. The dried leaves contain anthraquinone glycosides that are well-known cathartics. Derivatives of senna have been used medicinally; overdoses may be toxic. In six cases reported from New York and Pennsylvania in 1977, several persons became ill with severe diarrhea from drinking herbal teas that contained senna and buckthorn, *Rhamnus*. The labels on these commercially prepared herbal teas did not give information regarding the effects that might be expected from drinking them.

Catharanthus roseus, Periwinkle, Apocynaceae, has more than sixty different alkaloids, mostly indole and dihydroindole derivatives, some of which occur in plants of other species in the Apocynaceae. Drugs obtained from certain of these alkaloids are used in cancer therapy. Periwinkle has been smoked or used in a tea as a euphoriant and a hallucinogen.

Chamaemelum nobile [*Anthemis nobilis*], Roman, Russian, English, or Garden Chamomile, Asteraceae, and *Matricaria recutita* [*Chamomilla recutita, Matricaria chamomilla*], Sweet, Sweet False, German, or Hungarian Chamomile, Asteraceae, are Old World perennials commonly known as chamomile. They are popular in herbal teas, alone or in mixtures, and have been used since ancient times presumably for soothing therapeutic effects. Most packaged teas give no indication which of the two species of chamomile is included. Chamomile has been known to cause contact dermatitis and an anaphylactic reaction or severe hypersensitivity in persons allergic to ragweeds, asters, and chrysanthemums.

Cinnamomum camphora, Camphor, Lauraceae, contains an essential oil that consists in part of safrole; see *Sassafras albidum* below. When smoked it is a mild stimulant.

Corynanthe yohimbe, Yohimbe, Rubiaceae, a West African tree, contains the indole alkaloids yohimbine and yohimbinine. In Africa the plant is considered to be an aphrodisiac; in large doses it is hallucinogenic. It is said to be a stimulant when smoked or drunk in a tea.

Cytisus scoparius, Scotch Broom, and *Cytisus × racemosus*, Easter Broom, a hybrid of *C. canariensis* and *C. stenopetalus*, Fabaceae, contain the toxic quinolizidine alkaloids sparteine, cytisine, genisteine, and sarothamnine. Derivatives of these alkaloids have had medicinal uses. Brooms have been smoked for their sedative-hypnotic effects.

Datura stramonium, Thornapple or Jimsonweed, Solana-

ceae, contains the toxic indole alkaloids atropine and scopolamine. The plant, when smoked or made into a tea, can cause
hallucinations.

Ephedra viridis, Squaw Tea or Mormon Tea, Ephedraceae,
is a shrub of the southwestern United States. Indians and early
settlers drank infusions made from it, and probably from other
species of the genus as well, as a treatment for venereal diseases. Plants of the Asiatic and European species of *Ephedra*
contain the phenylethylamine alkaloids ephedrine and pseudo-
ephedrine, useful in the treatment of allergies including bronchial asthma and hay fever. Of the species native to North
America none has been reported to contain useful alkaloids.
Squaw Tea is drunk for its stimulating effects.

Equisetum hyemale and *E. arvense,* both called Horsetail or
Shavegrass, Equisetaceae, are spore-bearing plants related to
the ferns. They have been used in herbal medicines as a diuretic and a blood coagulant. Some plants of *Equisetum* contain thiaminase, an enzyme that promotes the breakdown of
thiamine. These are toxic to livestock, especially horses and
cattle. Their healthful use in herbal teas has been questioned.

Eschscholzia californica, California Poppy, Papaveraceae,
contains several toxic isoquinoline alkaloids. It has been
smoked for its mildly euphoriant effect.

Humulus lupulus, Hop, Cannabaceae, yields the substance
called lupulin from the glandular hairs of the female flowering
clusters. The principal ingredients of the bitter lupulin are a
volatile oil and a resin that, when ingested, are reputed to act
as a tonic that aids digestion. Used as a tea or smoked, it has
the effect of a sedative.

Hydrangea macrophylla and other species, Hydrangea,
Saxifragaceae, contain a cyanogenic glycoside. In a tea or
smoked it acts as a stimulant.

Ilex paraguariensis, Maté, Aquifoliaceae, is a South American shrub the leaves of which are used widely in Paraguay, Argentina, and Brazil for a tea called maté. The leaves contain
caffeine and tannin. In large doses the tea acts as a cathartic; it
also has diaphoretic and diuretic properties, increasing perspiration and urination.

Juniperus communis, Juniper, Cupressaceae, has berries

that contain a volatile oil. Although they have been used in folk medicine as a diuretic, they have irritant properties.

Lactuca virosa, Lettuce-opium or Wild Lettuce, Asteraceae, a European relative of edible lettuce, has a copious amount of bitter milky juice that contains lactucarine, an alkaloid. The milky juice, when dry and hard, is known as lactucarium and is reputed to be a mild sedative. One of its medicinal uses was in cough mixtures to replace opium. Lactucarium, when smoked as an opium substitute, gives a mild narcotic-analgesic effect.

Matricaria recutita, Sweet, Sweet False, German, or Hungarian Chamomile, Asteraceae; see *Chamaemelum nobile* above.

Myristica fragrans, Nutmeg, Myristicaceae, is a commercial spice long known for its effect on the central nervous system, probably the result of its active principles elemicin and myristicine. When used as a tea, it may produce hallucinogenic effects.

Nepeta cataria, Catnip, Lamiaceae, is an Old World perennial long used medicinally as a cold remedy and to relieve colic in infants. Its leaves and flowering stems contain volatile oils of which nepetalactone is an active component. Catnip is a euphoriant and is used in a tea or smoked. Cats have a fondness for the plant, and their apparent euphoric responses to it are well known.

Passiflora incarnata, Passion Flower, Passifloraceae, is a vine from the eastern United States. It contains harmine alkaloids, which are indole in structure and have been used medicinally as a sedative to relieve insomnia and in the treatment of convulsions and spasmodic disorders. When taken as tea, capsules, or cigarettes, the plant material acts as a mild stimulant.

Paullinia cupana, Guarana, Sapindaceae. Guarana is the common name of the plant as well as the name of the main constituent of Zoom, advertised as a natural organic stimulant and appetite depressant. Guarana is prepared from the crushed seeds of *Paullinia cupana,* a shrub from Brazil and Paraguay. The seeds contain caffeine and are used to make a stimulating beverage that is similar to coffee and tea.

Phoradendron, the mistletoes, Viscaceae, are sometimes

listed as ingredients in herbal teas; the material might be leaves of one of the species of *Phoradendron* or of *Viscum album,* which contain toxic amines.

Piper cubeba, Cubeb, Piperaceae, is a woody climber native to the East Indies and cultivated there. It contains a volatile oil and sesquiterpene alcohols and has been used medicinally as a diuretic, an antiseptic, an expectorant, and a carminative. The medicinal properties of Cubeb were known to Arabian physicians of the ninth and tenth centuries, but it was not until the nineteenth century that its medicinal properties became known in Europe. It is included in herbal tea mixtures, and sometimes information on the packets suggests its use as a diuretic.

Piper methysticum, Kava or Kava Kava, Piperaceae, is a shrub native to the South Pacific Islands. Natives there use its roots, which yield several pyrones, to prepare an intoxicating beverage. Used as an herbal tea or smoked it gives a mild hallucinogenic effect.

Rhamnus cathartica and *Rhamnus frangula,* Buckthorn, Rhamnaceae, contain anthraquinone glycosides that are catharthics. See *Cassia acutifolia* above.

Sassafras albidum, Sassafras, Lauraceae, a tree, was used medicinally by the Indians in the central and eastern United States and later by European settlers. The root bark of sassafras yields an oil that consists principally of safrole, which has aromatic and carminative properties, reducing the production of gas in the gastrointestinal tract. In 1960 it was reported that safrole showed evidence of causing cancer in rats.

In this same year, recognizing that safrole was potentially toxic, the Food and Drug Administration prohibited its use as a flavoring in foods. This toxicity was further clarified in 1974. Nevertheless, even after these rulings, sassafras not only continues to be available in herbal mixtures in health food stores but its use has undoubtedly increased with the public's current interest in so-called natural products.

Symphytum officinale, Common Comfrey, and *Symphytum × uplandicum,* Russian or Quaker Comfrey, Boraginaceae, are perennials with thick fleshy taproots and are among the best-known European herbs. During the time of the Crusades, Common Comfrey was used in the treatment of wounds, broken bones, and other injuries. Leaves and roots contain tannin, however, and eight pyrrolizidine alkaloids that cause liver damage. Experimentally it has been proved that toxic amounts of alkaloids are not absorbed through the skin. Therefore, the traditional use of Comfrey as an ointment for the relief of muscular pains and inflammations is not dangerous to patients' health.

Turnera diffusa, Damiana, Turneraceae, is a shrub native of the American tropics where it has had medicinal uses as a laxative and stimulant. It has a mildly stimulating effect when used in teas or smoked.

Valeriana officinalis, Valeriana, Valerianaceae. Another common name listed for this plant, Heliotrope, may cause this species to be confused with species of *Heliotropium* of the Boraginaceae. Valeriana plants contain the alkaloids chatinine and verline, oil of valerian (an essential oil), and the monoterpene valepotriate. Formerly Valeriana was used in medicine as a sedative and antispasmodic. It may be made into a tea, but because of its strong odor and taste it sometimes is taken in capsule form. Used either way it is reported to have a tranquilizing effect.

Photosensitization

Photosensitization is a condition of excessive sensitivity to sunlight. Substances activated by sunlight cause damage to the white areas of an animal's skin. There are two main kinds of photosensitivity. In the primary type the active compound is absorbed directly from the digestive tract or it may be injected into the animal as a therapeutic measure. The other type, hepatogenous photosensitivity, is the result of a damaged liver that does not remove the pigment phylloerythrin. In the peripheral circulation, phylloerythrin absorbs excessive energy from sunlight and causes damage to the skin.

In the primary type the photodynamic compound of Klamathweed, *Hypericum perforatum,* called hypericin, is a dianthrone derivative of eight carbon rings fused together. Hypericin is a very stable compound, resistant to heat and drying. The compound is not changed during digestion and is not destroyed in the liver. Minute amounts of hypericin cause he-molysis of red blood cells. Because necrosis of the white skin produces an intense itching, animals often rub the skin off the affected areas. Although cattle are affected the most, sheep and horses may be poisoned but not as readily. Primary photo-sensitization can also be caused by the pigment in Buckwheat, *Fagopyrum esculentum*. Injections in calves of phenothiazine or fluorescent dyes such as methylene blue likewise cause photosensitization.

Most photosensitization is of the hepatogenous type, the re-sult of liver damage. Chlorophyll pigments are degraded by bacteria in the digestive tract to form phylloerythrin. Nor-mally this compound is removed from the blood by the liver and secreted in the bile. Damage to the liver or bile ducts re-sults in accumulations of phylloerythrin in the blood. Any toxin causing liver damage can produce photosensitivity. In California, plant toxins capable of causing photosensitivity are found in species of *Tetradymia, Tribulus,* and *Lippia.* Some legumes such as alfalfa and Bur Clover, species of *Medicago,* have been implicated in photosensitization.

Although such reports are not known in California, sheep and cattle in New Zealand, Australia, and South Africa have been poisoned by fungal toxins that cause a photosensitization called facial eczema. The toxins are produced by the fungus *Pithomyces chartarum* [*Sporidesmium bakeri*], which grows on dead or damaged grass tissues.

In the fall and winter of 1957–1958 in Oklahoma and adja-cent states, the fungus *Periconia minutissima* was discovered growing on first-cutting alfalfa plants that were badly damaged by flooding and heavy rains. The spoiled hay caused wide-spread photosensitivity in cattle, but the fungus could not be proved as the causal agent.

A similar unexplained case of poisoning occurred in June 1966 near Goose Lake, Modoc County. As was done yearly, a herd of cows was turned onto a rangeland that included the dry-ing bottomland of several hundred acres of an abandoned reser-voir. This year proved different, however, since the lack of rainfall resulted in very little growth of forage on the range. Moreover, there had been a heavy frost very late in the spring that badly damaged the plants of rushes, *Juncus,* that were the dominant plants. The rushes formed a solid stand in the bottomland pasture, and the top third of the leaves of the rush plants exhibited much damaged tissue. By an oversight the animals were not seen until after they had been on this range for over a month. All of the white-faced cows had severe damage to unpigmented areas of their heads and bodies. Young calves in the herd were similarly affected. Although the fungus was not isolated from the lesions on the leaves, frost-damaged rush plants had been the only forage available to the animals.

Allelopathic Phytotoxins

Chemical compounds produced by vascular plants (seed plants, ferns, and their relatives) that can cause pathological effects on the growth of other vascular plants have been called allelopathic phytotoxins. Allele in this sense means another or similar kind of plant. Allelopathy is now considered to include all the biochemical reactions between plants; some may be positive or stimulatory reactions, others may be negative or inhibitory in their effects. These biologically active compounds are considered as phytotoxins only as they affect other plants. Toxicity to animals and to lower groups of plants is not considered under this term.

The concept of the toxicity of plant exudates was expressed as early as 1832 by de Candolle, who referred to it as a "soil sickness." In recent years much more attention has been directed toward the effects of one plant species preventing the growth of other species of plants.

The action of allelopathic plant toxins is the result of materials added to the environment that favor the growth of one species of plant over others. This addition of toxins to the environment is different from competition between species, where plants are removing some environmental factor such as water, mineral salts, or light.

There is no common chemical structure that characterizes these allelopathic phytotoxins. Their main action results from the toxic compounds present in the upper layers of the soil preventing the growth of seedlings. They occur in the soil from the secretions of the roots, decay of plant parts in the litter, or in water washed from the plants by rain or fog condensation. Some cases of seed dormancy are related to these phytotoxins. Seeds of *Datura* will not germinate until the alkaloids are washed from the seed coat. These alkaloids also prevent the growth of nearby seedlings of other species.

The substances detected as allelopathic phytotoxins include terpenes and phenolic compounds; the phenolic compounds are glycosides or phenolic acids. The structures of other substances remain unknown because the residues from some extractions show no inhibition of seedling growth. Many such phytotoxins are soluble in water and may be lost from the upper layers of soil by a very small amount of rain. Therefore, resistant seedlings may get a start on germination with the first rain; other seedlings can develop later after the leaching of the phytotoxic substances.

Chamise, *Adenostoma fasciculatum,* a shrub, develops pure stands in the chaparral. The first year following a fire, many annual species of herbs appear, growing in abundant numbers from seeds in the soil. This flush of herbs will not appear again in the area until another fire occurs. Clearly something inhibits their germination. On investigation the inhibiting chemical compounds have been found in the water dripping from the Chamise plants as rain or condensation from fog. These compounds accumulate in the litter under the shrubs during the drier periods but are lost rapidly by the leaching action of rain. The sudden growth of annuals can be achieved readily by mechanical removal of the Chamise shrubs in the first year, thereby duplicating the effects of a fire. Therefore, these allelopathic phytotoxins must be continually added every year to the upper layers of soil to achieve the dominance of the Chamise over other plants.

A study of the Eastwood Manzanita, *Arctostaphylos glandulosa* var. *zacaensis,* in the mountains of Santa Barbara County has shown that solid stands of this shrub are the result of a water-soluble group of phytotoxins that are phenolic compounds. Open areas occur in the solid stands of manzanita where water is washed over the soil surface from adjacent shrubs. These areas remain bare of any seedling growth. In similar open areas next to them but where the water does not wash from the manzanita shrubs, there are good stands of annual plants. The highest concentrations of these phenolic compounds are found in the fresh leaf litter. Mechanical removal of the manzanita shrubs does not result in immediate renewed growth of the annual herbs, however, as with the Chamise. Even after removal of the manzanita, seedling growth is almost completely inhibited the first year and well into the second year.

Purple Sage, *Salvia leucophylla,* shows a marked toxicity to herbaceous plants. No seedlings can be found in the interior of any mass of these shrubs. The ground at the edge of the shrubs shows a bare zone 3 to 6 ft in width, and stunted vegetation has been observed in years of average rainfall as far as 20 to 30 ft from the shrubs. In very wet years this effect is not seen. Cineole and camphor, terpene compounds, are released into the air surrounding the shrubs. The greatest amounts of these terpenes are released in late spring after the rains have stopped. Until the start of the fall rains these compounds are held on the particles of decaying plant litter in the surface layers of the soil and have been shown to be very toxic to sprouting seeds. Four to six weeks after the first rains these terpenes have disappeared from the soil surface, apparently destroyed by the soil bacteria and fungi.

The concept of gene splicing has caused significant interest in allelopathic phytotoxins. Crop plants appear to have naturally occurring phytotoxins, and breeding of cultivated varieties (cultivars) that have allelopathic phytotoxins could significantly reduce the major expense of weed control. Enzymatic degradation of the chromosomes of cultivated varieties or their wild relatives having allelopathic phytotoxins could be used to separate the genes. After identifying the gene or genes producing the toxins and attaching them to relatively harmless forms of bacteria or plant viruses, these desirable genes could then be introduced into crop plants. By this means cultivars of crop plants could be produced that would prevent or inhibit the germination and growth of weed seedlings.

Apart from the interaction between living plants caused by allelopathic phytotoxins, the effects of chemicals produced by decaying weeds or crop plants may markedly affect the growth of plants the following year, particularly when the cropping system involves little or no tillage. Inhibiting chemicals may be washed directly from the dead plants or may result from the aerobic or anaerobic decay of the plant material.

Black Mustard, *Brassica nigra,* is an abundant component of the grasslands of southern California and has been shown to be markedly toxic to the common annual grasses such as Wild Oat, *Avena fatua,* and species of *Bromus.* Water-soluble plant toxins are produced from the dead plant material of Black Mustard.

In crop rotations, residues of a previous crop may influence the following year's crop. Grain Sorghum, *Sorghum bicolor,* leaves a residue of plant material that significantly reduces the germination of wheat seedlings. This residue also inhibits the growth of two summer annual weeds, Barnyardgrass, *Echinochloa crus-galli,* and Jungle-rice, *Echinochloa colonum.*

The inhibition of plant growth is not always the result of phytotoxins. In California the bare zone between the chaparral shrubs and the grasslands, commonly just a narrow strip, is mechanical in origin, a pathway made by small animals.

Bloat

Bloat is an abnormal swelling of the rumen and reticulum in cattle, sheep, and horses caused by the trapped gases of fermentation. When the gas is free from the ingested food, it is called free gas bloat or secondary ruminal tympany. When the gas is in a persistent foam mixed with the fluids and food particles of the rumen, it is called frothy bloat or primary ruminal tympany.

Free gas bloat results from the physical blockage of the esophagus, preventing the normal escape of the gases through the mouth. This blockage is the result of a number of different causes. It is relieved by using a stomach tube or a trochar and tube that permit the trapped gases to escape, alleviating the pressure in the rumen.

Frothy bloat is not reduced to any great extent by the use of a stomach tube or a trochar and cannula, as these tubes usually are not wide enough to permit the foam to escape. A tube of at least 1 in. diameter is needed. Using a sharp knife an emergency incision can be made in the proper place in the lumbar region that permits an explosive release of the pressure, often saving the life of the animal.

Frothy bloat may occur in livestock on pastures or in feed-lots. Pasture bloat was formerly thought to be formed by various substances such as saponins, pectins, or derivatives of the decomposition of cellulose. Now, cytoplasmic proteins of the leaf cells of legumes, particularly alfalfa and clovers, are considered to be the responsible compounds. The cause of feedlot frothy bloat seems to be related to finely ground concentrates. The insoluble slime trapping the gases may be the result of bacterial growth on feed high in carbohydrate content. The foam is further stabilized by the finely ground particles of the feed concentrates.

In pasture bloat the mucoproteins of the saliva and the lipids or oils from the forage plants reduce the surface tension of the ruminal fluid. Synthetic surfactants or wetting agents such as poloxalene have given good protection against bloat. Because it is somewhat unpalatable, poloxalene should be given in an acceptable form such as in a grain mixture.

Bloat can be prevented by having pastures composed equally of clovers and grasses. Since the tannins present in trefoils, *Lotus,* are also helpful in reducing bloat, trefoils are commonly included in pasture plantings. Pastures with high percentages of clovers or alfalfa that could produce bloat can be used to make hay instead of being grazed.

Feedlot bloat is not so easily controlled. If given free choice, the animals will not eat enough of the roughage they need to prevent frothy bloat. Therefore 10% to 15% of grassy hay or grain straw, and not alfalfa, is included as roughage in the grain rations.

Polioencephalomalacia

Polioencephalomalacia, or cerebrocortical necrosis of the brain, occurs occasionally in cattle and sheep, particularly in young animals 6 to 12 months old, commonly when animals are changed from a dry rangeland to lush young green forage. The damage causes the sudden appearance of blindness, head-pressing, muscular tremors, and incoordination; eventually the condition is fatal. Edema and necrotic cells occur throughout the brain, especially in the cortex region. This brain damage, found on autopsy, has many different causes.

Thiamine deficiency, which produces the brain damage seen in polioencephalomalacia, is discussed under Bracken Fern, *Pteridium aquilinum.* Thiamine deficiency is also known to be caused by a high-sulfate diet. Thiaminase I, an enzyme produced in the rumen by the sulfates, causes the breakdown of thiamine. Thiaminase I can be found in the urine and feces of these animals; they respond favorably to injection or oral administration of thiamine.

Cattle on saline soils with a plant cover of Summer Cypress, *Kochia scoparia,* and particularly with a water supply of high-sulfate content, develop polioencephalomalacia. Photosensitization is the more common response, however. Cattle on pastures that also include Saltgrass, *Distichlis spicata,* are not subjected to such toxicity.

Cobalt deficiency in the diet will result in polioencephalomalacia as well; see the discussion under *Phalaris* in the Grass Family, Poaceae. Animals are protected from this injury by placing a large, slowly dissolving, cobalt pellet or "bullet" in the rumen. Animals with phalaris staggers do not respond to treatment with thiamine.

In Colorado and adjacent areas polioencephalomalacia of unknown origin is reported most frequently in cattle pastured during July and on feedlots in January. Sheep are less affected, but two cases have been seen, one on pasture in August and the other on a feedlot in January.

Although selenium toxicity has been postulated as one of the causes of polioencephalomalacia, an extensive investigation found no correlation between the selenium content of the forage and this disease.

Appendix A
Plants Causing Dermatitis

The plants listed here are recorded as causing dermatitis. This list includes both irritant contact dermatitis that occurs immediately in all persons and allergic contact dermatitis that occurs only in individuals susceptible to the particular toxin. Allergic contact dermatitis requires previous exposure and appears only after a delay of 24 to 48 hours. The list is adapted from Mitchell and Rook (1979) and other sources. Where only the genus is listed, all species of the genus are considered capable of causing dermatitis.

GYMNOSPERMS

GINKGOACEAE, Ginkgo Family
Ginkgo biloba, Ginkgo, Maidenhair Tree

ANGIOSPERMS: DICOTYLEDONS

ACERACEAE, Maple Family
Acer, Maple, Box Elder
ANACARDIACEAE, Sumac Family
Cotinus coggygria [*Rhus cotinus*], Smoke Tree, Smoke Bush, Venetian Sumac, Wig Tree
Schinus molle, Pepper Tree, Molle, California Pepper Tree, Peruvian Pepper Tree, Peruvian Mastic Tree
Schinus terebinthifolius, Brazilian Pepper Tree, Christmas Berry Tree
Toxicodendron diversilobum [*Rhus diversiloba*], Pacific Poison Oak, Western Poison Oak
APIACEAE, Parsley Family
Apium graveolens, Celery, Celeriac
Daucus carota, Wild Carrot, Queen Anne's Lace, Devil's Plague, Bird's-nest Plant
APOCYNACEAE, Dogbane Family
Allamanda cathartica, Yellow Allamanda, Golden Trumpet
Plumeria rubra, Frangipani, Temple Tree, Pagoda Tree, Amapola

Thevetia peruviana [*T. neriifolia*], Yellow Oleander, Lucky Nut, Be-still Tree, Flor Del Peru

Thevetia thevetioides, Giant Thevetia

ARALIACEAE, Ginseng Family

Hedera canariensis, Algerian Ivy, Canary Islands Ivy, Madeira Ivy

Hedera colchica, Persian Ivy

Hedera helix, English Ivy

Hedera nepalensis, Nepal Ivy

Polyscias balfouriana, Balfour Aralia

Polyscias guilfoylei, Geranium-leaf Aralia, Wild Coffee, Coffee Tree

ASTERACEAE, Aster Family

Achillea millefolium, Yarrow

Ambrosia [including *Franseria*], Ragweed, Burweed, Burbrush

Artemisia, Sagebrush, Mugwort, Wormwood

Aster, Aster

Calomeria [*Humea*], Crimson Shower

Chamomilla [*Anthemis*], Chamomile

Chrysanthemum, Chrysanthemum

Conyza, Horseweed

Encelia, Encelia, Brittlebush, Incienso

Gaillardia, Blanketflower

Iva [*Oxytenia*], Copperweed, Marshelder, Povertyweed

Lactuca, Lettuce

Lepidospartum, Scale Broom, Broomshrub

Senecio, Groundsel, Butterweed

Tagetes, Marigold

Tanacetum, Tansy

Xanthium, Cocklebur, Spiny Clotbur

BIGNONIACEAE, Bignonia Family

Campsis radicans, Trumpet Creeper, Trumpet Vine, Trumpet Honeysuckle, Cow-itch

CANNABACEAE, Hemp Family

Cannabis sativa, Hemp, Indian Hemp, Marijuana, Gallow Grass, Grass, Pot

Humulus lupulus, Hop, European Hop, Lupulo

CONVOLVULACEAE, Morning-glory Family
 Dichondra micrantha [has been confused with *D. repens*],
  Lawn Leaf, Dichondra
CRASSULACEAE, Stonecrop Family
 Crassula arborescens, Silver Jade Plant, Silver Dollar, Chi-
  nese Jade
 Crassula ovata [*C. argentea*], Jade Tree, Jade Plant, Baby
  Jade, Cauliflower Ears, Dollar Plant
 Sedum acre, Golden Carpet, Gold Moss
 Sedum album
 Sedum telephium, Live-forever, Orpine
EUPHORBIACEAE, Spurge Family
 Acalypha hispida, Chenille Plant, Red-hot Cattail, Red Cat-
  tail, Philippine Medusa, Foxtail
 Acalypha wilkesiana, Copperleaf, Painted Copperleaf,
  Fire-dragon, Beefsteak Plant, Match-me-if-you-can,
  Jacob's Coat
 Codiaeum variegatum, Croton [not to be confused with the
  genus *Croton,* native of California]
 Euphorbia cotinifolia, Red Spurge
 Euphorbia cyathophora [has been confused with *E. hetero-
  phylla*], Painted Leaf, Fire-on-the-mountain, Mexican
  Fire Plant, Fiddler's Spurge, Annual Poinsettia, Dwarf
  Poinsettia
 Euphorbia lactea, False Cactus, Candelabra Cactus, Can-
  delabra Spurge, Hatrack Cactus, Dragon Bones, Mottled
  Spurge
 Euphorbia lathyris, Caper Spurge, Mole Plant, Gopher
  Plant
 Euphorbia maculata, Prostrate Spotted Spurge, Eyebane,
  Milk Purslane, Wartweed
 Euphorbia marginata, Snow-on-the-mountain, Ghost Weed,
  Mountain Snow
 Euphorbia milii, Crown-of-thorns, Christ Plant, Christ
  Thorn
 Euphorbia peplus, Petty Spurge
 Euphorbia pulcherrima, Poinsettia, Christmas Flower,
  Painted Leaf, Mexican Flame Leaf, Easter Flower,
  Christmas Star, Lobster Plant

Euphorbia tirucalli, Milkbush, Malabar Tree, Indian Tree Spurge, Finger Tree, Pencil Tree, Rubber Euphorbia

Euphorbia trigona, African Milk Tree, Abyssinian Euphorbia

Jatropha curcas, Barbados Nut, Physic Nut, Poison Nut, Purging Nut

Pedilanthus tithymaloides, Slipper Flower, Redbird Cactus, Redbird Flower, Devil's Backbone, Japanese Poinsettia, Slipper Plant, Ribbon Cactus

Ricinus communis, Castor Bean, Castor Oil Plant, Wonder Plant, Palma Christi

Sapium sebiferum, Chinese Tallow Tree, Vegetable Tallow

Synadenium grantii, African Milk Bush

FABACEAE, Pea Family

Trifolium hybridum, Alsike Clover

HYDROPHYLLACEAE, Waterleaf Family

Phacelia brachyloba

Phacelia campanularia, Desert Bluebell

Phacelia crenulata

Phacelia grandiflora, Large-flowered Bluebell

Phacelia minor, California Bluebell

Turricula parryi [Nama parryi], Poodle-dog Bush

Wigandia caracasana

JUGLANDACEAE, Walnut Family

Juglans, Walnut

LAURACEAE, Laurel Family

Umbellularia californica, California Laurel, California Bay, California Olive, Oregon Myrtle, Pepperwood

LINACEAE, Flax Family

Linum usitatissimum, Flax

LOGANIACEAE, Logania Family

Gelsemium sempervirens, Carolina Jessamine, Yellow Jessamine, Carolina Jasmine, Yellow False Jasmine, Carolina Wild Woodbine, Wood Vine

MALVACEAE, Mallow Family

Alcea rosea [Althea rosea], Hollyhock

MORACEAE, Mulberry Family

Ficus carica, Fig

Ficus elastica, India Rubber Plant

Maclura pomifera, Osage Orange, Bow Wood, Hedge Apple

MYRTACEAE, Myrtle Family

Eucalyptus, Eucalypt, Eucalyptus

Melaleuca quinquenervia [has been confused with *M. leucadendron*], Cajeput, Paper-bark Tree, Punk Tree, Tea Tree

PLUMBAGINACEAE, Plumbago Family

Plumbago auriculata [*P. capensis*], Cape Plumbago

POLYGONACEAE, Buckwheat Family

Fagopyrum esculentum [*F. sagittatum*], Buckwheat

Rumex, Dock, Sorrel

PRIMULACEAE, Primrose Family

Anagallis arvensis, Pimpernel, Scarlet Pimpernel, Poor-man's-weatherglass, Shepherd's Clock

Primula obconica, German Primula, Poison Primula

PROTEACEAE, Protea Family

Grevillea banksii, Red-flowered Silky Oak, Dwarf Silky Oak

Grevillea robusta, Silky Oak, Silk Oak

RANUNCULACEAE, Buttercup Family

Anemone, Anemone, Windflower, Pasque Flower

Clematis, Clematis, Traveler's Joy, Old Man's Beard

Consolida ambigua [*Delphinium ajacis*], Rocket Larkspur

SAPINDACEAE, Soapberry Family

Cardiospermum halicabum, Balloon Vine, Blister Creeper

Sapindus drummondii, Western Soapberry, Drummond Soapberry

SIMAROUBACEAE, Quassia Family

Ailanthus altissima, Tree-of-heaven, Chinese Sumac

SOLANACEAE, Nightshade Family

Capsicum annuum, Chili Pepper

Lycopersicon lycopersicum [*L. esculentum*], Tomato

STERCULIACEAE, Sterculia Family

Brachychiton populneus, Kurrajong, Bottle Tree

THYMELAEACEAE, Mezereum Family

Dirca occidentalis, Western Leatherwood

TILIACEAE, Linden Family

Sparmannia africana, German Linden, African Linden, African Hemp

VITACEAE, Grape Family
> *Parthenocissus quinquefolia,* Virginia Creeper, Woodbine, American Ivy, Five-leaved Ivy
> *Parthenocissus tricuspidata,* Japanese Ivy, Boston Ivy

ZYGOPHYLLACEAE, Caltrop Family
> *Larrea tridentata* [*L. mexicana*], Creosote Bush

ANGIOSPERMS: MONOCOTYLEDONS

AGAVACEAE, Agave Family
> *Agave,* Agave, Century Plant
> *Furcraea,* Mauritius Hemp, Green Aloe, Cuban Hemp
> *Sansevieria,* Bow-string Hemp, Mother-in-law's Tongue, Devil's Tongue, Snake Plant, Good-luck Plant, Lucky Plant

AMARYLLIDACEAE, Amaryllis Family
> *Clivia,* Kaffir Lily
> *Crinum,* Crinum Lily, Spider Lily
> *Galanthus,* Snowdrop
> *Haemanthus,* Blood Lily, African Blood Lily
> *Hymenocallis,* Spider Lily, Basket Flower, Peruvian Daffodil, Sea Daffodil, Crown-beauty
> *Leucojum,* Snowflake
> *Narcissus,* Narcissus, Daffodil
> *Nerine,* Nerine

ARACEAE, Arum Family
> *Aglaonema commutatum*
> *Aglaonema costatum*
> *Aglaonema crispum* [*A. roebelinii*]
> *Aglaonema modestum,* Chinese Evergreen
> *Alocasia macrorrhiza,* Giant Elephant's Ear
> *Anthurium andreanum,* Florists' Anthurium, Flamingo Lily, Oilcloth Flower
> *Arum italicum,* Italian Arum
> *Arum maculatum,* Cuckoo Plant, Wake Robin, Lords-and-ladies
> *Caladium bicolor,* Fancy-leaved Caladium, Heart of Jesus
> *Colocasia esculenta* [*C. antiquorum*], Taro, Elephant's Ear, Kalo, Dasheen
> *Dieffenbachia maculata* [*D. picta*], Dumb Cane, Spotted Dumb Cane

Dieffenbachia seguine, Dumb Cane, Mother-in-law Plant

Dracunculus vulgaris

Epipremnum aureum [*Pothos aureus, Raphidophora aurea, Scindapsis aureus*], Pothos

Monstera deliciosa [*Philodendron pertusum*], Ceriman, Swiss-cheese Plant, Breadfruit Vine, Hurricane Plant, Fruit-salad Plant, Window Plant, Split-leaf Philodendron, Cut-leaf Philodendron

Philodendron bipennifolium [sometimes called *P. panduriforme* in error], Fiddle-leaf Philodendron, Horsehead Philodendron, Panda Plant

Philodendron bipinnatifidum

Philodendron domesticum [sometimes erroneously called *P. hastatum*], Spade-leaf Philodendron, Elephant's Ear Philodendron

Philodendron erubescens, Red-leaf Philodendron, Blushing Philodendron

Philodendron scandens, Heart-leaf Philodendron

Philodendron selloum, Selloum

Philodendron squamiferum

Philodendron verrucosum

Spathiphyllum

Xanthosma sagittifolium, Blue Elephant's Ear

Xanthosma violaceum, Blue Taro, Purple Elephant's Ear

Zantedeschia aethiopica, Calla Lily, Arum Lily, Florists' Calla, Garden Calla, Pig Lily, Trumpet Lily

COMMELINACEAE, Spiderwort Family

Commelina diffusa, Wandering Jew, Day-flower

Rhoeo spathacea [*R. discolor*], Purple-leaved Spiderwort, Oyster Plant, Boat Lily, Moses-in-a-boat, Moses-in-the-bulrushes, Two-men-in-a-boat

Setcreasea pallida 'Purple Heart' [*S. purpurea*], Purple Heart, Purple Queen, Purple Tradescantia

Tradescantia, Wandering Jew, Giant Inch Plant

IRIDACEAE, Iris Family

Iris foetidissima, Scarlet-seeded Iris, Stinking Iris, Gladwin, Stinking Gladwin

Iris germanica, Flag, Fleur-de-lis

Iris pseudacorus, Yellow Iris, Yellow Fleur-de-lis, Water Iris

LILIACEAE, Lily Family

Asparagus densiflorus 'Myers', Myers Asparagus; 'Sprengeri', Sprenger Asparagus, Emerald Fern, Emerald Feather

Asparagus officinalis, Garden Asparagus, Common Asparagus

Asparagus setaceus [*A. plumosus*], Asparagus Fern, Lace Fern

Convallaria majalis, Lily-of-the-valley, Mayflower, Conval Lily

Hyacinthus orientalis, Hyacinth, Dutch Hyacinth, Garden Hyacinth

Tulipa, Tulip

POACEAE, Grass Family

Cynodon dactylon, Bermudagrass, Devilgrass

Festuca, Fescue

Appendix B
Plants Causing Hay Fever and Asthma

Wind-borne spores and pollens of the plants listed here are reported as causing hay fever or asthma to humans. Considering the variations in human physiology, perhaps almost any plant can cause some kind of an allergic reaction to a particular individual. The list is adapted from Lewis and Elvin-Lewis (1977) and other sources.

ALGAE

CHLOROPHYTA, Green Algae
Chlorella
Chlorococcum

FUNGI

PHYCOMYCETES, Water Molds
Absidia
Cunninghamella
Mucor, Bread Mold
Plasmopara, Downy Mildew
Rhizopus, Bread Mold
Syncephalastrum
ASCOMYCETES, Sac Fungi
Chaetomium
Erysiphe, Powdery Mildew
Pleospora
Saccharomyces, Yeast
BASIDIOMYCETES, Club Fungi
Bullera, Jelly Fungi
Puccinia, Rusts
Sporobolomyces, Jelly Fungi
Tilletia, Smut
Uromyces, Rust
Ustilago, Smut

DEUTEROMYCETES, Imperfect Fungi
Alternaria
Aspergillus
Botrytis
Candida
Cladosporium [*Hormodendrum*]
Curvularia
Geotrichum
Helminthosporium
Monilia
Nigrospora
Penicillium
Trichoderma
Verticillium
and numerous other genera

GYMNOSPERMS

CUPRESSACEAE
Calocedrus, Incense Cedar
Chamaecyparis, False-cypress
Cupressus, Cypress
Juniperus, Juniper
Thuja, Arborvitae
GINKGOACEAE
Ginkgo biloba, Maidenhair Tree
PINACEAE
Pinus, Pine

ANGIOSPERMS: DICOTYLEDONS

ACERACEAE
Acer, Maple, Box Elder
AMARANTHACEAE
Amaranthus, Pigweed
ANACARDIACEAE
Pistacia, Pistachio
Schinus, Pepper Tree
APIACEAE
Anthriscus, Hedge Parsley
Heracleum, Hogweed, Cow Parsnip

ASTERACEAE
    Ambrosia [including *Franseria*], Ragweed, Burweed, Bur-brush
    Artemisia, Wormwood, Sagebrush
    Aster
    Callistephus, China Aster
    Chrysanthemum
    Dicoria
    Helianthus, Sunflower
    Hymenoclea, Burrobrush
    Iva, Marshelder, Povertyweed
    Solidago, Goldenrod
    Taraxacum, Dandelion
    Xanthium, Cocklebur

BETULACEAE
    Alnus, Alder
    Betula, Birch
    Carpinus, Hornbeam
    Corylus, Hazelnut
    Ostrya, Hop Hornbeam

BRASSICACEAE
    Brassica [*Sinapsis*], Charlock

CANNABACEAE
    Cannabis, Marijuana
    Humulus, Hop

CARICACEAE
    Carica, Papaya

CASUARINACEAE
    Casuarina, Beefwood

CHENOPODIACEAE
    Allenrolfea, Iodine Bush
    Atriplex, Saltbush
    Bassia
    Chenopodium, Goosefoot
    Eurotia, Winter Fat
    Grayia, Hop Sage
    Kochia, Summer Cypress
    Salicornia, Glasswort

Salsola, Russian Thistle
Sarcobatus, Greasewood
Suaeda, Sea-blite
DATISCACEAE
Datisca, Durango Root
EUPHORBIACEAE
Mercurialis, Mercury
FABACEAE
Acacia
Prosopis, Mesquite
FAGACEAE
Castanea, Chestnut
Castanopsis, Chinquapin
Fagus, Beech
Quercus, Oak
GARRYACEAE
Garrya, Silk-tassel Bush
HAMAMELIDACEAE
Liquidambar, Sweet Gum
JUGLANDACEAE
Carya, Pecan, Hickory
Juglans, Walnut
LAMIACEAE
Leonotis, Lion's Tail
MORACEAE
Broussonetia, Paper Mulberry
Maclura, Osage Orange
Morus, Mulberry
MYRICACEAE
Comptonia, Sweet Fern
Myrica, Wax Myrtle
MYRTACEAE
Eucalyptus, Eucalypt, Eucalyptus
OLEACEAE
Fraxinus, Ash
Ligustrum, Privet
Olea, Olive
Syringa, Lilac

PLANTAGINACEAE
  Plantago, Plantain
PLATANACEAE
  Platanus, Sycamore
POLYGONACEAE
  Fagopyrum, Buckwheat
  Rheum, Rhubarb
  Rumex, Sorrel, Dock
RANUNCULACEAE
  Ranunculus, Buttercup
ROSACEAE
  Rosa, Rose
  Spiraea
SALICACEAE
  Populus, Poplar
  Salix, Willow
SCROPHULARIACEAE
  Leucophyllum, Ceniza
  Verbascum, Mullein
SIMAROUBACEAE
  Ailanthus, Tree-of-heaven
TILIACEAE
  Tilia, Basswood
ULMACEAE
  Celtis, Hackberry
  Ulmus, Elm
URTICACEAE
  Parietaria, Pellitory
  Urtica, Nettle
VISCACEAE [LORANTHACEAE]
  Arceuthobium, Dwarf Mistletoe
  Phoradendron, Mistletoe

ANGIOSPERMS: MONOCOTYLEDONS

ARECACEAE
  Phoenix, Date Palm
CYPERACEAE
  Carex, Sedge
  Eleocharis, Spikerush

Eriophorum, Cottongrass
Scirpus, Bulrush

JUNCACEAE

Juncus, Rush
Luzula, Wood Rush

POACEAE

Agropyron, Wheatgrass
Agrostis, Bentgrass
Aira, Hairgrass
Ammophila, Beachgrass
Anthoxanthum, Vernalgrass
Aristida, Threeawn
Arundo, Giant Reed
Avena, Oats
Bouteloua, Grama
Briza, Quaking-grass
Bromus, Brome, Chess
Cynodon, Bermudagrass
Dactylis, Orchardgrass
Danthonia, Oatgrass
Digitaria, Crabgrass
Distichlis, Saltgrass
Echinochloa, Barnyardgrass
Elymus, Wildrye
Eragrostis, Lovegrass
Festuca, Fescue
Gastridium, Nitgrass
Hilaria, Galleta
Holcus, Velvetgrass
Hordeum, Barley
Koeleria, Junegrass
Lamarckia, Goldentop
Lolium, Ryegrass
Melica, Melic
Oryza, Rice
Oryzopsis, Ricegrass
Panicum, Millet
Paspalum, Knotgrass, Dallisgrass
Pennisetum, Fountaingrass

Phalaris, Canarygrass
Phleum, Timothy
Poa, Bluegrass
Polypogon, Rabbitfootgrass
Secale, Rye
Sitanion, Squirreltail
Sorghum, Johnsongrass, Sudangrass
Sporobolus, Dropseed
Stipa, Needlegrass
Triticum, Wheat
Zea, Corn
TYPHACEAE
Typha, Cattail

Appendix C
Plants Accumulating Nitrates

Listed here are plants found to accumulate free nitrates in quantities capable of causing death or distress to cattle. The list is adapted from Kingsbury (1964) and from analyses made by the Division of Chemistry, California Department of Food and Agriculture.

AMARANTHACEAE
Amaranthus blitoides, Prostrate Pigweed
Amaranthus graecizans, Tumbling Pigweed
Amaranthus retroflexus, Rough Pigweed

APIACEAE
Ammi majus, Bishop's Weed
Apium graveolens, Celery
Conium maculatum, Poison Hemlock
Daucus carota, Carrot
Pastinaca sativa, Parsnip

ASTERACEAE
Bidens frondosa, Beggar-ticks
Carduus nutans, Musk Thistle
Cirsium arvense, Canada Thistle
Gnaphalium purpureum [*Gamochaeta purpurea*], Purple Cudweed
Haplopappus venetus, Coast Goldenbush
Helianthus annuus, Common Sunflower
Helianthus tuberosus, Jerusalem Artichoke
Hemizonia pungens, Common Spikeweed
Lactuca sativa, Lettuce
Lactuca serriola, Prickly Lettuce
Rafinesquia californica, California Chicory
Silybum marianum, Milk Thistle
Solidago, Goldenrod

Sonchus asper, Prickly Sowthistle
Sonchus oleraceus, Common Sowthistle
Verbesina encelioides, Crownbeard

BORAGINACEAE

Amsinckia douglasiana, Douglas' Fiddleneck
Amsinckia intermedia, Common Fiddleneck
Plagiobothrys, Popcorn Flower

BRASSICACEAE

Brassica napus, Rape, Rutabaga
Brassica oleracea, Broccoli, Kale
Brassica rapa, Turnip
Descurainia pinnata, Tansymustard
Raphanus sativus, Radish
Thelypodium lasiophyllum, California Mustard

CAPPARACEAE

Cleome serrulata, Rocky Mountain Bee Plant

CAPRIFOLIACEAE

Sambucus mexicana, Southwestern Elderberry

CARYOPHYLLACEAE

Stellaria media, Chickweed

CHENOPODIACEAE

Atriplex serenana, Bracted Saltbush
Bassia hyssopifolia, Five-hook Bassia
Beta vulgaris, Sugar Beet, Mangel
Chenopodium album, Lamb's Quarters
Chenopodium ambrosioides, Mexican Tea
Chenopodium californicum, Soap Plant
Chenopodium murale, Nettleleaf Goosefoot
Kochia scoparia, Summer Cypress
Salsola australis, Common Russian Thistle

CONVOLVULACEAE

Convolvulus arvensis, Field Bindweed

CUCURBITACEAE

Cucurbita maxima, Hubbard Squash
Cucumis sativus, Cucumber

EUPHORBIACEAE

Euphorbia maculata, Prostrate Spotted Spurge

FABACEAE

Glycine max, Soybean
Melilotus officinalis, Yellow Sweetclover
Parkinsonia aculeata, Jerusalem Thorn

LINACEAE

Linum usitatissimum, Flax

MALVACEAE

Malva parviflora, Cheeseweed

POACEAE

Avena sativa, Oats
Bromus willdenovii, Rescuegrass
Echinochloa crus-galli, Barnyardgrass
Eleusine indica, Goosegrass
Hordeum vulgare, Barley
Panicum capillare, Witchgrass
Secale cereale, Rye
Sorghum bicolor, Grain Sorghum
Sorghum halepense, Johnsongrass
Sorghum sudanense, Sudangrass
Triticum aestivum, Common Wheat
Zea mays, Corn

POLYGONACEAE

Polygonum, Smartweed
Rumex, Dock

PORTULACACEAE

Montia perfoliata, Miner's Lettuce

SOLANACEAE

Datura, Jimsonweed
Solanum carolinense, Carolina Horsenettle
Solanum nigrum, European Black Nightshade

ZYGOPHYLLACEAE

Tribulus terrestris, Puncturevine

General References

Abrams, L., and R. S. Ferris. 1923–1960. *Illustrated flora of the Pacific States.* 4 vols. Stanford: Stanford University Press.

Arora, D. 1979. *Mushrooms demystified: A comprehensive guide to the fleshy fungi of the central California coast.* Berkeley: Ten Speed Press.

Bailey Hortorium, Staff of. 1976. *Hortus Third: A concise dictionary of plants cultivated in the United States and Canada.* Initially compiled by L. H. Bailey and E. Z. Bailey. New York: Macmillan.

Bailey, L. H. 1949. *Manual of cultivated plants.* Rev. ed. New York: Macmillan.

Blood, D. C., J. A. Henderson, and O. M. Radostits. 1979. *Veterinary medicine.* 5th ed. Baltimore: Williams & Wilkins.

Clarke, M. L., D. G. Harvey, and D. J. Humphreys. 1981. *Veterinary toxicology.* London: Baillière Tindall.

Cooper, M. R., and A. W. Johnson. 1984. *Poisonous plants in Britain and their effects on animals and man.* Reference Book 161, Ministry of Agric. Fisheries and Food. London: Her Majesty's Stationery Office.

Duffy, T. J., and P. P. Vergeer. 1977. *California toxic fungi.* San Francisco: Mycological Society of San Francisco.

Everist, S. L. 1981. *Poisonous plants of Australia.* Rev. ed. Sydney: Angus & Robertson.

Fowler, M. H. 1980. *Plant poisoning in small companion animals.* St. Louis: Ralston Purina Co.

Frohne, D., and H. J. Pfänder. 1984. *A colour atlas of poisonous plants.* Translated from the second German edition by N. G. Bisset. London: Wolfe.

Hardin, J. W., and J. M. Arena. 1974. *Human poisoning from native and cultivated plants.* 2nd ed. Durham, N.C.: Duke University Press.

Hulbert, L. C., and F. W. Oehme. 1984. *Plants poisonous to livestock: Selected plants of the United States and Canada of importance to veterinarians.* Manhattan: Kansas State University Printing Service.

Jones, T. C., and R. D. Hunt. 1983. *Veterinary pathology.* 4th ed. Philadelphia: Lea & Febiger

Keeler, R. F., K. R. Van Kampen, and L. F. James, eds. 1978. *Effects of poisonous plants on livestock.* Symposium on poisonous plants, Utah State University, 1977. New York: Academic Press.

Keeler, R. F., and A. T. Tu, eds. 1983. *Handbook of natural toxins.* Vol. 1: *Plant and fungal toxins.* New York: Marcel Dekker.

Kinghorn, A. D., ed. 1979. *Toxic plants*. New York: Columbia University Press.

Kingsbury, J. M. 1964. *Poisonous plants of the United States and Canada*. Englewood Cliffs, N.J.: Prentice-Hall.

Lampe, K. F., and R. Fagerström. 1968. *Plant toxicity and dermatitis*. Baltimore: Williams & Wilkins.

Lampe, K. F., and M. A. McCann. 1985. *AMA handbook of poisonous and injurious plants*. Chicago: Chicago Review Press.

Lewis, W. H., and M. P. F. Elvin-Lewis. 1977. *Medical botany: Plants affecting man's health*. New York: Wiley-Interscience.

Lincoff, G. 1981. *The Audubon Society field guide to North American mushrooms*. New York: Knopf.

Lincoff, G., and D. H. Mitchell. 1977. *Toxic and hallucinogenic mushroom poisoning: A handbook for physicians and mushroom hunters*. New York: Van Nostrand.

Mitchell, J., and A. Rook. 1979. *Botanical dermatology: Plants and plant products injurious to the skin*. Vancouver: Greenglass, Ltd.

Morton, J. F. 1971. *Plants poisonous to people*. Miami: Hurricane House.

Munz, P. A. 1968. *A California flora and supplement*. Berkeley and Los Angeles: University of California Press.

Munz, P. A. 1974. *A flora of southern California*. Berkeley and Los Angeles: University of California Press.

Parker, K. F. 1958. *Arizona ranch, farm, and garden weeds*. Cir. 265. Tucson: Agri. Ext. Serv., University of Arizona.

Pier, A. C. 1973. An overview of the mycotoxicoses of domestic animals. *J. Amer. Vet. Med. Assoc.* 163:1259–1261.

Rice, E. L. 1984. *Allelopathy*. 2nd ed. New York: Academic Press.

Robbins, W. W., M. K. Bellue, and W. S. Ball. 1951. *Weeds of California*. Sacramento: Documents & Publications, State of California.

Rosenthal, G. A., and D. H. Jansen, eds. 1979. *Herbivores: Their interaction with secondary plant metabolites*. New York: Academic Press.

Sampson, A. W., and H. E. Malmsten. 1942. *Stock-poisoning plants of California*. Rev. ed. Bull. 593. Berkeley: Agri. Expt. Sta., University of California.

Schmutz, E. M., and L. B. Hamilton. 1979. *Plants that poison*. Flagstaff, Ariz.: Northland Press.

Schmutz, E. M., B. N. Freeman, and R. E. Reed. 1968. *Livestock poisoning in Arizona*. Tucson: University of Arizona Press.

Schultes, R. E., and A. Hofmann. 1980. *The botany and chemistry of hallucinogens*. 2nd ed. Springfield, Ill.: Charles C Thomas.

Scimeca, J. M., and F. W. Oehme. 1985. Postmortem guide to common poisonous plants of livestock. *Vet. Hum. Toxicol.* 27:189–199.

Tyler, V. E., L. R. Brady, and J. E. Robbers. 1977. *Pharmacognosy.* 7th ed. Philadelphia: Lea & Febiger.

Vahrmeijer, J. 1981. *Poisonous plants of southern Africa that cause stock losses.* Cape Town: Tafelberg Pub., Ltd.

Vergeer, P. P. 1983. *Poisonous mushrooms.* In D. H. Howard, ed., *Fungi pathogenic for humans and animals.* New York: Marcel Dekker.

Watt, J. M., and M. G. Breyer-Brandwijk. 1962. *The medicinal and poisonous plants of southern and eastern Africa.* Edinburgh: E. & S. Livingston, Ltd.

Index of Common Names

PLANT FAMILIES, GENERA, AND SPECIES

Page number for figures are in italics.

Acacia, Catclaw, 143
Aconite, 2, 212–13, 214
Agaric, Fly, 28, 36–37
Agaricus, California, 42
Agave, 264
Agave Family, Agavaceae, 264, 374
Air Plant, 126
Akee, 327
Alder, 379
    Black, 224
Alfalfa, 121, 164–66, 172, 181,
    324, 361, 366
Algae, 11–14, 327, 377
    Green, 377
Allamanda, Yellow, 5, 76, 369
Almond, 225
    Bitter, 225–26, 319
    Indian, 252
Aloe, Barbados, 280, 281, 326, 355
    Bitter, 280
    Curaçao, 280, 355
    Green, 374
    Medicinal, 280
Amapola, 369
Amaranth Family, Amaranthaceae,
    64, 378, 384
Amaryllis Family, Amaryllidaceae,
    265, 268, 280, 328, 344, 374
Ammi, Greater, 69, 384
    Toothpick, 70; pl. 4a
Andromeda, 131
Anemone, 212, 216, 373, 375
    Japanese, 216
    Poppy-flowered, 216
Angelica Tree, 79
Angel's Trumpet, 232, 233–34,
    235, 238, 336
    Red, 4, 234; pl. 13a
    White, 234
Anise, 355
Anthurium, Florists', 270, 374
Apple, 225
    Hedge, 373

Kangaroo, 247–48; pl. 14c
    Mad, 239
Apricot, 225, 226, 319
Aralia, Balfour, 370
    Geranium-leaf, 370
Arborvitae, 378
Arrowgrass, 7, 278–80
    Common, 278
    Marsh, 278, 279
    Seaside, 278
    Three-ribbed, 278
Arrowgrass Family, Juncaginaceae,
    278
Artichoke, Jerusalem, 384
Arum, Italian, 270, 374
Arum Family, Araceae, 269–70,
    273, 315, 355, 374
Ash, 380
    Prickly, 79
    Wafer, 227
Asparagus, Common, 376
    Garden, 281, 376
Aster, 356, 370, 379
    China, 379
    Woody, 309
Aster Family, Asteraceae, 83, 309,
    349, 370, 379, 384
Aubergine, 248
Aucuba, Japanese, 124
Avocado, 183–84
Azalea, 129, 131
    Mock, 76
    Western, 132; pl. 8c

Baby Jade, 371
Baileya, Desert, 84
Balloon Vine, 228, 273
Balm, Field, 181
Balsamroot, 85
    Arrow-leaved, 85
Baneberry, 214
    Red, 212, 214, 215
Barbados Nut, 139, 372

Barbados Pride, 157
Barberry, 101
Barberry Family, Berberidaceae, 101, 122, 337, 342
Barley, 18, 21, 90, 107, 293, 310, 330, 382, 386
Barnyardgrass, 365, 382, 386
Bassia, Fivehook, 118, 315, 379, 385
Basswood, 381
Bay, California, 184, 349, 372
Bay Leaf, 185
Beachgrass, 382
Bead, Indian, 142
    Seminole, 142
Bean, Black, 158
    Broad, 142, 172
    Castor, 2, 5, 140–41, 260, 339, 347–48, 372; pl. 9b, d
    Coffee, 2
    European, 172
    Fava, 172
    Field, 172
    Horse, 172
    Jack, 142
    Kidney, 142
    Lima, 168
    Love, 142
    Lucky, 142
    Mescal, 170–71
    Prayer, 142
    Precor, 142
    Sieva, 168
    Windsor, 172
Bearberry, 224, 355
Beauty-of-the-night, 197
Beech, 380
    American, 174
    European, 174
Beech Family, Fagaceae, 174
Beefsteak Plant, 371
Beefwood, 379
Bee Plant, Rocky Mountain, 385
Beet, 118
    Forage, 118
    Sugar, 307, 385
Beggar-ticks, 384
Belladonna, 232, 246, 336
    Cape, 268
Bellflower Family, Campanulaceae, 185
Bentgrass, 295, 382

Bermudagrass, 9, 18, 22–23, 86, 249, 274, 293, 316, 376, 382
Be-still Tree, 78, 370
Betel, 338
Betty, Sweet, 117
Bignonia Family, Bignoniaceae, 370
Bindweed, Field, 320, 385
Birch, 379
Birch Family, Betulaceae, 379
Bird-of-paradise Bush, 156, 157
Bird-of-paradise Flower, 156
Bird's-nest Plant, 369
Bishop's Weed, 69, 70, 384
Bittersweet, 117, 247
    American, 117
    Climbing, 117
Bitterweed, 88
    Fragrant, 89
    Western, 89
Bladder Flower, White, 81
Blanketflower, 370
Bleedingheart, Western, 177
Blister Creeper, 373
Bloodberry, 206
Blood Flower, 82
Bluebell, California, 372
    Desert, 372
    English, 284
    Large-flowered, 372
Bluegrass, 21, 383
    Kentucky, 295
Blue-green Algae, 11–13, 327
Blue Witch, 251
Blusher, The, 43–44
Bolete, 24, 44–45
Bootlace Mushroom, 44
Borage Family, Boraginaceae, 102, 318, 332, 385
Bottle Tree, 252, 373
Bouncing Bet, 117, 324
Bow Wood, 373
Box, 108
    African, 241
    Victorian, 207
Boxwood, 108
    African, 195
Boxwood Family, Buxaceae, 108
Bracken, 56
Bracken Fern, 6, 56–59, 57; pl. 2a
Brain Fungus, 31
Brake, 56
Breadfruit Vine, 271, 375

Bread Mold, 377
Bristlegrass, 293, 301
    Yellow, 301
Brittlebush, 370
Broccoli, 385
Brome, 382
Bromegrass, 21, 293
Broom, 158
    Easter, 338, 356
    French, 158
    Scale, 89–90, 370
    Scotch, 158, 338, 356
    Spanish, 337
Broomshrub, 370
Broomweed, 87
Buckeye, 178
    California, 175, 178–79
Buckeye Family, Hippocastanaceae,
    178
Buckthorn, 102, 103, 223, 326,
    356, 359
    Alder, 224
    Alder-leaved, 224
    Common, 224
    Purging, 224
Buckthorn Family, Rhamnaceae,
    223, 325, 326
Buckwheat, 208, 361, 373, 381
Buckwheat Family, Polygonaceae,
    208, 325, 373, 381, 386
Bulrush, 382
Bunny Ears, 109
Burbrush, 370, 379
Burclover, California, 165
Burdock, 354
Burning Bush, 120
Burrobrush, 379
Burweed, 370, 379
Bushman's Poison, 76, 264
Buttercup, 212, 221–22, 381
    Bermuda, 199–200
    Bur, 222–23
    Creeping, 222
Buttercup Family, Ranunculaceae,
    212, 324, 329, 342, 373, 381
Butterfly Flower, 81–82
Butternut, 181
Butterweed, 370
    Shrubby, 92–93
Buttonbush, 226–27
Buttonwillow, 226–27

Cabbage, 106, 325, 327
    Cow, 288
    Skunk, 288
Cacao, 337, 346
Cactus, Beavertail, 109
    Candelabra, 135–36, 371
    False, 135–36, 371
    Hatrack, 135–36, 371
    Redbird, 140, 372
    Ribbon, 140, 372
    Unguentine, 280
Cactus Family, Cactaceae, 109
Cajeput, 196–97, 373
Caladium, Fancy-leaved, 270, 374
Calamus, 355
Calfkill, 129
Calico Bush, 129
Calla, Florists', 272, 375
    Garden, 272, 375
Caltrop Family, Zygophyllaceae,
    262, 374, 386
Calycanthus Family, Caly-
    canthaceae, 111
Camas, 290, 291
Camphor Tree, 182–83, 314–15,
    356
Canarygrass, 383
    Mediterranean, *229–301*
Cape Lilac, 191
Caper Family, Capparaceae, 113,
    324, 385
Cardinal Flower, 185
Carrot, 69
    Wild, 74, 331, 369, 384
Cassava, 318
Cassine, 79
Castor Oil Plant, 140, 372
Casuarina Family, Casuarinaceae,
    379
Catalina, 135
Catclaw, 143
    Devil's, 143
    Gregg's, 143
Catnip, 358
Cattail, 383
    Red, 371
    Red-hot, 371
Cattail Family, Typhaceae, 383
Cauliflower Ears, 125, 371
Cedar, 190
    Incense, 378
    White, 191

Celeriac, 369
Celery, 69, 316, 369, 384
    pink rot of, 69, 321
Ceniza, 381
Centaury, June, 177
Century Plant, 264, 374
Ceriman, 271, 375
Cestrum, 236
    Day, 236
    Day-blooming, 236
    Orange-flowered, 236; pl. 13b
    Red, 236
Chalice Vine, 245–46
Chamise, 363, 364
Chamomile, 86, 90, 356, 370
    English, 356
    Garden, 86, 356
    German, 356, 358
    Hungarian, 356, 358
    Roman, 356
    Russian, 86, 356
    Sweet, 356
    Sweet False, 90, 356, 358
Chanterelle, 35, 47
Charas, 112
Chard, 118
Charlock, 320, 379
Chat, 330
Cheeseweed, 189–90, 386
Chenille Plant, 371
Cherry, 319
    Flowering, 225
    Holly-leaved, 225
    Jerusalem, 249–50; pl. 14d
    Sour, 225
    Sweet, 225
    Western Choke, 225; pl. 11b
Chess, 382
Chestnut, 380
    Horse, 178
    Moreton Bay, 158
    Red Horse, 178
Chicalote, 201
Chicken-of-the-woods, 49
Chickweed, 385
Chicory, California, 384
China Ball Tree, 191
Chinaberry Tree, 191, 351
China Tree, 191
Chincherinchee, 286
Chinese Evergreen, 270, 374
Chinese Jade, 371

Chinese Tallow Tree, 372
Chinquapin, 380
Chokecherry, Western, 6, 225;
    pl. 11b
Cholla, 109
Christ Plant, 137, 371
Christ Thorn, 137, 371
Christmas Berry, 225
Christmas Berry Tree, 65, 369
Christmas Flower, 138, 371
Christmas Star, 138, 371
Chrysanthemum, 356, 370, 379
Cineraria, 92
Citrus Family, Rutaceae, 227–28,
    320
Clotbur, Spiny, 99–100, 101, 370
Clover, 360
    Alsike, 171, 372
    Berseem, 172
    Bur, 361
    Ladino, 171
    Red, 171, 320
    Subterranean, 171–72, 200,
      277
    White, 171
Club Fungi, 377
Club Moss, 329
Club Moss Family, Lycopodiaceae,
    329
Clustered Woodlover, 39, 50
Coal-oil-brush, 97
Coca, 336–37
Cocoa, 337, 346
Coca Family, Erythroxylaceae, 336
Cockle, Corn, 116–17, 323–24
    Purple, 116
Cocklebur, 7, 99–101, 100, 370,
    379; pl. 6b
Coffee, 159, 246, 346, 352, 358
    Wild, 370
Coffee Berry, American, 159
Coffee Tree, 370
    Kentucky, 159, 338
Cola, 353
Cola Nut, 251
Comfrey, 105–6, 231, 354
    Common, 105, 359–60
    Prickly, 105
    Quaker, 105, 359–60
    Russian, 105, 359–60
Conkers, 178
Conquerors, 178

Copa-de-oro, 245–46
Copperleaf, 371
    Painted, 371
Copperweed, 89, 370
Coral-bead Plant, 142
Coralberry, 214
Coral Tree, 158–59
Coriaria Family, Coriariaceae, 123
Corn, 16, 303–4, 307, 310, 383,
    386
    Indian, 303
    Wild Indian, 288
Corynocarpus Family, Corynocar-
    paceae, 124
Cotoneaster, 225
Cotton, 188–89
Cottongrass, 382
Couchgrass, 9
Cow Cabbage, 288
Cow-itch, 370
Cow Poison, 217, 218
Coyotillo, 223
Crabgrass, 382
Crab's Eye, 142
Crazyweed, 143, 167
Creeping Charlie, 181
Creosote Bush, 262, 374
Cress, Indian, 254
    Winter, 106
Crimson Shower, 370
Crocus, Autumn, 281–82
    Fall, 281
Crofton Weed, 87
Croton, 371
Crowfoot, 221–22
    Celeryleaf, 222–23
    Cursed, 222–23
Crownbeard, 98–99, 385
    Golden, 98
Crown-beauty, 374
Crown Fungus, 32
Crown-of-thorns, 137–38, 371; pl. 8d
Cruel Plant, 81
Cubeb, 359
Cuckoo Plant, 270, 374
Cucumber, 385
Cudweed, Purple, 384
Cup-of-gold Vine, 245–46
Cycad Family, Cycadaceae, 61
Cyclamen, 211
Cymopterus, 74
Cypress, 378

Monterey, 61
    Summer, 118, 210–11, 367,
        379, 385
Cypress Family, Cupressaceae, 61,
    349, 378

Daffodil, 268, 269, 374
    Peruvian, 374
    Sea, 374
Daisy, African, 90–91
    Butter, 98
    Freeway, 91
    Trailing African, 91
Dallisgrass, 18, 21–22, 382
Damiana, 360
Dandelion, 379
Daphne, February, 253
    Garland, 253
    Winter, 253; pl. 15b
Darnel, 297–98
Dasheen, 270, 374
Datisca Family, Datiscaceae, 127, 380
Datura, Sacred, 238
    Tree, 233–34
Day-flower, 375
Deadly Hemlock. See Poison
    Hemlock
Death Angel, 30
Deathcamas, 175, 289–92, 345
    Chaparral, 290
    Meadow, 5, 7, 290–91; pl. 16b
Death Cap, 29–30; pl. 1c
Desert Velvet, 91
Destroying Angel, 30
Devilgrass, 293, 376
Devil's Backbone, 126, 140, 372
Devil's Catclaw, 143
Devil's Plague, 369
Devil's Tongue, 374
Devil's Walking Stick, 79
Dinoflagellate, 13–14
Dittany, 227
Divine Coca, 336
Dock, 210, 373, 381, 386
Dodder, 121
Dogbane, Spreading, 76
Dogbane Family, Apocynaceae, 75,
    369
Dog Brush, 97
Dogwood Family, Cornaceae, 124
Dollar Plant, 125, 371
Doll's Eyes, 214

Dove Weed, 133
Downy Mildew, 377
Dragon Bones, 135, 371
Dropseed, 383
Dumb Cane, 270, 273–74, 374, 375
    Spotted, 270, 273–74, 374;
    pl. 15c
Durango Root, *127*–28, 380; pl. 8b
Dusty Miller, 92
Dwarf Bay, 253

Earthballs, 24, 53
Easter Flower, 138, 371
Eggplant, 248–49
Elder, 114
    Box, 369, 378
Elderberry, 114–16
    Southwestern, *115*, 385
Elephant's Ear, 270, *271*, 374
    Blue, 272, 375
    Giant, 270, 375
    Purple, 272, 375
Elfin Saddle, California, 32
Elm, 381
Elm Family, Ulmaceae, 381
Emerald Feather, 376
Encelia, 370
Ephedra Family, Ephedraceae, 330
Ergot, *18*–23, 326, 327, 329, 340
Eucalypt, 195–96, 373, 380
Eucalyptus, 195–96, 373, 380
Euphorbia, Abyssinian, 372
    Rubber, 139, 372
Eyebane, 371
Eyebright, 353

Fairy Ring Mushroom, 34
False-cypress, 378
Fence Flower, 157
Fern, Asparagus, 376
    Bracken, 6, 56–59, *57*, 367
    Emerald, 316
    Lace, 376
Fescue, 21, 376, 382
    Chewings, 294–95
    Reed, 294
    Tall, 294
Fetterbush, 131
Fiddleneck, 6, *102*, 103, 307, 385;
    pl. 7a
    Douglas', 102, 385
Field Balm, 181

Fig, 193–94, 372
    Indian, 109
Figwort Family, Scrophulariaceae,
    230, 381
Filaree, 177–78
Finger Tree, 139, 372
Fire-dragon, 371
Fire-on-the-mountain, 134, 371
Fire Plant, Mexican, 134, 371
Firethorn, 225, 226
Fireweed, 102
    Rancher's, 103
Fitweed, 6, 175–77, *176*
Flag, 275, 375
    Sweet, 355
Flax, 185, 372, 386
    Prairie, 185
Flax Family, Linaceae, 185, 372
Flax Olive, 253
Fleur-de-lis, 275, 375
    Yellow, 275, 375
Flixweed, 107–8
Flor del Peru, 78, 370
Fountaingrass, 383
Four-o'clock, *197*–98
Four-o'clock Family, Nyctaginaceae,
    197
Foxglove, Common, 5, 230–31,
    354; pl. 12c
Foxtail, 371
    Bristly, 301
    Yellow, 301
Frangipani, 369
Friar's Cap, 213
Fruit-salad Plant, 271, 375
Fuller's Herb, 117
Fumitory Family, Fumariaceae, 175,
    342
Fungus, Brain, 31
    Crown, 32
    Oak Root, 44
    Panther, 37
    Veiny Cup, 32

Galleta, 382
    Big, 296
    James', 18, 22
    Woolly, 296
Gallow Grass, 111
Garland Flower, 253
Gas Plant, 227
Gentian Family, Gentianaceae, 177

Geranium Family, Geraniaceae, 177
Ghost Weed, 136, 371
Giant Inch Plant, 375
Giant Kelp, 14
Gidee-gidee, 142
Gill-over-the-ground, 181
Ginkgo Family, Ginkgoaceae, 369, 378
Ginseng Family, Araliaceae, 79, 370
Gladwin, Stinking, 275, 375
Glasswort, 379
Goat Head, 262
Gold Cup, 245–46
Goldenbush, Coast, 384
Golden Buttons, 96
Golden Carpet, 371
Golden Chain, 159, 338
Golden Dewdrop, 256–57
Golden Ear-drops, 177
Golden Rain, 338
Goldenrod, 379, 384
    Basin, 96
Golden Seal, 353
Golden Shower, 157, 326
Goldentop, 382
Golden Trumpet, 76, 369
Gold Moss, 371
Good-luck Plant, 374
Gooseberry, Barbados, 244
    Cape, 244
Goosefoot, 306, 379
    Nettleleaf, 385
Goosefoot Family, Chenopodiaceae, 117, 337, 379, 385
Goosegrass, 386
Gopher Plant, 136, 371
Gordoloba Yerba, 334, 353–54
Gota Kola, 353
Gotu Kola, 353
Gourd Family, Cucurbitaceae, 126, 385
Grama, 382
Grape, Holly, 101
    Oregon, 101
Grape Family, Vitaceae, 260, 374
Grapefruit, 227
Grass, 11, 370
Grass, Gallow, 111, 370
Grass Family, Poaceae, 292–93, 376, 382, 386
Greasewood, 121, 380
    Black, 7, 121; pl. 7d

Gregg's Catclaw, 143
Grisette, 44
Groundcherry, 242–45
    Lanceleaf, 244, 245
Goundsel, 91–92, 370
    Common, 94–96, 95
Guarana, 358
Guildfordgrass, 277
Gum, Blue, 196
    Manna, 196
    Sugar, 196
    Tasmanian Blue, 196
Gum Tree, 195–96, 373, 380
Gumweed, 87
Gyromitra, Hooded, 32

Hackberry, 381
Hairgrass, 382
Halogeton, 6, 118–20, 119; pl. 7c
Harding-grass, 298–301
Harebell, 284
Harmel, Common, 262
Hartshorn, 224
Hawkweed, 320
Haymaker's Mushroom, 39
Hazelnut, 379
Headache Tree, 184, 349
Heart-of-Jesus, 270, 374
Heath Family, Ericaceae, 128–29, 350
Heliotrope, 104, 360
    Alkali, 104
    Common, 105
    European, 105
    Seaside, 104
Hellebore, 221, 280, 288
    California False, 288
    Corsican, 221
    Del Norte False, 288
    European White False, 289
    False, 345
    Fringed False, 288
    Green False, 288, 289
    Stinking, 221
    Swamp, 288
Helmet Flower, 213
Helvella, Cabbage-leaf, 32
Hemlock, Deadly, 72
    Poison, 2, 5, 68, 69, 72–74, 335, 384; pl. 3a,b
    Spotted, 69, 72
    Water. See Waterhemlock

Hemp, 111–13, *112*, 370;
  pl. 8*a*
    African, 373
    American, 76
    Bow-string, 374
    Cuban, 374
    Dogbane, 76
    Indian, 76, 111, 370
    Mauritius, 374
Hemp Family, Cannabaceae, 111,
  370, 379
Henbane, 2, 311, 336
    Black, 240, 332
Herb Christopher, 214
Hercules' Club, 79
Hickory, 380
Hoary-cress, Heart-podded, 107
Hogweed, 378
    Giant, 69
Holly, 79
    Box-leaved, 79
    Burford, 79, *80*
    Chinese, 79, *80*
    English, 80
    European, 80
    Japanese, 80
Holly Family, Aquifoliaceae, 79,
  346, 357
Holly Grape, 101
Hollyhock, 372
Honeybush, 192
Honey Flower, Cape, 192–93
    Tall Cape, 192–93
Honey Mushroom, 38, 44
Honeysuckle, Tatarian, 114
Honeysuckle Family, Caprifoliaceae,
  114, 385
Hop, 113, 357, 370, 379
Hopbush, 228–29
Hop Hornbeam, 379
Hopseed, 228–29
Hopseed Bush, 228–29
Hop Tree, 227
Horsebrush, Gray, 97
    Littleleaf, 7, 97–98; pl. 6*d*
    Mojave, 98
    Smooth, 97
    Spineless, 97–98; pl. 6*c*
Horsenettle, Carolina, 386
    White, 247
Horseradish, 106, 205, 325
Horsetail, 59, 329, 357
    Common, 59
    Field, 59, 339
Horsetail Family, Equisetaceae, 59,
  329, 357
Horseweed, 370
Hound's Tongue, 104
Hurricane Plant, 271, 375
Hyacinth, 285–86, 376
    Dutch, 285, 376
    Garden, 285, 376
    Star, 287
Hyacinth-of-Peru, 287
Hydrangea, 357
    Bigleaf, 229
    French, 229
    Garden, 229

Imperfect Fungi, 378
Incienso, 370
Indian Rubber Plant, 372
Indian Tree Spurge, 139, 372
Inky Cap, 28, 32–33, 314
Iodine Bush, 379
Iris, 275, 351
    Butterfly, 276
    Scarlet-seeded, 275, 375
    Stinking, 275, 375
    Water, 275, 375
    Wild, 276
    Yellow, 275–76, 375
Iris Family, Iridaceae, 274, 375
Ivy, Algerian, 81, 370
    American, 260, 374
    Boston, 261, 374
    Canary Islands, 81, 370
    Cape, 94
    English, 80–81, 370
    Five-leaved, 260, 374
    German, 94
    Ground, 181–82, 190
    Japanese, 261–62, 374
    Madeira, 81, 370
    Mountain, 129
    Nepal, 370
    Parlor, 94
    Persian, 81, 370
    Poison, 65
    Water, 94
Ivy Bush, 129

Jacinth, Peruvian, 287
Jackassclover, 113–14

Jack O'Lantern Mushroom, 35
Jacob's Coat, 371
Jade Plant, 125, 371
Jade Tree, 371
Jalap, 122, 198
    Wild, 122
Jamberry, 244
Jamestown Weed, 239
Japanese Lantern, 243
Jasmine, Carolina, 186, 372
    Yellow False, 186
Jelly Fungi, 377
Jerusalem Thorn, 386
Jessamine, Carolina, 186–88, *187*, 340, 372
    Night-blooming, 236–37
    Willow-leaved, 236–37
    Yellow, 186, 372
Jetbead, 225
Jimsonweed, 5, 232, 235, 239, 332, 356–57, 386; pl. 13*c*
Johnny-jump-up, 259
Johnsongrass, 302–3, 383, 386
Junegrass, 382
Jungle-rice, 365
Juniper, 357–58, 378

Kale, 325, 385
Kalmia, Bog, 129
Kalo, 270, 374
Karaka, 124
Kava, 359
Kava-Kava, 359
Kelp, Giant, 14
Khat, 330
Kikuyugrass, 298
Klamathweed, 6, 179–80, 360–61; pl. 10*b, d*
Knapweed, Russian, 86
Knotgrass, 18, 22, 382
Kola, 353
Kotolo, 82
Kotu Cola, 353
Kurrajong, 252, 373
Kyllinga, 274

Labrador Tea, Western, 130
Laburnum, 159
Lamb's Quarters, 385
Lantana, Creeping, 258
    West Indian, 257–58
Lantern Plant, Chinese, 243

Larkspur, 213, 217–20
    Alkali, 218
    Anderson's, 218
    Annual, 216–17
    Candle, 219
    Chinese, 219
    Coast, 218
    Gypsum, 218
    Menzies', 218–*19*
    Mountain, 218
    Parry's, 218
    Poison, 218
    Rocket, 216–17, 373
    Royal, 218; pl. 12*b*
    Smooth, 218
    Western, 6, 218
Laurel, Alpine, 129
    Black, 130
    Bog, 129
    California, 184–85, 372
    Cherry, 225
    Grecian, 185
    Japanese, 124
    Mountain, 129
    New Zealand, 124
    Pale, 129
    Portugal, 225
    Sierra, 130
    Spurge, 253
    Texas Mountain, 170–71
Laurel Family, Lauraceae, 182, 372
Laurel-wicky, Dwarf, 129
Lawn Leaf, 371
Lawn Mower's Mushroom, 39
Leatherwood, Western, 253–54, 350, 373
Lentil, 142
Lepiota, Ragged, 49
Lettuce, 370, 384
    Bitter, 89
    Miner's, 386
    Prickly, 358
    Wild, 89, 358
Lettuce-opium, 89, 358
Leucothoë, Drooping, 130
Licorice, Indian, 142
Life Plant, 126
Lilac, 380
Lily, African, 266
    Arum, 272, 375
    Barbados, 268
    Belladonna, 268

Lily (*continued*)
  Blood, 374
  Blue African, 266
  Blue Flax, 283–84
  Boat, 375
  Calla, 272, 375
  Checkered, 284
  Climbing, 284
  Conval, 282, 376
  Corn, 7, 288; pl. 16*a*
  Crinum, 374
  Cuban, 287
  Flamingo, 270, 374
  Gloriosa, 284–*85*
  Glory, 284
  Kaffir, 374
  Natal, 276
  Pig, 272, 375
  Spider, 374
  Trumpet, 272, 375
Lily Family, Liliaceae, 265, 280, 325, 328, 345, 376
Lily-of-the-Nile, 266
Lily-of-the-valley, 282–*83*, 376
Lily-of-the-valley Shrub, 131
Lime, 227
  Rangpur, 227
Linden, African, 383
  German, 373
Lion's Tail, 380
Live-for-ever, 371
Lobelia, Scarlet, 185–86
Lobelia Family, Lobeliaceae, 185
Lobster Plant, 138, 371
Loco, Douglas, 147
  Gray, *144,* 149
  Horse, 146–*47*; pl. 9*c*
  Morton's, 152
  Sheep, *144,* 148
  Spotted, 6, 148
Locoweed, 143–44, 146, 167–68
  Threadleaf, *144,* 153
Locust, Black, 169–70, 347–48; pl. 9*f*
Logania Family, Loganiaceae, 186, 372
Loosestrife Family, Lythraceae, 188
Loquat, 225
Lords-and-ladies, 270, 374
Lovegrass, 382
Lucerne, 164
Lucky Nut, 78, 370

Lucky Plant, 374
Lupine, 161–64, 311, 314, 337, 338
  Broadleaf, 6, 161; pl. 9*c*
  Dense-flowered, 164
  Douglas' Annual, 164
  False, 339
  Forage, 337
  Grassland, 161
  Intermountain Low, 162
  Loosely-flowered Annual, 164
  Plumas, 162
  Pursh's Silky, 162
  Small-flowered Annual, 164
  Tailcup, 161
  Velvet, 161–62
  Watson's Bush, 164
  Western, 162
  Woodland, 162
  Woolly-leaf, 162
Lupulo, 113, 370
Lychee, 228

Madder Family, Rubiaceae, 226, 344
Maguey, 264
Mahogany, Mountain, 225
Mahogany Family, Meliaceae, 190
Maidenhair Tree, 369, 378
Maize, 16, 303
Malabar Tree, 139, 372
Mallow, Alkali, 190
  Bristly, 190
  Red-flowered, 190
  Small-flowered, 189
  Wheel, 190
Mallow Family, Malvaceae, 188, 372, 386
Mandrake, European, 232, 336
Mangel, 118, 385
Mango, 65
Mannagrass, Fowl, 295
Manna Gum, 196
Manzanita, Eastwood, 363–64
Maple, 369, 378
  Red, 64
  Scarlet, 64
  Swamp, 64
Maple Family, Aceraceae, 64, 369, 378
Marigold, 96, 370
  Colorado Desert, 84
  Desert, 84–85

French, 96
Marsh, 212, 216
Wild, 96
Woolly, 84
Marijuana, 111–13, *112,* 127, 239,
 351, 370, 379; pl. 8*a*
Marshelder, 370, 379
Marvel-of-Peru, 197
Marygold, 96
Mastic Tree, Peruvian, 65, 369
Match-me-if-you-can, 371
Matchweed, 87
Maté, 346, 357
Matrimony Vine, 240–41
Maui Pamakani, 87
Mayflower, 282, 376
Meadow Foam Family, Limnantha-
 ceae, 324
Meadowgrass, Fowl, 295
Meadow Mushroom, 41
Melianthus Family, Melianthaceae,
 192
Melic, 382
Melon, Paddy, 126–27
Mercury, Annual, 139, 380
Mescal Button, 109–10; pl. 7*b*
Mesquite, 380
Mexican Devil, 87
Mexican Fire Plant, 134, 135
Mexican Flame Leaf, 138, 371
Mexican Flame Vine, 138, 371
Mezereum, 253
Mezereum Family, Thymelaeaceae,
 252, 373
Mignonette Family, Resedaceae, 324
Milkbush, 139, 372
Milk Bush, African, 372
Milk Caps, 48, 52
Milk Purslane, 371
Milk Tree, African, 372
Milkvetch, Anderson, 151
 Balloon, 155
 Basalt, *144,* 155
 Bolander, 151–52
 Bush, 154
 Cleveland, 153
 Field, 150–51
 Freckled, 6, 148–49
 Gibbs, 153–54
 Klamath, 152
 Lambkill, 146
 Layne, 154

Naked, 154
Preuss', 145
Rattle-box, *144,* 145, 308, 309
Rogue River, 150
Short-toothed Canada, 152
Sickle, *144,* 153
Sylvan, 155
Torrey, 152
Two-keeled, 151
Webber, 155
Milkweed, 81–83
 Indian, 82
 Narrowleaf, 82
 Purple, 82
 Showy, 82, 83; pl. 5*b*
 Whorled, 6, 82, 83; pl. 5*a*
Milkweed Family, Asclepiadaceae,
 81, 322
Millet, 382
Mint Family, Lamiaceae, 181, 380
Miracle Leaf, 126
Mistletoe, 358–59
 Common, 259
 Dwarf, 381
 European, 259, 260, 327, 330
 Hairy, 259
Mistletoe Family, Viscaceae, 259
Mole Plant, 136, 371
Molle, 65, 369
Monkshood, 212–13, 217
 Garden, 213
 Western, *213;* pl. 12*a*
Moonseed Family, Menispermaceae,
 193
Morel, 24, 50, 54
 Black, 50
 False, 27, 31–32
Moringa Family, Moringaceae, 324
Morning-glory, 122–*23,* 340
 Woolly, 123
Morning-glory Family, Convolvula-
 ceae, 121, 371
Morning-noon-and-night, 235
Moses-in-a-boat, 375
Moses-in-the-bulrushes, 375
Mother-in-law Plant, 270, 273–74,
 375
Mother-in-law's Tongue, 374
Moth Plant, White, 81
Mountain Snow, 136, 371
Mower's Mushroom, 39
Mugwort, 370

Mulberry, Paper, 380
Mulberry Family, Moraceae, 193, 372, 380
Mullein, 381
Mushroom, Bootlace, 44
    Fairy Ring, 34
    Haymaker's, 39
    Honey, 44
    Jack O'Lantern, 35
    Lawn Mower's, 39
    Meadow, 41
    Mower's, 39
    Panther, 37
    Parasol, 49
        Green-gilled, 45–46
        Green-spored, 45–46
    Salmon Coral, 52
    Shoestring, 44
    Sweating, 34
Mustard, Black, 365
    California, 307, 385
    Field, 385
    Tansy, 107, 385
    Wild, 107, 320
Mustard Family, Brassicaceae, 106, 254, 324–25, 379, 385
Myoporum Family, Myoporaceae, 194
Myrsine Family, Myrsinaceae, 195
Myrtle, Cape, 195
    Oregon, 184, 372
    Wax, 380
Myrtle Family, Myrtaceae, 195, 373, 380
Mysteria, 281

Naked Lady, 267, 268
Nap-at-noon, 286
Narcissus, 268, 374
Nasturtium, Garden, 254
    Tall, 254
Nasturtium Family, Tropaeolaceae, 254, 324
Nectarine, 225
Needlegrass, 293, 295, 383
Nerine, 374
Nettle, 255–56, 381
    Brewer's, 255
    Burning, 255
    California, 255
    Creek, 255
    Hoary, 255

    Lyall's, 255
    Roman, 256
    Serra, 255
    Small, 255
    Stinging, 4, 255–56
    Western, 254–55
Nettle Family, Urticaceae, 254, 381
Ngaio, 194–95
Nightshade, 246–47
    American Black, 246–47
    Climbing, 247
    Deadly, 232, 331, 332, 336; pl. 12d
    Douglas', 247
    European Black, 249, 346, 386
    Hairy, 250; pl. 15a
    Jasmine, 247
    Silverleaf, 247
    Stinking, 240
    Woolly, 248
Nightshade Family, Solanaceae, 232, 373, 386
Nitgrass, 382
Nutmeg, 195, 358
Nutmeg Family, Myristicaceae, 195, 358

Oak, 7, 174–75, 314, 317, 380
    Blue, 175
    California Black, 175
    Dwarf Silky, 211, 373
    Pacific Poison, 5, 65–68, 66, 136, 179, 317, 350, 369; pl. 2c, d
    Red-flowered Silky, 211, 373
    Silk, 211, 373
    Silky, 211, 373
    Western Poison. See Oak, Pacific Poison
Oak Root Fungus, 44
Oat, 312, 316, 382, 386
    Wild, 21, 365
Oatgrass, 382
Oilcloth Flower, 270, 374
Old-man-of-the-spring, 94
Old Man's Beard, 216, 373
Oleander, 5, 77–78; pl. 4c
    Yellow, 78; pl. 4d
Olive, 380
    California, 184, 372
Olive Family, Oleaceae, 198, 380
Onion, 266
    False Sea, 286
    German, 286

Pregnant, 286
Sea, 286, 287–88
Swamp, 267
Wild, 266–67, 289
Oniongrass, 277
Orache, Red, 118
Orange, 227
Bergamot, 227
Mock, 207
Osage, 373, 380
Sour, 227
Trifoliate, 227–28
Orange Glow Vine, 92
Orchardgrass, 21, 295, 382
Orchid Family, Orchidaceae, 320, 332
Oxeye, 88
Oyster Plant, 375

Paddy Melon, 126–27
Pagoda Tree, 369
Painted Leaf, 134, 371
Palm, Areca, 338
Date, 381
Japanese Sago, 61
Sago, 61
Palma Christi, 140, 372
Palm-Beach-bells, 125
Palm Family, Arecaceae, 338, 381
Panaeolus, Bell-cap, 39
Belted, 40
Girdled, 40
Panda Plant, 271, 375
Pansy, Wild, 259
Panther Cap, 37
Panther Fungus, 37
Panther Mushroom, 37; pl. 1a
Papaya, 379
Papaya Family, Caricaceae, 324, 379
Paper-bark Tree, 196, 373
Paper Flower, 91
Parasol Mushroom, 31, 49
Green-gilled, 45–46
Green-spored, 45–46
Parsley, Hedge, 378
Poison, 72
Spring, 74
Water, 68, 350
Parsley Family, Apiaceae, 68–69, 316, 320, 321, 335, 349, 350, 369
Parsnip, 68, 69, 384

Cow, 378
Cutleaf Water, 68, 70
Water, 68
Pasque Flower, 216, 373
Passion Flower, 358
Passion Flower Family, Passifloraceae, 358
Paxillus, Poison, 51
Pea, Austrian Winter, 160
Everlasting, 160
Garden, 142
Glory, 170
Grass, 160
Indian, 160
Jequirity, 142–43
Rosary, 4, 142–43, 347–48; pl. 9a
Rough, 160
Singletary, 160
Sweet, 160
Tangier, 160
Wild, 159, 327
Peach, 225
Pea Family, Fabaceae, 141–42, 318, 325, 329, 332, 337, 347, 351, 372, 380, 386
Pear, Prickly, 109
Pecan, 380
Pedalium Family, Pedaliaceae, 204
Pellitory, 381
Pencil Tree, 139, 372
Pennyroyal, 182, 315
Pepper, Baby, 206
Black, 335
Chili, 236, 373
White, 335
Peppercorn, pink, 65
Pepper Family, Piperaceae, 335, 338
Pepper Tree, 65, 369, 378
Brazilian, 65, 369
California, 65, 369
Peruvian, 65, 369
Pepperwood, 184, 372
Periwinkle, 77, 356; pl. 4b
Rose, 340
Peyote, 109–10, 171, 265, 330; pl. 7b
False, 110
Peyotl, 110
Pheasant's-eye, Summer, 214–16
Philippine Medusa, 371
Philodendron, Blushing, 271, 375
Cut-leaf, 271, 375
Elephant's Ear, 271, 375

Philodendron (*continued*)
    Fiddle-leaf, 271, 375
    Heart-leaf, 271–72, 375
    Horsehead, 271, 375
    Red-leaf, 271, 375
    Spade-leaf, 271, 375
    Split-leaf, 271, 375
Pholiota, Scaly, 52
Physic Nut, 139, 372
Pie Plant, 208
Pieris, Japanese, 131
    Mountain, 131
Pigeonberry, 204, 256–57
Pig's Ears, 124
Pigweed, 378
    Prostrate, 384
    Redroot, 64
    Rough, 64, 384
    Tumbling, 384
Pimpernel, Scarlet, 210–11, 373
Pine, 378
    Ponderosa, 61–62
    Western Yellow, 61
Pine Family, Pinaceae, 61–62, 378
Pink, Hedge, 117
    Indian, 185
Pink Berries, 65
Pink Crown, 32
Pink Family, Caryophyllaceae, 116
Pistachio, 378
Pittosporum, 206–8
    Japanese, 207
    Willow, 207
Pittosporum Family, Pittosporaceae, 206
Plane Tree Family, Platanaceae, 381
Plantain, 381
Plantain Family, Plantaginaceae, 381
Plum, 225, 319
    Cherry, 225
    Japanese, 225
Plumbago, Cape, 208, 373
Plumbago Family, Plumbaginaceae, 208, 373
Poha, 244
Poinciana, Dwarf, 157
    Paradise, 156
Poinsettia, 132, 138
    Annual, 134, 371
    Dwarf, 134, 371
    Japanese, 140, 372
Poison Hemlock, 2, 5, *68*, 69, 72–74, 335, 384; pl. 3*a,b*

Poison Nut, 139, 372
Poison Pie, 47–48
Poison Sego, 289
Poison Weed, 217
Poke, 204
    Indian, 288
Pokeberry, 204
Pokeweed, 5, 204–6, 348; pl. 11*a*
Pokeweed Family, Phytolaccaceae, 204
Polypody Family, Polypodiaceae, 56–59, 60
Polypore, 24
Poodle-dog Bush, 179, 372
Poor-man's-weatherglass, 210, 373
Popcorn Flower, 103, 385
Poplar, 381
Poppy, California, 201–2, 342, 357
    Celandine, 201
    Mexican, 201
    Opium, 1, 201, 202–4, *203*, 329, 342, 343; pl. 11*d*
    Prickly, 201, 342, 355; pl. 11*c*
    Prickly Mexican, 201
    Thorn, 201
Poppy Family, Papaveraceae, 200–201, 329, 337, 342
Pot, 11, 370
Potato, 329
    Irish, 250–51
    Sweet, 122, 352
    White, 250–51
Potato Creeper, Giant, 251
Potato Family, Solanaceae, 232, 329, 335–36, 345, 373, 386
Potato Vine, 247
Pothos, 270, 375
Povertyweed, 370, 379
Powdery Mildew, 377
Pride-of-India, 191
Pride-of-Madeira, 104
Primrose Family, Primulaceae, 210, 373
Primula, German, 373
    Poison, 373
Prince's Plume, 309
    Golden, 108
    Green, 108
    Panamint, 108
Privet, 198–99, 380
    Japanese, *198;* pl. 10*c*
Protea Family, Proteaceae, 211, 351, 373

Psilocybe, Dung, 40
Puffballs, 24
    Hard-skinned, 53
    Pigskin Poison, 53
Puncturevine, 262–64, *263*, 386
Punk Tree, 196, 373
Purging Nut, 139, 372
Purple Heart, 375
Purple Queen, 375
Purslane, 210
Purslane Family, Portulacaceae, 210, 386
Pyracantha, 225, 226

Quaking-grass, 382
Quassia Family, Simaroubaceae, 231, 373, 381
Queen Anne's Lace, 369

Rabbitbrush, 86
    Rubber, 86
    Spring, 97
Rabbitfootgrass, 383
Radish, 385
Ragweed, 83, 356, 370, 379
    Giant, 83
    Short, 83
    Western, 83, *84*
Ragwort, 93
    Tansy, 7, 60, 93–94, 334; pl. *6a*
Ranger's Button, 74–75
Ranunculus, Persian, 221–23
    Turban, 221–22
Rape, 325, 385
Rat Brush, 97
Rattlebox, 170
Rattleweed, 143
Red-bean Vine, 142
Redbird Flower, 140, 372
Reed, Giant, 382
Rescuegrass, 386
Resinweed, 87
Rhododendron, West Coast, 131–32
Rhubarb, 2, 326, 381
    Garden, 208–*9*, 315
Rice, 382
Ricegrass, 382
Rose, 381
    Christmas, 221
    Desert, 76
    Lenten, 221
Rose-bay, California, 131–32

Rose Family, Rosaceae, 225–26, 318, 381
Rouge Plant, 206
Rubberweed, Bitter, 89
Rue, 227
    African, 262
    Syrian, 262
Rue Family, Rutaceae, 227–28, 320
Runaway Robin, 181
Rush, 277, 362, 382
    Brown-headed, 277–78
    Common Scouring, 59
    Scouring, 59
    Smooth Scouring, 59
    Wood, 382
Rush Family, Juncaceae, 277, 382
Russula, Emetic, 52
Rust, 377
Rusty Leaf, 130–31
Rutabaga, 385
Rye, 20–21, 383, 386
Ryegrass, 21, 382
    Darnel, 297–98
    Italian, 297
    Perennial, 60, 297

Sac Fungi, 377
Saffron, Meadow, 281
Sage, Purple, 364
    Red, *257*–58
    Yellow, 257
Sagebrush, 370, 379
St. Anthony's Fire, 20
St. Johnswort, Common, 179
St. Johnswort Family, Hypericaceae, 179
Salmon Coral Mushroom, 52
Saltbush, 379
    Bracted, 385
    Fourwing, 118
    Nuttall's, 118
Saltgrass, 367, 382
Sanicle, Poison, 74
Sassafras, 184, 359
Saxifrage Family, Saxifragaceae, 229
Scourwort, 117
Sea-blite, 380
Sedge, 381
Sedge Family, Cyperaceae, 274, 381
Selloum, 272, 375
Seminole Bead, 142
Senna, 157

Senna (*continued*)
    Alexandria, 355–56
    Tinnevelly, 355–56
Sesame, 204
Shavegrass, 357
Shepherd's Clock, 210, 373
Shoestring Mushroom, 44
Sickener, The, 52
Silk-tassel Bush, 380
Silk-tassel Family, Garryaceae, 380
Silkweed, 81–82
Silver Dollar, 125, 371
Skoke, 204
Skunk Cabbage, 288
Skyflower, 256–57
Sleepygrass, 303
Slipper Flower, 140, 372
Slipper Plant, 140, 372
Smartweed, 386
Smoke Bush, 369
Smoke Tree, 369
Smut, 23, 377
Snake Berry, 214
Snake Plant, 374
Snake's Head, 284
Snakeweed, 72
    Broom, 87
    Threadleaf, 87
Sneezeweed, Bitter, 88
    Common, 88
    Fine-leaved, 88
    Orange, 88–89, pl. 5*d*
    Western, 88
Snowberry, 116, 373
Snowdrop, 374
Snowflake, 374
    Summer, 286
Snow-on-the-mountain, 136–*37*, 371
Soapbark Tree, 323–24
Soapberry, 323
    Drummond, 229, 373
    Western, 229, 373
Soapberry Family, Sapindaceae, 228, 318, 373
Soap Plant, 385
Soapwort, 117
Soldier's Cap, 213
Sorghum, 301–2
    Grain, 301–2, 365, 386
Sorrel, 210, 373, 381
    Lady's, 199

Sowthistle, Common, 385
    Prickly, 385
Soybean, 328, 347, 386
Spicebush, Western, 111
Spiderwort, Purple-leaved, 375
Spiderwort Family, Commelinaceae, 375
Spikerush, 381
Spikeweed, Common, 384
Spindle Tree, European, 117
Spotted Hemlock. *See* Poison Hemlock
Sprouting Leaf, 126
Spurge, 133–38
    Candelabra, 135, 371
    Caper, 136, 371
    Cypress, 134–35
    Fiddler's, 134, 371
    Indian Tree, 139, 372
    Leafy, 135
    Mottled, 135, 371
    Petty, 138, 371
    Prostrate Spotted, 371, 386
    Red, 134, 371
Spurge Family, Euphorbiaceae, 4, 132, 253, 350, 351, 371, 380, 386
Spurge Olive, 253
Squash, Hubbard, 385
    Zucchini, 126
Squaw Bush, *66*
Squill, 280, 287–88
    Red, 287–88
Squirrel Food, 289
Squirreltail, 383
Staff-tree Family, Celastraceae, 117, 330
Stargrass, 9
Star-of-Bethlehem, 286
Starthistle, Yellow, 6, 85–86; pl. 5*c*
Sterculia Family, Sterculiaceae, 251, 337, 346, 373
Stinging Tree, 256
Stinkgrass, 293–94
Stinking Willie, 93
Stinkweed, 239
    Poison, 72
Stinkwort, 239
Stonecrop Family, Crassulaceae, 124, 371
Subclover, Subterranean Clover, 171–72, 200, 277

Sudangrass, 302–3, 304, 383, 386
Sulfur Shelf, 49
Sulfur Tuft, 50
Sumac, Chinese, 231–32, 373
    Poison, 65
    Venetian, 369
Sumac Family, Anacardiaceae, 65,
    317, 350, 369, 378
Sunflower, 379, 384
    False, 88
    Swamp, 88
Swan Plant, 82
Sweating Mushroom, 34
Sweet Betty, 117
Sweetclover, 166, 167, 320
    Annual Yellow, 166
    White, 166
    Yellow, 166, 386
Sweet Fern, 380
Sweet Gum, 380
Swiss-cheese Plant, 271, 375
Sycamore, 381
Sycamore Family, Platanaceae, 381

Tanglehead, 293, 295–96; pl. 15c
Tansy, 96, 97, 370
Tansymustard, 107, 385
Taro, 270, 374
    Blue, 272, 375
Tarweed, 102, 103
Tea, 346
    Chamomile, 86, 90, 356, 358
    China, 352
    Herbal, 352–60
    Labrador, 130
    Mexican, 385
    Mormon, 357
    Paraguay, 346
    Squaw, 357
    Trapper's, 130
    Western Labrador, 130
Tea Family, Theaceae, 346
Tea Tree, 196, 373
Temple Tree, 369
Thevetia, Giant, 78, 370
Thimble Cap, 54
    Wrinkled, 54
Thistle, Barnaby's, 85
    Canada, 384
    Flowering, 201
    Jamaica, 201
    Milk, 306, 384

Musk, 384
    Russian, 380, 385
    Spanish, 99
Thornapple, 238–39, 336, 356–57
    Chinese, 238
    Common, 239
    Desert, 238
    Downy, 238
Thoroughwort, White, 87
Threeawn, 293, 382
Timothy, 383
Titoki, 228
Tobacco, 186, 232, 241–42, 329,
    331, 339
    Coyote, 242
    Desert, 242
    Flowering, 241–42
    Indian, 186, 335, 353
    Jasmine, 241–42
    Tree, 5, 6, 242, 243, 339;
        pl. 13d
Tobira, 207
Tolguacha, 238–39; pl. 14a,b
Tomatillo, 244
Tomato, 241, 329, 373
    Cherry, 244
    Gooseberry, 244
    Husk, 244
    Mexican Husk, 244
    Strawberry, 243
Tonga, 235
Toyon, 225
Tradescantia, Purple, 375
Traveler's Joy, 216, 373
Tree-of-Heaven, 231–32, 373, 381
Trefoil, 366
    Birdsfoot, 160–61
    Narrow Birdsfoot, 160–61
Tricholoma, Soapy, 54
Truffles, 24
Trumpet Creeper, 370
Trumpet Honeysuckle, 370
Trumpet Plant, 245–46
Trumpet Vine, 370
Tulip, 287, 376
    Cape, 274, 287
    Cape Blue, 276
    Guinea-hen, 284
Tulp, 274, 287
    Blue, 276
Tung Oil Tree, 132–33
Turkey Mullein, 133

Turnera Family, Turneraceae, 360
Turnip, 385
Turpentineweed, 87
Two-men-in-a-boat, 375

Umbrella Tree, Australian, 80
    Queensland, 80
    Texas, 191
Uva-ursi, 355

Valeriana, 360
Valerian Family, Valerianaceae, 360
Vegetable Tallow, 372
Veiny Cup Fungus, 32
Velvetgrass, 296–97, 382
Velvet Rosettes, 91
Verbena, Lemon, 256
Vernalgrass, Sweet, 320, 382
Vervain Family, Verbenaceae, 256
Vetch, Green, 160
    Spanish, 160
Vetchling, 159–60, 327
Violet, English, 258
    Florists', 258
    Garden, 258
    Sweet, 258–59
Violet Family, Violaceae, 258
Virginia Creeper, 260–*61*, 374

Wake Robin, 270, 374
Walnut, 180–81, 372, 380
    Black, 180–81
    Eastern Black, 181
    English, 181
    Persian, 181
Walnut Family, Juglandaceae, 180,
    372, 380
Wandering Jew, 375
Wartweed, 371
Waterhemlock, 70–72, 73, 316,
    349–50
    Western, 5, 6, *68*, 71; pl. 3*c, d*
Waterleaf Family, Hydrophyllaceae,
    179, 372
Water Molds, 377
Waxberry, 116
Waxy Caps, 48
Wheat, 21, 308, 310, 316, 365,
    383, 386
Wheatgrass, 382
White Heads, 74–75
    Swamp, 74–75

Wig Tree, 369
Wildrye, 382
Willow, 381
Willow Family, Salicaceae, 381
Windflower, 216, 373
Window Plant, 271, 375
Wine Plant, 208
Wintercherry, 243, 244
Winter Fat, 379
Wintersweet, 76
Wisteria, Chinese, 173–74
    Japanese, 142, *173*
    Silky, 173
Wisteria Tree, Scarlet, 170; pl. 10*a*
Witchgrass, 386
Witch Hazel Family, Hamamelida-
    ceae, 380
Witch's Hat, 48
Wolfbane, Garden, 213
Wolf's Milk, 135
Wonder Bulb, 281
Wonder Flower, 286
    African, 286
Wonder Plant, 140, 372
Woodbine, 114, 260, 374
    Carolina Wild, 186, 372
Wood Rose, Hawaiian Baby, 123
Wood Rush, 382
Wood Vine, 186, 372
Woodsorrel, 199
    Creeping, 199
Woodsorrel Family, Oxalidaceae,
    199, 315
Wormwood, 355, 370, 379

Yam, 122
Yarrow, 320, 349, 370
Yaupon, 79
Yeast, 377
Yellow Star, 88
Yesterday-and-today, 235
Yesterday-today-tomorrow, 235
Yew, English, 5, 62–63, 358
    Irish, 63; pl. 2*b*
    Japanese, 62–63
    Western, 63
Yew Family, Taxaceae, 62
Yohimbe, 356

Zacatechichi, 355
Zigadene, 289–92

Index of Scientific Names

PLANT FAMILIES, GENERA, AND SPECIES

Page numbers for figures are in italics.

Abrus precatorius, 4, 142–43, 169, 348; pl. 9*a*
Absidia, 377
Acacia, 317, 380
    greggii, 143
Acalypha hispida, 371
    wilkesiana, 371
Acer, 369, 378
    rubrum, 64
Aceraceae, 64, 369, 378
Achillea millefolium, 320, 349, 370
Acokanthera, 322
    oblongifolia, 76
    oppositifolia, 76, 264
Aconitum, 212–13, 217, 219, 345
    columbianum, *213*; pl. 12*a*
    napellus, 213–14
Acorus calamus, 355
Actaea rubra, 212
    rubra subsp. *arguta*, 214–*15*
    spicata, 212, 214
Adenium obesum, 76
Adenostoma fasciculatum, 363, 364
Adonis aestivalis, 214–16
Aesculus, 178–79, 320
    californica, 175, 178–79
    × *carnea*, 178
    hippocastanum, 178
Agapanthus africanus, 266
    orientalis, 266
    umbellatus, 266
Agaricus, 41, 49, 317
    bisporus, 41
    californicus, 42
    campestris, 41, 53
    hondensis, 42
    meleagris, 42
    praeclaresquamosus, 42
    placomyces, 42
    xanthodermis, 43; pl. 1*d*
Agavaceae, 264, 374
Agave, 264–65, 374
    americana, 264–65

    atrovirens, 265
    attenuata, 264
    tequilana, 265
Ageratina adenophora, 87
Aglaonema commutatum, 270, 374
    costatum, 270, 374
    crispum, 270, 374
    modestum, 270, 374
    roebelinii, 270, 374
Agropyron, 382
Agrostemma githago, 116–17, 323–24
Agrostis, 382
Ailanthus altissima, 231–32, 373, 381
Aira, 382
Alcea rosea, 372
Alectryon excelsus, 228
Aleurites fordii, 132
Allamanda cathartica, 76, 369
Allenrolfea, 379
Allium canadense, 266
    cepa, 266
    vallidum, 267
Alnus, 379
Alocasia macrorrhiza, 270, 374
Aloë, 355
    barbadensis, 280, 281, 326, 355
    perfoliata var. *vera*, 280
    vera, 280, 355
Aloysia triphylla, 256
Alternaria, 378
Althea rosea, 372
Amanita, 27, 28, 29, 30, 35, 43–44, 327
    chlorinosma, 43
    citrina, 340
    cokeri, 43
    mappa, 340
    muscaria, 5, 28, 33, 35, *36*, 37; pl. 1*b*
    ocreata, 5, 30

Amanita (continued)
    pantherina, 5, 33, 35, 37, 44;
      pl. 1*a*
    phalloides, 5, 25–26, 29–30;
      pl. 1*c*
    rubescens, 43–44
    solitaria, 43
    vaginata, 44
Amaranthaceae, 64, 378, 384
Amaranthus, 378
    blitoides, 384
    graecizans, 384
    retroflexus, 64, 384
Amaryllidaceae, 265, 268, 280,
    328, 344, 374
Amaryllis belladonna, 267, 268
Ambrosia, 83, 370, 379
    artemisiifolia, 83
    elatior, 83
    psilostachya, 83, *84*
    trifida, 83
Ammi majus, 69, 384
    visnaga, 70; pl. 4*a*
Ammophila, 382
Amsinckia, 102, 105, 332, 333
    douglasiana, 6, 102, 307, 385
    intermedia, 102, 103, 307,
      385; pl. 7*a*
Anabaena, 11–13
    flos-aquae, 11, *12*
Anacardiaceae, 65, 317, 350, 369,
    378
Anacystis cyanea, 11
Anagallis arvensis, 210–11, 373
Anemone, 222, 373, 375
    blanda, 212, 216
    × *hybrida,* 216
Anthemis nobilis, 86, 356, 370
Anthoxanthum odoratum, 320, 382
Anthriscus, 378
Anthurium andreanum, 270, 374
Aphanizomenon, 11
Apiaceae, *68–69,* 316, 320, 321,
    335, 349, 350, 353, 369, 378,
    384
Apium, 321
    graveolens, 69, 369, 384
Apocynaceae, 75, 322, 329, 369
Apocynum androsaemifolium, 76–77
    cannabinum, 76
Aquifoliaceae, 79, 346, 357
Araceae, 269–70, 273, 315, 355,
    374

Araliaceae, 79, 370
Aralia spinosa, 79
Araujia sericofera, 81
Arceuthobium, 259, 381
Arctium lappa, 354
Arctostaphylos glandulosa var.
    zacaensis, 363–64
    uva-ursi, 355
Arctotis, 90
Areca catechu, 338
Arecaceae, 338, 381
Argemone, 342
    mexicana, 201, 355
    munita, 201; pl. 11*c*
    platyceras, 201
Argyreia nervosa, 123
Ariocarpus, 110
Aristida, 293, 382
Armillaria mellea, 38, 44
Armoracia rusticana, 106, 325
Artemisia, 98, 370, 379
    absinthium, 349, 355
Arum italicum, 270, 374
    maculatum, 270, 374
Arundo, 382
Asclepiadaceae, 81, 322
Asclepias, 81–82, 322
    cordifolia, 82
    curassavica, 82
    eriocarpa, 82
    fascicularis, 6, 82; pl. 5*a*
    fremontii, 82
    fruticosa, 82
    physocarpa, 82
    speciosa, 82; pl. 5*b*
Asparagus densiflorus, 376
    officinalis, 281, 376
    plumosus, 376
    setaceus, 376
Aspergillus, 16, 167, 316, 378
    flavus, 17
    parasiticus, 17
Aster, 309, 370, 379
Asteraceae, 83, 309, 349, 370, 379,
    384
Astragalus, 143–56, 167, 309
    accidens var. *hendersonii,* 150
    agnicidus, 146
    agrestis, 150–51
    andersonii, 151
    asymmetricus, 146, *147;* pl. 9*c*
    bicristatus, 151
    bolanderi, 151–52

californicus, 152
calycosus var. *calycosus*, 152
canadensis var. *brevidens*, 152–53
clevelandii, 153
crotalariae, *144*, 145, 308, 309
curvicarpus var. *curvicarpus*, *144*, 153
dasyglottis, 150
douglasii, 147–48, 149
    var. *douglasii*, 147
    var. *parishii*, 147
filipes, *144*, 153
gibbsii, 153–54
hornii, *144*, 148
layneae, 154
lentiginosus, 6, 148–49
    var. *borreganus*, 148
leucophyllus, 146
menziesii, 149
miser, 149
mortoni, 152
mortonii, 152
nuttallii, *144*, 149
pachypus, 154
preussii, 145–46, 309
serenoi var. *serenoi*, 154
umbracticus, 155
webberi, 155
whitedii, 153
whitneyi, 155
Atriplex, 309, 379
    canescens, 118
    nuttallii, 118
    rosea, 118
    serenana, 385
Atropa belladonna, 232–33, 240, 331, 332, 336; pl. 12*d*
Aucuba japonica, 124
Avena, 382
    fatua, 365
    sativa, 386

Baileya, 84–85
    multiradiata, 84
    pauciradiata, 84
    pleniradiata, 84
Balsamorhiza sagittata, 85
Barbarea vulgaris, 106
Bassia hyssopifolia, 118, 315, 379, 385
Berberidaceae, 101, 122, 337, 342
Berberis vulgaris, 101

Berula erecta, *68*, 70
    thunbergii, 70
Beta vulgaris, 118, 307, 385
Betula, 379
Betulaceae, 379
Bidens frondosa, 384
Bignoniaceae, 370
Blighia sapida, 327
Boletus, 44–45
    erythropus, 45
    piperitus, 45
    satanus, 45
Boraginaceae, 102, 318, 332, 385
Botrytis, 378
Bouteloua, 382
Brachychiton populneus, 252, 373
Brassaia actinophylla, 80
Brassica, 106, 327, 379
    campestris, 325
    hirta, 325
    kaber, 107, 320
    napus, 325, 385
    nigra, 325, 365
    oleracea, 325, 385
    rapa, 385
Brassicaceae, 106, 254, 324–25, 379, 385
Briza, 382
Bromus, 293, 365, 382
    willdenovii, 386
Broussonetia, 380
Brugmansia, 232, 233–35, 238, 336
    × *candida*, 234
    sanguinea, 4, 234–35; pl. 13*a*
    suaveolens, 234
Brunfelsia pauciflora, 235
Bryophyllum calycinum, 126
    pinnatum, 126
Bullera, 377
Buxaceae, 108
Buxus sempervirens, 108–9
Byttneriaceae, 337, 346

Cactaceae, 109
Caesalpinia gilliesii, *156*–57
    pulcherrima, 157
Caladium bicolor, 270, 374
Calea zacatechichi, 355
Callistephus, 379
Calocedrus, 378
Calomeria, 370
Calotropis, 322

Caltha, 222
 palustris, 212, 216
Calycanthaceae, 111
Calycanthus occidentalis, 111
Camassia, 290
Camellia sinensis, 346, 352
Campanulaceae, 185
Campsis radicans, 370
Canavalia ensiformis, 142
Candida, 378
Cannabaceae, 111, 370, 379
Cannabis, 379
 indica, 111
 sativa, 111–13, *112,* 351, 370;
  pl. 8*a*
Cantharellus cibarius, 35, 47
 floccosus, 47
Capparaceae, 113, 324, 385
Caprifoliaceae, 114, 385
Capsicum annuum, 236, 373
Cardaria draba, 107
Cardiospermum grandiflorum, 228
 halicabum, 373
Carduus nutans, 384
Carex, 381
Carica, 379
Caricaceae, 324, 379
Carpinus, 379
Carya, 380
Caryophyllaceae, 116, 385
Cassia, 158, 326
 acutifolia, 355–56
 angustifolia, 355–56
 didymobotrya, 157
 fistulosa, 157, 326
 nairobensis, 157
Castanea, 380
Castanopsis, 380
Castanospermum australe, 158
Casuarina, 59, 379
Casuarinaceae, 379
Catha edulis, 330
Catharanthus roseus, 77, 340–41,
 356; pl. 4*b*
Cedrella, 190
Celastraceae, 117, 330
Celastrus scandens, 117
Celtis, 381
Centaurea repens, 86
 solstitialis, 6, 85–86; pl. 5*c*
Centaurium beyrichii, 177
 calycosum, 177
 floribundum, 177

Cephaelis acuminata, 344
 ipecacuanha, 344
Cephalanthus occidentalis, 226–27
Ceratocephalus testiculatus, 222
Cercocarpus, 225
Cestrum aurantiacum, 236; pl. 13*b*
 diurnum, 236
 elegans, 236
 fasciculatum, 236
 nocturnum, 236–37
 parqui, 237
 purpureum, 236
Chaetomium, 377
Chamaecyparis, 378
 thyoides, 349
Chamaemelum nobile, 86, 90, 356
Chamomilla recutita, 89, 356, 370
Chelidonium, 342
 majus, 201
Chenopodiaceae, 117, 337, 379,
 385
Chenopodium, 306, 379
 album, 385
 ambrosioides, 385
 californicum, 385
 murale, 385
Chlorella, 377
Chlorococcum, 377
Chlorophyllum molybdites, 6,
 45–46
Chrysanthemum, 349, 370
Chrysothamnus nauseosus, 86
Cicuta, 69, 70–72, 316, 349–50
 douglasii, 5, 6, *68,* 71; pl.
  3*c,d*
 maculata, 71
  var. *angustifolia,* 71
  var. *bolanderi,* 71
  var. *maculata,* 71
Cinchona calisaya, 341–42
Cinnamomum camphora, 182–83,
 314–15, 356
Cirsium arvense, 384
Citrus aurantifolia, 227
 aurantium, 227
 aurantium 'Bouquet', 227
 × *limonia,* 227
 × *paradisi,* 227
Cladosporium, 378
Claveria formosa, 52
Claviceps, 16, 18–23, 329
 cineraria, 18, 22
 cynodontis, 23

paspali, 18, 21–22
purpurea, *18*, 21, 23, 326,
   327, 340
Clematis, 212, 216, 222, 373
Cleome serrulata, 385
Clitocybe, 28, 33–34
   dealbata, 6, 34–35
Clivia, 268, 374
Clostridium, 15
Cocculus laurifolius, 193
Codiaeum variegatum, 371
Coelosphaerium, 11
Coffea, 346
Cola, 346
   acuminata, 251, 353
   nitida, 251, 353
Colchicum autumnale, 281, *282*
Colocasia antiquorum, 270, 374
   esculenta, 270, *271*, 374
Commelinaceae, 375
Commelina diffusa, 375
Compositae. *See* Asteraceae
Comptonia, 380
Conium maculatum, 5, *68*, 69,
   72–74, 335, 384; pl. 3*a,b*
Conocybe, 29
   filaris, 35, 304
Consolida ambigua, 216–17, 373
Convallaria, 322
   majalis, 282–*83*, 286, 376
Convolvulaceae, 121, 371, 385
Convolvulus arvensis, 320, 385
Conyza, 370
Coprinus atramentarius, 28, 32–33
Coriaria japonica, 123
   nepalensis, 123
   ruscifolia, 123
Coriariaceae, 123
Cornaceae, 124
Cortinarius, 46–47
   camphoratus, 47
   orellanus, 46
   purpurascens, 47
   traganus, 47
   vibritalis, 47
Corydalis caseana, 6, 175–77,
   176, 342
Corylus, 379
Corynanthe yohimbe, 356
Corynocarpaceae, 124
Corynocarpus laevigatus, 124
Coryphantha, 110
Cotinus coggygria, 369

Cotyledon cacalioides, 125
   orbiculata, 124–25
   wallichii, 125
Crassula arborescens, 125, 371
   argentea, 125, 371
   ovata, 125, 371
Crassulaceae, 124, 371
Crinum, 268, 378
Crocus sativus, 281
Crotalaria, 170, 333
Croton, 132, 371
   tiglium, 350
Cruciferae. *See* Brassicaceae
Cucumis myriocarpus, 126–27
   sativus, 385
Cucurbitaceae, 126, 385
Cucurbita maxima, 385
   pepo 'Zucchini', 126
Cunninghamella, 377
Cupressaceae, 61, 349, 378
Cupressus, 349, 378
   macrocarpa, 61
Curvularia, 378
Cuscuta, 121
Cuscutaceae. *See* Convolvulaceae
Cycadaceae, 61
Cycas revoluta, 61
Cyclamen europaeum, 211
   purpurascens, 211
Cymopterus watsonii, 74
Cynodon dactylon, 9, 18, 23, 293,
   376, 382
Cynoglossum officinale, 104
Cyperaceae, 274, 381
Cyperus brevifolius, 274
Cytisus, 158
   canariensis, 356
   monspessulanus, 158
   × *racemosus*, 338, 356
   scoparius, 158, 338, 356
   stenopetalus, 356

Dactylis, 382
Danthonia, 382
Daphne, 252–53, 320, 350
   cneorum, 253
   mezereum, 253, 350
   odora, 253
   'Marginata', 253; pl. 15*b*
Datiscaceae, 127, 380
Datisca glomerata, *127*–28, 380; pl. 8*b*
Datura, 238, 336, 363, 386
   × *candida*, 234

Datura (*continued*)
    discolor, 238
    ferox, 238
    inoxia, 238
    metel, 336
    meteloides, 238–39; pl. 14*a, b*
    stramonium, 5, 239, 332,
      356–57; pl. 13*c*
    suaveolens, 234
    tatula, 239
    wrightii, 238
Daubentonia punicea, 170
Daucus carota, 69, 331, 369, 384
Delphinium, 213, 217–20, 345
    ajacis, 216, 373
    andersonii, 218
    × *belladonna*, 219
    californicum, 218
    elatum, 219
    glaucum, 219
    grandiflorum, 219
    gypsophilum, 218
    hesperium, 6, 218
    menziesii, 218, *219*
    parryi, 218
    patens, 218
    recurvatum, 218
    semibarbatum, 219
    trolliifolium, 218
    variegatum, 218; pl. 12*b*
Descurainia pinnata, 107, 385
    sophia, 107–8
Dianella tasmanica, 283–84
Dicentra, 177, 342
    chrysantha, 177
    formosa, 177
Dichondra micrantha, 371
    repens, 371
Dichroa febrifuga, 240
Dicoria, 379
Dictamnus albus, 227
Dieffenbachia, 273–74, 315, 374
    maculata, 270; pl. 15*c*
    picta, 270, 374
    seguine, 270, 273–74, 375
Dietes bicolor, 276
    vegeta, 276–77
Digitalis purpurea, 5, 230–31, 322,
    354; pl. 12*c*
Digitaria, 382
Dimorphotheca, 90
Dirca occidentalis, 253, 351, 373
    palustris, 253–54

Disciotis, 31, 50
    venosa, 32
Distichlis spicata, 367, 382
Dodonaea viscosa, 228–29
Dracunculus vulgaris, 270, 375
Dugaldia hoopesii, 88
Duranta plumieri, 256
    repens, 256

Echinochloa, 382
    colonum, 365
    crus-galli, 365, 386
Echinopsilon hyssopifolium, 118
Echium, 102, 332
    fastuosum, 104
Eleocharis, 381
Eleusine indica, 386
Elymus, 382
Encelia, 370
Endoconidium temulentum, 297–98
Endymion non-scriptus, 284
Entoloma rhodopolium, 6, 47
Ephedra, 330, 357
    viridis, 357
Ephedraceae, 330
Epipremnum aureum, 270, 357, 375
Equisetaceae, 59, 329, 357
Equisetum, 59, 357
    arvense, 59, 339, 357
    hyemale, 59, 357
    laevigatum, 59
Eragrostis, 382
    cilanensis, 293–94
    megastachya, 293
Eremocarpus setigerus, 133
Ericaceae, 128–29, 350
Eriobotrya japonica, 225
Eriophorum, 382
Erodium, 177–78
Erysiphe, 377
Erythrina, 158–59
    caffra, 158
    coralloides, 158
    crista-galli, 158
    lysistemon, 158
Erythroxylaceae, 336
Erythroxylon coca, 336
Eschscholzia californica, 201–2,
    342, 357
Eucalyptus, 170, 195–96, 373, 380
    cladocalyx, 196
    globulus, 196
    viminalis, 196

Euonymus europaea, 117
Eupatorium adenophorum, 87
  rugosum, 317
Euphorbia, 132, 133–39, 350
  cotinifolia, 134, 371
  cyathophora, 134, 371
  cyparissias, 134–35
  esula, 135
  heterophylla, 134, 135
  lactea, 135–36, 371
  lathyris, 136, 371
  maculata, 371, 386
  marginata, 136, *137,* 371
  milii, 137–38; pl. 8*d*
  peplus, 138, 371
  pulcherrima, 138, 371
  tirucalli, 139, 372
  trigona, 372
Euphorbiaceae, 4, 132, 253, 350,
  351, 371, 380, 386
Euphrasia officinalis, 353
Eurotia, 379

Fabaceae, 141–42, 318, 325, 329,
  332, 337, 347, 351, 372, 380,
  386
Fagaceae, 7, 173, 380
Fagopyrum, 381
  esculentum, 208, 361, 373
  sagittatum, 208, 373
Fagus, 380
  grandiflora, 174
  sylvatica, 174
Festuca, 376, 382
  arundinacea, 294
  rubra var. *commutata,* 294–95
Ficus carica, 193–94, 372
  elastica, 372
Frangula alnus, 224
Franseria, 370, 379
Fraxinus, 380
Fritillaria meleagris, 284
Fumariaceae, 175, 342
Furcraea, 265, 374
Fusarium, 16

Gaillardia, 370
Galanthus, 268, 374
Galega officinalis, 99
Galerina, 29
  autumnalis, 30, 31
  marginata, 31
Galipea officinalis, 341

Gamochaeta purpurea, 384
Garrya, 380
Garryaceae, 380
Gastridium, 382
Gelsemium sempervirens, 186–88,
  187, 340, 372
Gentianaceae, 177
Geotrichum, 378
Geraniaceae, 177
Ginkgoaceae, 369, 378
Ginkgo biloba, 369, 378
Glechoma hederacea, 181–82
Gleotrichia, 11, 13
Gloriosa rothschildiana, 284–*85*
  superba, 284
Glyceria striata, 295
Glycine max, 328, 347, 386
Gnaphalium, 353
  purpureum, 384
Gnidia polystachya, 254, 351
Gomphocarpus fruticosus, 82
  physocarpus, 82
Gomphus floccosus, 47
Gonyaulax acatenella, 13
  catenella, 13
  polyedra, 13
Gossypium, 188–89
Gramineae. *See* Poaceae
Grayia, 379
Grevillea banksii, 211, 373
  robusta, 211, 373
Grifola sulphurea, 49
Grindelia squarrosa, 87
Gutierrezia microcephala, 87
  sarothrae, 87
Gymnocladus dioica, 159, 338
Gymnopilus, 37
  junonius, 38
  spectabilis, 38–39
Gyromitra, 27, 31, 50
  californica, 31
  esculenta, 6, 31–32
  infula, 32

Haemanthus, 268, 374
Halogeton glomeratus, 6, 118–20,
  119, 315; pl. 7*c*
Hamamelidaceae, 380
Haplopappus heterophyllus, 317
  venetus, 384
Hebeloma crustuliniforme, 6,
  47–48
  mesophaeum, 48

Hebeloma (*continued*)
  sinapizans, 48
Hedera canariensis, 81, 370
  colchica, 82, 370
  helix, 80–81, 370
  nepalensis, 81, 370
Heimia salicifolia, 188, 338
Helenium amarum, 88
  autumnale, 88
  hoopesii, 88–89; pl. 5*d*
  tenuifolium, 88
Helianthus, 379
  annuus, 384
  tuberosus, 384
Heliotropium, 102, 170, 332, 333, 360
  curassavicum, 104
  europaeum, 105
Helleborus, 221, 288, 322
  foetidus, 221
  lividus subsp. *corsicus*, 221
  niger, *221*
  orientalis, 221
Helminthosporium, 378
  biseptatum, 277
Helvella, 31, 50
  acetabulum, 32
  californica, 31
Hemizonia pungens, 384
Heracleum, 378
  mantegazzianum, 69
Hesperocnide tenella, 254–55
Heteromeles arbutifolia, 225
Heteropogon contortus, 293, 295–96; pl. 16*c*
Hieracium, 320
Hilaria, 382
  jamesii, 18, 22
  rigida, 296
Hippeastrum equestre, 268
  puniceum, 268
Hippocastanaceae, 178
Holcus lanatus, 296–97, 382
Homeria, 274–75, 287
  breyniana, 274
  collina, 274
  ochroleuca, 274
Hordeum vulgare, 293, 330, 382, 386
Hormodendrum, 378
Humea, 370
Humulus lupulus, 113, 357, 370, 379

Hyacinthus orientalis, 268, 285, 376
Hydrangea macrophylla, 229, 357
Hydrastis canadensis, 353
Hydrocotyle asiatica, 353
Hydrophyllaceae, 179, 372
Hygrophorus conicus, 48
Hymenocallis, 268, 374
Hymenoclea, 379
Hymenoxys odorata, 89
Hyoscyamus niger, 232, 240, 336
Hypericum perforatum, 6, 179–80, 360–61; pl. 10*b*, *d*
Hypericaceae, 179
Hypolepidaceae, 56

Ilex aquifolium, 79
  cornuta 'Burfordii', 79–*80*
  crenata, 79
  paraguariensis, 346, 357
  vomitoria, 79
Inocybe, 28, 33–34
  geophylla, 34
  pudica, 34
  sororia, 35
Ipomoea batatas, 122, 352
  purga, 122, 198
  tricolor, 122–23, 340
  violacea, 122–23
Iridaceae, 274, 375
Iris, 275
  foetidissima, 275, 351, 375
  germanica, 275, 351, 375
  missouriensis, 276
  pseudacorus, 275–76, 351, 375
Iva, 370, 379
  acerosa, 89

Jatropha curcas, 139, 372
Juglandaceae, 180, 372, 380
Juglans, 180–81, 372, 380
  cineraria, 181
  nigra, 181
  regia, 181
Juncaceae, 277, 382
Juncaginaceae, 278, 382
Juncus, 362, 382
  phaeocephalus, 277
Juniperus, 378
  communis, 357–58
  sabina, 347

Kalanchoë daigremontiana, 126
   lanceolata, 125
   pinnata, 126
   rotundifolia, 125
Kalmia angustifolia, 129
   latifolia, 129
   polifolia, 129
Karwinskia humboldtiana, 223
Kochia scoparia, 118, 210–11,
   367, 379, 385
Koeleria, 382
Koelreuteria paniculata, 228
Kyllinga brevifolia, 274

Labiatae. *See* Lamiaceae
Laburnum anagyroides, 159, 338
Lactarius, 48–49, 52
   chrysorheus, 48
   insulsus, 48
   resimus, 48
   rufus, 49
   trivialis, 49
   uvidus, 49
   zonarius, 49
Lactuca, 370
   sativa, 384
   serriola, 384
   virosa, 89, 358
Laetiporus sulphureus, 49
Lamarckia, 382
Lamiaceae, 181, 380
Lantana camara, 257–58
   montevidensis, 258
Laportea, 256
Larrea divaricata, 262
   mexicana, 262, 374
   tridentata, 262, 374
Lathyrus, 159–60, 327
   clymenum, 160
   hirsutus, 160
   latifolius, 160
   odoratus, 160
   sativus, 160
   tingitanus, 160
Lauraceae, 182, 372
Laurus nobilis, 185
Ledum glandulosum, 130
Leguminosae. *See* Fabaceae
Lens culinaris, 142
Leonotis, 380
Lepidospartum squamatum, 89–90,
   370
Lepiota, 29, 49

   castanea, 31
   clypeolaria, 49
   cristata, 49
   felina, 31
   helveola, 31
   josserandii, 31
   lutea, 50
   molybdites, 45
   morganii, 45
   naucina, 49
   rhacodes, 46
Leucoagaricus naucinus, 49
Leucocoprinus birnbaumii, 50
   luteus, 50
Leucojum, 268, 374
Leucophyllum, 381
Leucothoë axillaris, 130
   catesbaei, 130
   davisiae, 130
   fontanesiana, 130
   grayana, 350
Ligustrum, 198–99, 380
   japonicum, *198;* pl. 10c
   lucidum, 198
   ovalifolium, 198
   vulgare, 198
Liliaceae, 265, 280, 325, 328, 345,
   376
Limnanthaceae, 324
Linaceae, 185, 372, 386
Linum lewisii, 185
   perenne subsp. *lewisii*, 185
   usitatissimum, 185, 372, 386
Lippia, 361
   citriodora, 256
Liquidambar, 380
Litchi chinensis, 228
Lobelia cardinalis subsp. *graminea*,
   185–86
   inflata, 186, 335, 353
   laxiflora, 186
   siphilitica, 186
   splendens, 185
Lobeliaceae, 185
Loganiaceae, 186, 372
Lolium, 382
   multiflorum, 297
   perenne, 60, 297
   temulentum, 297–98
Lonicera periclymenum, 114
   tatarica, 114
Lophophora, 265
   diffusa, 110

Lophophora (*continued*)
    williamsii, 109–10, 330; pl. 7*b*
Loranthaceae. *See* Viscaceae
Lotus, 366
    corniculatus, 160–61
    tenuis, 160–61
Lupinus, 161–64, 171, 242
    × *alpestris*, 162
    arbustus, 161
    argenteus, 161
    caudatus, 161–63, 338
    densiflorus, 164
    latifolius, 6, 161, 162, 163;
        pl. 9*c*
    laxiflorus, 161, 163, 338
    leucophyllus, 162
    longifolius, 164
    macounii, 162
    nanus, 164
    onustus, 162
    polycarpus, 164
    pusillus subsp. *intermontanus*,
        162
    sericeus, 162, 163, 338
    sparsiflorus, 164
Luzula, 382
Lycium barbarum, 240–41
    ferocissimum, 241
    halimifolium, 240
Lycoperdon, 53
Lycopersicon esculentum, 241, 373
    lycopersicum, 241, 373
Lyngbya, 11
Lythraceae, 188

Machaeranthera orcuttii, 309
Maclura pomifera, 373, 380
Macrocystis pyrifera, 14
Mahonia aquifolium, 101
Malus domestica, 225
Malva parviflora, 189–90, 386
Malvaceae, 188, 372, 386
Mammillaria, 110
Mandragora officinarum, 232, 336
Mangifera indica, 65
Manihot esculenta, 2, 318
Marasmius oreades, 34
Matricaria chamomilla, 90, 356
    recutita, 86, 90, 356, 358
Medicago hispida, 165
    polymorpha, 165
    sativa, 121, 164–66, 172, 324,
        361, 366

Melaleuca leucadendron, 196, 373
    quinquenervia, 196–97, 373
Melia azedarach, *191*–92, 351
Meliaceae, 190
Melianthaceae, 192
Melianthus comosus, 192
    major, 192
Melica, 382
Melilotus, 166, 320
    albus, 166–67
    indicus, 166–67
    officinalis, 166–67, 386
Menispermaceae, 193
Mentha pulegium, 182, 315
Menziesia ferruginea, 130–31
Mercurialis annua, 139, 380
Microcystis aeruginosa, 11, 12, 13
Mirabilis jalapa, 122, *197*–98
Modiola caroliniana, 190
Monilia, 378
Monstera deliciosa, 271, 375
Montia perfoliata, 386
Moraceae, 111, 193, 320, 372, 380
Moraea polystachya, 276–77
Morchella, 50, 54
    angusticeps, 50
    conica, 50
    elata, 50
Moringaceae, 324
Morus, 380
Mucor, 167, 377
Myoporaceae, 194
Myoporum laetum, 194–95, 315
Myrica, 380
Myricaceae, 380
Myrsinaceae, 195
Myristica fragrans, 195, 358
Myrsinaceae, 195
Myrsine africana, 195
Myrtaceae, 195, 373, 380

Naematoloma fasciculare, 6, 39,
    50–51
    popperianum, 39
Nama parryi, 179, 372
Narcissus pseudonarcissus, 268,
    269
Nepeta cataria, 358
    glechoma, 181
    hederacea, 181
Nerine, 268, 374
Nerium, 322
    indicum, 77

oleander, 5, 77–78; pl. 4*c*
Nicotiana, 232, 241–42, 328, 335
    alata, 241
    attenuata, 242
    glauca, 5, 6, 242, *243*, 339;
        pl. 13*d*
    rusticana, 241
    × *sanderae*, 242
    sylvestris, 242
    tabacum, 241
    trigonophylla, 242
Nigrospora, 378
Nodularia, 11
Nostoc, 11
Nyctaginaceae, 197

Oenanthe sarmentosa, *68*, 350
Olea, 380
Oleaceae, 198, 380
Omphalotus, 28, 33, 35
    olearius, 35
Opuntia, 109
    basilaris, 109
    ficus-indica, 109
    microdasys, 109
Orchidaceae, 320, 332
Ornithogalum, 322
    arabicum, 286
    caudatum, 286
    thyrsoides, 286
    umbellatum, 286
Oryza, 382
Oryzopsis, 382
Oscillatoria, 11
Osteospermum ecklonis, 90–91
    fruticosum, 91
Ostrya, 379
Oxalidaceae, 199, 315
Oxalis, 199
    cernua, 199
    corniculata, 199
    pes-caprae, 199–200
Oxytenia acerosa, 89, 370
Oxytropis, 146, 167–68

Panaeolina, 37
    foenisecii, 39, 51
Panaeolus, 28, 37, 53
    campanulatus, 40
    foenisecii, 39
    retirugis, 51
    sphinctrinus, 39–40
    subbalteatus, 40

Panellus stipticus, 51
Panicum, 382
    capillare, 386
Panus stipticus, 51
Papaveraceae, 200–201, 329, 337,
    342
Papaver somniferum, 177, 201,
    202–4, *203*, 329–30, 342, 343;
    pl. 11*d*
Parietaria, 381
Parkinsonia aculeata, 386
Parthenocissus quinquefolia, 260–
    61, 262, 374
    tricuspidata, 261–62, 374
Paspalum, 382
    dilatatum, 18, 21
    distichum, 18, 22
    paspalodes, 18, 22
Passifloraceae, 358
Passiflora incarnata, 358
Pastinaca sativa, *68*, 69, 384
Paullinia cupana, 358
    tomentosa, 228
Paxillus involutus, 6, 51
Pedaliaceae, 204
Pedilanthus tithymaloides, 140, 372
Peganum harmala, 262
Penicillium, 16, 167, 378
    crustosum, 180
Pennisetum, 383
    clandestinum, 298
Periconia minutissima, 361
Pernettya mucronata, 131
Persea americana, 183–84
Phacelia, 179
    brachyloba, 372
    campanularia, 372
    crenulata, 372
    grandiflora, 372
    minor, 372
Phalaris, 367, 383
    aquatica, 298–301, 340
    minor, 299–301, 340
    tuberosa var. *stenoptera*, 298
Phaseolus limensis, 168
    lunatus, 168
    vulgaris, 142
Phialea temulenta, 297
Philodendron bipennifolium, 271, 375
    bipinnatifidum, 271, 375; pl. 15*d*
    domesticum, 271, 375
    erubescens, 271, 375
    hastatum, 271, 375

Philodendron (*continued*)
    panduriforme, 271, 375
    pertusum, 271, 375
    scandens, 271–72, 375
    selloum, 272, 375
    squamiferum, 272, 375
    verrucosum, 272, 375
Phleum, 383
Phoenix, 381
Pholiota aurivella, 51–52
    spectabilis, 38
    squarrosa, 52
Phomopsis, 16
Phoradendron, 259–60, 358–59, 381
    flavescens, 259
    tomentosum subsp. *macro- phyllum*, 259
    villosum subsp. *villosum*, 259, 260
Physalis, 242–45
    alkekengi, 243
    angulata var. *lanceifolia*, 244–45
    franchetii, 243
    ixocarpa, 244
    lanceifolia, 243
    peruviana, 244
    philadelphica, 244
Physianthus albens, 81
Phytolacca americana, 5, 204–6, 348; pl. 11*a*
    decandra, 204
Phytolaccaceae, 204
Pieris floribunda, 131
    japonica, 131
Pimelea prostrata, 252, 254, 351
Pinaceae, 61–62, 378
Pinus, 378
    ponderosa, 61–62
Piperaceae, 335, 338
Piper betle, 338
    cubeba, 359
    methysticum, 359
    nigrum, 335
Piptadenia, 340
Pistacia, 378
Pisum sativum, 142
Pithomyces, 16
    chartarum, 361
Pittosporaceae, 206
Pittosporum, 206–8
    crassifolium, 206
    erioloma, 206

    eugenioides, 206
    phillyraeoides, 207
    tenuifolium, 207
    tobira, 207–8
    undulatum, 207
Plagiobothrys, 102, 103, 385
Plantaginaceae, 381
Plantago, 381
Plasmopara, 377
Platanaceae, 381
Platanus, 381
Pleospora, 377
Plumbago auriculata, 208, 373
    capensis, 208, 373
Plumeria rubra, 369
Poa, 383
Poaceae, 292–93, 328, 332, 376, 382, 386
Podophyllum peltatum, 122
Poinciana gilliesii, 156
    pulcherrima, 157
Poincirus trifoliata, 227
Polygonaceae, 208, 325, 373, 381, 386
Polygonum, 386
Polypodiaceae, 56–59, 60
Polypogon, 383
Polyporus sulphureus, 49
Polyscias balfouriana, 370
    guilfoylei, 370
Populus, 381
Portulacaceae, 210, 386
Portulaca oleracea, 210
Pothos aureus, 270, 375
Primulaceae, 210, 373
Primula obconica, 373
Prosopis, 380
Proteaceae, 210, 351, 373
Prunus, 225, 319
    virginiana var. *demissa*, 6; pl. 11*b*
Psathyrella foenisecii, 39
Psathyrotes annua, 91
Psilocybe, 28, 37
    beocystis, 40
    coprophila, 40
    cyanescens, 40
    pelliculosa, 41
Psilostrophe cooperi, 91
Psoralea, 320
Ptelea crenulata, 227
    trifoliata, 227
Pteridaceae, 56

Pteridium aquilinum, 60, 367
    var. *pubescens*, 6, 56–59, 57;
    pl. 2*a*
Puccinia, 377

Quercus, 7, 174–75, 317, 380
    douglasii, 175
    kelloggii, 175
Quillaja saponaria, 323–24

Rafinesquia californica, 384
Ramaria formosa, 52
Ranunculaceae, 212, 324, 329, 342,
    373, 381
Ranunculus, 212, 221–22, 324, 381
    asiaticus, 221–22
    repens, 222
    sceleratus, 222, 223
        var. *multifidus*, 222
    testiculatus, 222–23
Raphanus sativus, 385
Raphidophora aurea, 270, 375
Rauvolfia, 75
Remijia, 341
Resedaceae, 324
Rhamnaceae, 223, 325, 326
Rhamnus, 223–25, 326, 356
    alnifolia, 224
    cathartica, 224, 359
    frangula, 224, 359
    purshiana, 224, 225, 326
Rheum, 381
    officinale, 326
    rhabarbarum, 208–9, 315, 326
    rhaponticum, 208, 326
Rhizoctonia, 16
Rhizopus, 377
Rhododendron, 129, 131–32
    japonicum, 131, 350
    macrophyllum, 131–32
    occidentale, 132; pl. 8*c*
Rhodotypos scandens, 225
Rhoeo discolor, 375
    spathacea, 375
Rhus cotinus, 369
    diversiloba, 65, 369
    trilobata, 66
Rhynchosia pyramidalis, 168–69
Ricinus communis, 2, 5, 140–41,
    260, 339, 347–48, 372; pl. 9*b, d*
Rivea corymbosa, 123
Rivina humilis, 206
Rivularia, 11

Robinia pseudoacacia, 169–70,
    347–48; pl. 9*f*
Romulea longifolia, 277
Rosa, 381
Rosaceae, 225–26, 318, 381
Rubiaceae, 226, 329, 341, 342, 344
Rumex, 210, 373, 381, 386
Russula, 52–53
    albonigra, 52
    densifolia, 53
    emetica, 52
    foetentula, 52
    fragilis, 52
    laurocerasi, 53
    nigricans, 53
    rosacea, 53
    subfoetens, 53
    veternosa, 53
Rutaceae, 227–28, 320
Ruta chalepensis, 227
    graveolens, 227

Saccharomyces, 377
Salicaceae, 381
Salicornia, 379
Salix, 381
Salmonella, 15
Salsola, 379
    australis, 385
Salvia leucophylla, 364
Sambucus, 114–15
    mexicana, 115, 385
Sanguinaria, 342
Sanicula bipinnata, 74
Sansevieria, 374
Sapindaceae, 228, 318, 373
Sapindus drummondii, 228, 229,
    373
    saponaria, 323
    utilis, 228
Sapium sebiferum, 372
Saponaria officinalis, 117, 324
Sarcobatus vermiculatus, 7, 121,
    380; pl. 7*d*
Sarcosphaera, 31
    crassa, 32
Sassafras albidum, 184, 356, 359
    variicolor, 184
Saxifragaceae, 229
Schefflera actinophylla, 80
Schinus molle, 65, 369, 378
    terebinthifolius, 65, 369, 378
Scilla, 322

Scilla (*continued*)
　　maritima, 287
　　non-scripta, 284
　　nutans, 284
　　peruviana, 287
Scindapsis aureus, 270, 375
Scirpus, 382
Scleroderma aurantium, 6, 53
　　citrinum, 6, 53
Sclerotinia sclerotiorum, 69, 321
Scrophulariaceae, 230, 381
Secale cereale, 383, 386
Sedum acre, 371
　　album, 371
　　telephium, 371
Senecio, 91–92, 170, 332, 333, 335, 370
　　bicolor subsp. *cineraria*, 92
　　cineraria, 92
　　confusus, 92
　　douglasii, 92–93
　　　　var. *longilobus*, 354
　　jacobaea, 7, 60, 93–94, 334; pl. 6*a*
　　longilobus, 334, 354
　　mikanioides, 94
　　vulgaris, 94–96, 95
Sesamum indicum, 204
　　orientale, 204
Sesbania punicea, 170; pl. 10*a*
　　tripetii, 170
Setaria, 293, 301
　　glauca, 301
　　lutescens, 301
Setcreasea pallida 'Purple Heart', 375
　　purpurea, 375
Sida leprosa var. *hederacea*, 190
Silybum marianum, 306, 384
Simaroubaceae, 231, 373, 381
Sinapsis, 379
Sitanion, 383
Sium suave, 68
Solanaceae, 232, 329, 335–36, 345, 373, 386
Solandra guttata, 245
　　hartwegii, 245
　　maxima, 245–46
Solanum, 232, 246–51, 345
　　americanum, 246

　　auriculatum, 248
　　aviculare, 248
　　californicum, 251
　　carolinense, 386
　　douglasii, 247
　　dulcamara, 247
　　elaeagnifolium, 247
　　jasminoides, 247
　　laciniatum, 247–48; pl. 14*c*
　　mauritianum, 248
　　melongena, 248–49
　　nigrum, 249, 386
　　nodiflorum, 246
　　pseudocapsicum, 249–50; pl.14*d*
　　sarrachoides, 250; pl. 15*a*
　　seaforthianum, 250
　　tuberosum, 250–51
　　umbelliferum, 251
　　wendlandii, 251
Solidago, 379, 384
　　spectabilis, 96
Sonchus asper, 385
　　oleraceus, 385
Sophora secundiflora, 170–71
Sorghum, 318, 383
　　bicolor, 301–2, 365, 386
　　halepense, 302, 386
　　sudanense, 302–3, 386
　　vulgare, 301
Sparmannia africana, 373
Spartium junceum, 337–38
Spathiphyllum, 272, 375
Sphenosciadium capitellatum, 74–75
Spiraea, 381
Sporidesmium bakeri, 361
Sporobolomyces, 377
Sporobolus, 383
Stachybotrys, 16
Stanleya, 309
　　elata, 108
　　pinnata, 108
　　viridiflora, 108
Staphylococcus, 15
Stellaria media, 385
Sterculiaceae, 251, 337, 346, 373
Sterculia foetida, 252
Stipa, 293, 295, 383
　　robusta, 303
　　vaseyi, 303

Strophanthus, 322
Stropharia ambigua, 53
  coronilla, 53
Strychnos nux-vomica, 186
Suaeda, 380
Swietenia, 190
Symphoricarpos albus, 116
Symphytum, 102, 105–6, 332
  asperum, 105
  officinale, 354, 359–60
  peregrinum, 105
  × *uplandicum*, 105, 354,
    359–60
Synadenium grantii, 372
Syncephalastrum, 377
Syringa, 380

Tagetes, 96, 370
  minuta, 96
  patula, 96
Tanacetum vulgare, 96, 97, 349, 370
Taraxacum, 379
Taxaceae, 62
Taxus baccata, 5, 62–63, 338
    'Stricta', 63; pl. 2*b*
  brevifolia, 63
  cuspidata, 62
Tetradymia, 361
  canescens, 97–98; pl. 6*c*
  glabrata, 7, 97–98; pl. 6*d*
  stenolepis, 98
Theaceae, 346
Thelypodium lasiophyllum, 307, 385
Theobroma cacao, 337, 346
Thermopsis, 339
Thevetia neriifolia, 78, 322, 370
  peruviana, 78, 322, 370; pl. 4*d*
  thevetioides, 78, 322, 370
Thuja, 378
  occidentalis, 349
Thymelaeaceae, 252, 350, 373
Tilia, 381
Tiliaceae, 373, 381
Tilletia, 377
Toxicodendron diversilobum, 5,
    65–68, *66*, 179, 317, 369;
    pl. 2*c, d*
Tradescantia, 375
Tribulus terrestris, 262–64, *263*,
    361, 386

Trichocereus, 110
Trichoderma, 378
Tricholoma, 53–54
  pardinum, 6, 54
  saponaceum, 54
  virgatum, 54
Trifolium, 171–72
  alexandrinum, 172
  hybridum, 171, 372
  pratense, 171, 320
  repens, 171
  subterraneum, 171–72, 200
Triglochin concinnum, 7, 278–80
    var. *debile*, 278, 280
  maritimum, 278, 279
  palustre, 278
  striatum, 278
Tripleurospermum maritimum
  subsp. *inodorum*, 90
Triticum aestivum, 383, 386
Tropaeolaceae, 254, 324
Tropaeolum majus, 254
Tulipa, 287, 376
Turbina corymbosa, 123
Turneraceae, 360
Turnera diffusa, 360
Turricula parryi, 179, 372
Tylocodon cacalioides, 125
  wallichii, 125
Typha, 383
Typhaceae, 383

Ulmaceae, 381
Ulmus, 381
Umbelliferae. *See* Apiaceae
Umbellularia californica, 184–85,
    349, 372
Ungnadia speciosa, 228
Urginea maritima, 280, 287–88
Uromyces, 377
Urtica, 255–56, 381
  breweri, 255
  californica, 255
  holosericea, 255
  lyallii, 255
  pilulifera, 256
  serra, 255
  urens, 255
Urticaceae, 254, 381
Ustilago, 377

Valerianaceae, 360
Valeriana officinalis, 360
Veratrum, 280, 284, 288–89, 291, 345
    album, 289
    californicum, 7, 288; pl. 16*a*
    fimbriatum, 288
    insolitum, 288
    viride, 288, 289
Verbascum, 381
Verbenaceae, 256
Verbesina encelioides, 98–99, 385
Verpa bohemica, 54–55
Verticillium, 378
Vicia, 327
    faba, 142, 172
    narbonensis, 172
Vinca rosea, 340
Violaceae, 258
Viola odorata, 258–59
    tricolor, 259
Viscaceae, 259, 381
Viscum album, 259, 260, 327, 330, 359, 381
Vitaceae, 260, 374

Wigandia caracasana, 372
Wislizenia refracta, 113–14
Wisteria, 173–74
    floribunda, 142, *173*
    sinensis, 173
    venusta, 173

Xanthium, 370, 379
    spinosum, 99–*100*
    strumarium, 7, 99–101, *100*; pl. 6*b*
Xanthosma sagittifolium, 272, 375
    violaceum, 272, 375
Xylorrhiza, 309

Zantedeschia aethiopica, 272, 375
Zea mays, 303–4, 307, 383, 386
Zigadenus, 175, 289–92, 345
    fremontii, *290*
    venenosus, 5, 7, 290–91, 292; pl. 16*b*
Zygadenus, 289
Zygophyllaceae, 262, 374, 386

General Index

Abrin (lectin), 142, 347–48
Absinthe, 349, 355
Acetaldehyde, 32, 314
Acetone, 90, 346
Acetylcholine, 255
Acokantherin (cardiac glycoside), 76
Aconitine (alkaloid), 214, 219
Acorn calves, 314
Acorns, 7, 173–75
Adonidin (cardiac glycoside), 215
Aesculin (coumarin glycoside), 178, 320
Aflatoxins, 15, 17
Agapanthogenin (steroid saponin), 266
Agrilus hyperici, 180
Ailanthin, 231
Ajacine (diterpene alkaloid), 217, 219
Ajaconine (diterpene alkaloid), 217, 219
Alcohol, 316–17, 348
Alcohol, reaction with mushrooms, 28, 32, 33, 50, 52, 314
Aldehyde, 256, 314
Aliphatic alcohol, 71, 349
Aliphatic nitro compounds, 145, 149, 150
Alkali disease, 145, 310
Alkaloids, 328–46
Allelopathic phytotoxins, 362–65
Aloin (anthraquinone glycosides), 280–81
Amanitin poisoning, 27, 28–31
Amatoxin (polypeptide), 27, 28, 29
Amine, 304
    toxic, 19, 260, 327, 359
Amine alkaloid, 330
Amino acid, 305, 326–28, 342
    toxic, 61, 160, 327
Amino alcohol, 332, 336
Ammonia, 305–6, 327, 333
    toxicity, 92
Amoeba, 344
Amygdalin (cyanogenic glycoside), 225–26, 319
Anabasine (pyridine-piperidine alkaloid), 242, 339

Anagyrine (quinolizidine alkaloid), 162–64, 171, 314, 338
Analgesic, 203, 355, 358
Andromedotoxins (diterpenes), 128, 130, 131, 350
Anemia, 51, 64, 172
Anemonin, 212, 324
Angostura bark, 341
Anguina agrostis (nematode), 295
Antabuse, 32, 33
Anthraquinone glycosides, 17, 157, 280, 318, 325–26, 355, 359
Anthrone, 326
Antibiotic, 15, 254
Anticoagulant, 320
Anti-estrogen, 62
Antihistamine therapy, 274
Antimalarial, 230
Antimetabolite, 172
Antineuralgic, 188
Antiseptic, 184
Antispasmodic agent, 73, 188, 281, 360
Aphrodisiac, 356
Apocynamarin (cardiac glycoside), 76
Arbutin (glycoside), 355
Arecoline (pyridine alkaloid), 338–39
Argemone, oil of, 201
Arrow poison, 76, 134, 193, 322
Atropine, 232–33, 234, 240, 335, 336, 354, 357

Barbaloin, 280
Barberry poisoning, 102
Beetle, Klamathweed Flea, 180
Benzo-alpha-pyrone, 320
Benzocain, 67
Benzylisoquinoline alkaloids, 342–43
Benzylisothiocyanate, 254
Benzyl mustard oil glycoside, 254
Berbamine (isoquinoline alkaloid), 101
Berberine (isoquinoline alkaloid), 101, 193, 201
Berberis (drug), 101

Bermudagrass poisoning, 293
Bermudagrass tremors, 22–23
Bhang, 112
Biforine, 273
Bighead, 98, 171, 263
Black drinks, 79
Blind staggers, 145, 310
Bloat, 164, 165, 365–66
Botulism, 13
Bromide poisoning, 312
Bronchodilators, 70
Bufadienolide (cardiac glycoside), 192, 275
Bufotenine (indole alkaloid), 340
Bufo vulgaris, 340
Buxine (alkaloid), 109

Caffeine (purine alkaloid), 79, 251, 346, 352, 353, 357, 358
Calcinosis, 237
Calves, deformed, 72–73, 163, 242, 311, 314, 338
Calycanthidine (alkaloid), 111
Calycanthine (alkaloid), 111
Camphor (ketone), 314–15, 364
Cancer, 340–41, 356, 359. *See also* Carcinogen, Cocarcinogen
Cannabinol, tetrahydro- (resin), 112
Cannabis, 112
Capsaicin, 236
Carboline (indole alkaloid), 262
Carboxylatractyloside (glycoside), 99
Carboxylic acid, 315, 316, 327
Carcinogen, 17, 56, 58–59. *See also* Cancer, Cocarcinogen
Cardiac glycosides, 75–78, 192, 212, 215, 221, 229, 275, 282, 284, 286, 287, 316, 321, 322, 337–38
Carminative, 184, 359
Cascara bark, 224–25, 326
Cascara sagrada, 224, 225, 326
Castanospermine (indolizidine alkaloid), 158
Castor oil, 141, 201
Cathartic, 76, 101, 109, 117, 122, 141, 157, 195, 198, 201, 204, 211, 223–25, 265, 281, 325–26, 349, 350, 355, 357, 359
Cathine (protoalkaloid), 330
Cerebrin (cardiac glycoside), 78
Chat, 330

Chatinine (alkaloid), 360
Charas, 112
Chelidonine (isoquinoline alkaloid), 116
Chewing disease, 85
Chlorine odor, 43
Cholesterol, 249, 323
Chondoinine (alkaloid), 193
Chrysolina quadrigemina, 180
Cicutoxin (terpenoid resin), 71, 349–50
Cinchona bark, 341–42
Cinchonidine (quinoline-quinuclidine alkaloid), 341
Cinchonine (quinoline-quinuclidine alkaloid), 341
Cineole (terpene), 364
Citral (aldehyde), 256
Citrus, oil of, 349
Clams, 14
Cobalt deficiency, 300, 367
Cocaine (tropane alkaloid), 336–37
Cocarcinogen, 132, 350–51
Cocculine (alkaloid), 193
Coclaurine (alkaloid), 193
Cocoa, 337, 346
Codeine (morphinan or isoquinoline alkaloid), 1, 202–4, 342–44
Colchicine (alkaloid), 281, 284
Coniceine (alkaloid), 72
Coniine (piperidine alkaloid), 72–73, 335
Consolidine (alkaloid), 104
Convallamarin (cardiac glycoside), 282
Convallarin (cardiac glycoside), 282
Convallotoxin (cardiac glycoside), 282
Coprine poisoning, 28, 32, 314
Cotyledonosis, 124–25
Coumarin glycosides, 17, 166–67, 171, 178, 227, 316, 319–21
    hydroxylated derivatives, 253, 320
Coumarol (coumarin glycoside), 166–67, 320
Coumestan (estrogen), 165
Coumestrol (estrogen), 165, 172, 351
Cracker heels, 150
Crooked calf disease, 163, 242, 311, 314, 338
Crooked calves, 73–74

Croton oil, 350
Cryogenine (quinolizidine alkaloid), 188, 338
Cucurbitacins, 126
Cyanogenic glycosides, 2, 90, 114, 143, 161, 168, 185, 225–26, 230, 279, 304, 318–19, 357. *See also* Hydrocyanic acid
Cyanogenic lipids, 228, 318
Cyclamin (glycoside), 210, 211
Cyclopamine (steroid alkaloid), 288
Cynoglossine (alkaloid), 104
Cytisine (quinolizidine alkaloid), 158, 159, 171, 338, 356

Daphnetoxin (resin), 253, 254, 350–51
Daphnin (coumarin derivative), 320
Daphnine (alkaloid), 254
Dehydrogenase, 172
Delphinine (diterpene alkaloid), 219
2,6-Diaminopurine, 172
Dianthrone, 180, 326, 360
Dicoumarol, 166–67, 320
Digitalis (cardiac glycoside), 1, 221, 230, 260, 322
Dihydroxycoumarin (coumarin glycoside), 166, 178, 253, 318–19
Diosgenin (saponin), 263
Diphenylnitrosamine, 307
Disulfiram, 32
Diterpene, 128, 131, 217, 219, 253, 345, 350, 351
Diterpene acetate, 254
Diterpene alcohol, 132
Diterpene alkaloids, 217, 219, 345
Diuretic, 89, 259, 265, 337, 346, 355, 357, 358, 359
DNA, 321
Drug Abuse and Control Act, Comprehensive, 38
Dugaldin (sesquiterpene lactone), 89

Elemicin, 358
Embelic acid, 195
Embelin, 195
Emetic, 101, 109, 201, 259, 344
Emetine (isoquinoline alkaloid), 344
Emodin anthraquinone glycosides, 157, 224, 326
Entamoeba histolytica, 344
Enzyme, proteolytic, 274

Enzyme inhibitors, 347
Ephedrine (protoalkaloid), 330, 357
Ergine (indole alkaloid), 123
Ergosine (indole alkaloid), 122
Ergosinine (indole alkaloid), 122
Ergot alkaloids, 19, 20, 122, 123, 340
Ergotamine (polypeptide), 19
Estrogen, 62, 164, 165, 172
Ethyl alcohol, 316
Eucalyptol, 196
Euphorbol, 140
Euphoriant, 35, 110, 122, 355, 356, 357, 358
Expectorant, 359

Facial eczema, 361
Fats, 348
Favism, 172
Febrifuge, 341
Fescue foot, 294
Ficin (proteolytic enzyme), 194
Fish poison, 128, 133, 134, 139, 157, 178, 207, 229, 265, 323
Flavanoids, 351
Folic acid, 57
Foot rot, 149
Formaldehyde, 167, 319, 320
Formic acid, 256
Fuochromones, 70
Furanocoumarin, 320
Furanoid sesquiterpene ketone, 194, 315
Furocoumarin, 69, 74, 227, 320–21, 352
Furofuran, 16–17

Galegine, 99
Galitoxin (resinoid), 83
Gallic acid, 317
Ganja, 112
Gastrointestinal irritants, 28, 39, 41–55, 317
Gelsemicine (indole alkaloid), 187
Gelsimine (indole alkaloid), 187, 340
Genistein (estrogen), 172
Genisteine (quinolizidine alkaloid), 356
Githagenin (saponin), 116
Gitogenin (saponin), 263
Glucose-6-phosphate dehydrogenase, 172

Glucoside, 317
Glucosinolates, 106
Glucotropaeolin (glycoside), 254
Glycoalkaloid, 243, 246, 248, 249, 250, 344
    steroid, 240, 329
Glycolipids, 347
Glycoproteins, 347
Glycoside, 317–26
Glycosinolates, 324–25
Goitrogenic glycosides, 328
Goitrogens, 325, 327–28
Gossypol (phenol), 189, 317
Gramine (indole alkaloid), 340
Grass tetany, 7, 120, 312–14
Grayanotoxins (resin), 350
Gyromitrin, 31–32, 55
Gyromitrin poisoning, 24, 27, 31–32, 50

Hallucinations, hallucinogens, 28, 37–41, 110, 112, 122, 169, 171, 188, 195, 233, 235, 246, 262, 265, 330, 336, 337, 338, 340, 351, 354, 355, 356, 357, 358, 359
Harmaline (indole alkaloid), 262
Harmine (indole alkaloid), 262, 358
Hashish, 112
Hay, moldy alfalfa, 121, 361
Hayfever plants. *See* Appendix B
Hederin (saponin glycoside), 81
Helenalin (sesquiterpene lactone), 88
Hemagglutinins, 142, 168, 169, 347
Hemolytic saponins, 265, 266
Heroin (morphinan or isoquinoline alkaloid), 202, 342–44
Histamine, 255
Hordenine (protoalkaloid), 330
Hydrangin (cyanogenic glycoside), 230
Hydrocyanic acid, 90, 91, 116, 143, 161, 168, 178, 183, 185, 196, 211, 225–26, 228, 230, 252, 277–80, 292, 297, 302–4, 318–19. *See also* Cyanogenic glycoside
    in invertebrate animals, 319
Hydroquinone (glycoside), 99
5-Hydroxy-1,4-naphthoquinone, 181
4-Hydroxycoumarin, 320

5-Hydroxytrypamine, 255
Hymenovin, 89
Hyoscine (tropane alkaloid), 234, 336
Hyoscyamine (tropane alkaloid), 232–34, 239–40, 246, 336
Hypericin (dianthrone), 180, 360–61
Hypnotic, 203, 343, 356
Hypotensive, 289

Ibotenic acid and muscimol poisoning, 28, 35, 43
Ilicin, 79
Imperaline (alkaloid), 284
Indole alkaloids, 37, 38, 123, 187, 262, 323, 340–41, 356–57, 358
Indolizidine alkaloid, 158
Indomethacin, 274
Infertility, 171
Insecticide, 204, 289, 339, 345, 349
Ipecac, Brazilian, 344
    Cartagena, 344
Ipecacuanha (isoquinoline alkaloids), 344
Ipomearone (sesquiterpene), 122
Isoergine (indole alkaloid), 123
Isoflavanoids, 351–52
Isoflavone, 172
Isoprene, 348
Isoquinoline alkaloids, 101, 110, 116, 176, 177, 201, 202, 342–44, 355, 357
Isothiocyanate, 106, 254, 324–25

Jalap, 122
Juglone, 181

Karakin (glycoside), 124
Ketone, 182, 194, 314–15
Khat, 330
Khellin (fuochromone), 70
Klamathweed Flea Beetle, 180
Kola nut, 353

Lactone, 17, 84, 88, 89, 91, 124, 316, 322
Lactucarine (alkaloid), 358
Lactucarium, 89, 358
Laetrile, 226, 319
Lantadene, 258
Lathyrism, 160

Laxative, 223–25, 360. *See also* Cathartic
Lead poisoning, 129, 310–11
Lectin, 2, 140, 142, 168, 169, 260, 347–48
Leucoderma, 321
Ligustrin (glycoside), 199
Limonene (monoterpene), 349
Linamarase, 90, 185
Linamarin (cyanogenic glycoside), 90, 91, 161, 185
Linseed oil, 185
Lipids, 348–52, 366
cyanogenic, 228, 318
Lobeline (piperidine alkaloid), 186, 335, 353
Locoine, 145, 146
Lotaustralin (cyanogenic glycoside), 161
LSD, 20, 122, 340
Lupinine (quinolizidine alkaloid), 337
Lupulin (volatile oil and resin), 357
Lycorine (isoquinoline alkaloid), 268
Lysergic acid, 19–20, 122–23, 340

Mace, 195
Magnesium deficiency, 7, 120, 312–14
Malaria, antimalarial action, 230
Malvalic acid, 190
Manganese, 314
Manioc, 2
Mechanical injury, 293, 295–96, 301
Medicagenic acid, 165–66
Melanin, 321
Meloidogyne arenaria, 96
hapla, 96
incognita, 141
Mescal, 265
Mescal de pulque, 265
Mescaline (protoalkaloid), 109–10, 265, 330, 355
Metabolites, 15, 16, 17
Meteloidine (tropane alkaloid), 234
Methemoglobin, 306
8-Methoxypsoralen, 194, 321
Methyl alcohol, 319, 339
Methyl bromide, 312
3-Methyl-2-butenylguanidine, 99

3,3'-Methylenebis (4-hydroxycoumarin), 166–67
Methylene blue, 361
Methyl salicylate, 259
Mezerein (diterpene resin), 350–51
Miserotoxin (aliphatic nitro compound), 149
Mitogens, 142, 205, 348
Mitotic poison, 281
MMH, 27, 31
Mollusks, 14
Molybdenum poisoning, 311–12
Monomethylhydrazine, 27, 31
Monoterpene, 96, 332, 348–49, 360
Morphine (morphinan or isoquinoline alkaloids), 1, 202–4, 328, 329, 342–44
Mucoproteins, 366
Muscarine poisoning (alkaloid), 28, 33–35
Muscimol poisoning, 28, 35–36
Mushrooms, poisonous, 23–55
Mussels, 4, 14
Mustard gas, 106, 325
Mustard oil glycosides, 106, 324–25
Mustard seed, 201, 325
Mycological Society of San Francisco, 26
Mycothecium, 16
Mycotoxins, 15–17, 180
Myristicine (alkaloid), 355, 358

Naphthoquinone, 181, 208
Narcotic, 111, 202–4, 210, 235, 241, 342, 343, 344, 355, 358
Necic acid, 332
Nematode, grass seed, 294–95
root knot, 96, 141
root lesion, 96, 141
Nepetalactone (volatile oil), 358
Neriifolin (cardiac glycoside), 78
Neurotoxin (tetranorterpenoid), 192
Ngaione (furanoid sesquiterpene ketone), 194
Nicotine (pyridine-piperidine alkaloid), 72, 186, 232, 237, 242, 328, 329, 335, 339
Nicotinic acid, 339
Nitrate poisoning, 7, 103–4, 264, 302–4, 305–7

Nitrates, plants capable of accumulating toxic amounts of, 64, 99, 102, 107, 116, 117, 263, 292, 384–86. *See also* Appendix C
Nitrile, 106, 324
Nitrite, 304, 305, 306
Nitro compounds, aliphatic, 145, 149–50
    3-nitropropanol, 149
    3-nitropropionic acid, 149, 150
Nitrosamine poisoning, 304, 307
Nitrosohemoglobin, 304
Noratropine (tropane alkaloid), 246
Norbelladine (isoquinoline alkaloid), 344
Norhyoscyamine (tropane alkaloid), 234, 246
Nucleic acid, 227
    deoxyribose, 321
Nutmeg, 195, 358

Oenanthetoxin (diterpene resin), 350
Oestrogen. *See* Estrogen
Oil, 348
Oil, essential, 96, 109, 194, 196, 348–49, 356
Oil, volatile, 182, 210, 256, 259, 265, 315, 349, 357, 358, 359
Oilseed varieties of Opium Poppy, 329
Oleandrin (cardiac glycoside), 77
Opium, 201, 202–4, 342–44, 358
Orellanine, 47
Ouabain (cardiac glycoside), 76, 322
Oxalate, 64, 118, 119–20, 121, 209, 210, 265, 273, 274, 315–16
    acid potassium, 118, 200, 315
    calcium, 120, 261, 269–70, 273–74, 286, 315–16
Oxalic acid, 265, 315
Oxyacanthine (isoquinoline alkaloid), 101

Papaverine (benzylisoquinoline alkaloid), 343
Paralyzed tongue, 107–8
Parathion, antidote for, 233
Pellotine (alkaloid), 110
Penitrem A, 180
Peptide, 122, 326, 328
Phalaris staggers, 299–301, 367; pl. 16d

Phallotoxin (polypeptide), 27
Phaseolunatin (cyanogenic glycoside), 90, 168
Phasin (lectin), 169
Phenol, 41–43, 66, 189, 317, 350, 363–64
Phenothiazine, 361
Phenylalanine, 320, 328, 342, 343, 344
Phenylethylamine, 110, 260, 328, 330, 342, 357
Phenylethylamine alkaloids, 109–10, 357
Phoratoxin (lectin), 260
Phorbol (phenolic resin), 132, 253, 350–51
Photodermatitis, 69, 227
Photophobia, 75
Photosensitization, 12, 16, 70, 75, 98, 121, 165, 171, 180, 194, 195, 208, 258, 263, 293, 321, 360–62, 367
Phylloerythrin, 360–61
Phytoalexins (isoflavanoids), 352
Phytoestrogen. *See* Estrogen
Phytohemagglutinin, 142, 168, 169, 347
Phytolaccatoxin (resin), 205
Phytolaccigenin (triterpene saponin), 205
Phytolaccine (alkaloid), 205
Phytotoxin, 132, 347–48
    allelopathic, 362–65
Picrotoxin, 124, 193
Pinkeye, increase in cases of, 149
Pink rot of Celery, 69, 321
Piperidine alkaloids, 335
Piperine alkaloids, 335
Placenta, transmission through, 103, 163–64
    retention of, 62
Plumbagin (naphthoquinone), 208
Pneumonia, increase in cases of, 149
Polioencephalomalacia, 60, 121, 301, 366–67
Poloxalene, 366
Polyhydroxyphenol, 317
Polypeptide, 11, 19, 308, 326, 327, 340
    amatoxin, 28, 327
Polyphenol, 189
Potato eruption, 251

Pratylenchus, 96
    alleni, 141
Pressor action, 330
N-Propyl-disulfide, 266
Prostratin (diterpene acetate), 254,
    351
Proteins, 326–27
    cytoplasmic, 366
Prothrombin, 167
Protoalkaloids, 328, 330
Protoanemonin glycosides, 212,
    214, 216, 221, 222, 324
Protopine (isoquinoline alkaloid),
    201
Prunasin (cyanogenic glycoside),
    225
Prussic acid. *See* Hydrocyanic acid
Pseudocyanogenic glycosides, 319
Pseudoephedrin (protoalkaloid), 357
Psilocin (indole alkaloid), 28, 37–
    39, 340
Psilocybin (indole alkaloid), 28,
    37–39, 340
Psilocybin–psilocin poisoning, 37–
    41, 281
Psilotropin (sesquiterpene lactone),
    91
Psoralen (furocoumarin), 194, 227,
    320–21
Psoriasis, 321
Psychoactivity, 20, 38, 112, 122–
    23, 338
Pulegone (ketone volatile oil), 182,
    315
Pulque, 265
Purapurine (glycoalkaloid), 248
Purgative, 114, 139, 157, 224, 252,
    259
Purine, 172
Purine alkaloids, 346
Pyrethrin, 204, 349
Pyridine alkaloids, 186, 335, 338–
    39
Pyridine–piperidine alkaloids, 339
Pyrimidine base, 227
Pyrone, 359
Pyrrolidine alkaloids, 331–32, 335
Pyrrolizidine alkaloids, 17, 92–95,
    102, 103, 104, 105–6, 170, 328,
    332–34, 354, 360
Pyruvate, 60

Quinidine (quinoline-quinuclidine
    alkaloid), 341
Quinine (quinoline-quinuclidine al-
    kaloid), 341–42
Quinoline alkaloids, 341, 344
Quinoline-quinuclidine alkaloids,
    341–42
Quinolizidine alkaloids, 171, 188,
    337–38, 356
Quinone, 325
Quinuclidine alkaloids, 341

Ragweed pollen, 86, 90
Ranunculin (protoanemonin
    glycoside), 212, 221, 222, 324
Reserpine, 75
Resin, 76, 83, 101, 109, 111–12,
    128, 131, 139, 173, 205, 229,
    253, 254, 262, 276, 286, 349–
    51, 357
Rhodojaponins (resins), 131, 350
Ricin (lectin), 140–41, 260, 339,
    347–48
Ricinine (pyridine alkaloid), 339
Robatin (glycoside), 169
Robin (lectin), 169, 347–48
Rodent poison, 288, 320
Ruminal engorgement, 298
Ruscogenin, 263
Ryegrass staggers, 297

Safrole, 184, 356, 359
Salicylic acid, 259
Sapogenins, 263, 265, 322
Saponin, 87, 90, 116, 124, 132,
    164, 165, 166, 170, 174, 193,
    205, 207–8, 210, 227–28, 229,
    230, 231, 237, 256, 265, 266,
    282, 318, 322, 324, 345, 366
Saponin glycosides, 81, 321, 322–
    23, 351
Saporubic acid (saponin), 116
Saporubin (saponin), 116
Sarothamnine (quinolizidine al-
    kaloid), 356
Sassafras, oil of, 184, 359
Scallops, 14
Scopolamine (tropane alkaloid),
    232, 234–35, 240, 336, 357
Sedative, 73, 89, 289, 356, 357,
    358, 360
Selenium, 17, 87, 108, 118, 143,
    145, 307–10, 367

Sempervirine (indole alkaloid), 340
Sesamin, 204
Sesbanine (alkaloid), 170
Sesquiterpene, 56, 122, 189, 194, 315, 348, 349, 352
Sesquiterpene alcohol, 359
Sesquiterpene lactone, 84, 88, 89, 91
Shellfish, 13–14
Shivers, 189–90
Silicates, 59
Solanidine (steroid glycoalkaloid), 246, 250, 345
Solanine (steroid glycoalkaloid), 240, 241, 243, 246, 249, 250, 345
Solanocapsine (steroid alkaloid), 249
Solasodamine (steroid glycoalkaloid), 248
Solasodine (steroid glycoalkaloid), 246, 249
Solasonine (steroid glycoalkaloid), 246, 248
Solauricine (steroid glycoalkaloid), 248
Sparteine (quinolizidine alkaloid), 158, 337–38, 356
Spewing sickness, 89
Staggers, 189–90
    blind, 145, 310, 312
    phalaris, 299–301; pl. 16d
    ryegrass, 297
Sterculic acid, 190, 251, 373
Steroid, 266, 291, 322, 329
Steroid alkaloids, 249, 345
Steroid glycosides, 230, 240, 321, 351
Steroid saponins, 266, 322, 345
G-Strophanthin (cardiac glycoside), 76
Strychnine, 111, 186, 188
Sulfate, high content of, 121, 367
Syringin (glycoside), 199

Tannin, 7, 174–75, 317, 352, 355, 357, 360, 366
Tapioca, 2
Taxine (alkaloid), 63
Tenulin (sesquiterpene lactone), 88
Tequila, 265
Terpene, 62, 140, 196, 348–52, 363, 364

monoterpene, 332, 348–49, 360
diterpene, 217, 219, 253, 345, 349, 351
    acetate, 254
    alcohol, 132
sesquiterpene, 194, 348, 349, 352, 359
triterpene, 124, 192, 351–52
    pentacyclic, 258
Tetany, grass, 7, 120, 312–14
Tetradymol, 98
Tetrahydrocannabinol (resin), 112, 351
Tetrahydroharmine (indole alkaloid), 262
Tetrahydroisoquinoline alkaloids, 110
Theobromine (purine alkaloid), 251, 346
Theophylline (purine alkaloid), 346
Thevetin (cardiac glycoside), 78
Thiaminase, 56, 57, 58, 60, 357, 367
Thiamine, 56, 57, 60, 121, 357, 367
Thiocyanate, 106, 324–25
Thioglucosidase, 324
Thiophene, 96
Thujone (monoterpene), 96–97, 349, 355
Toad, 340
Tomatidine (steroid glycoalkaloid), 241
Tomatine (steroid glycoalkaloid), 241
Tonga, 235
Toxalbumin, 2, 132, 139, 140, 339, 347–48
Tranquilizing effect, 360
Tremetol (alcohol), 316–17
Trigonelline (alkaloid), 198
Triterpene, 124, 192, 322, 351–52
    pentacyclic, 258
Triterpene saponins, 124, 205, 210, 322
Tropane alkaloids, 232, 233, 234, 238, 240, 335–37
3-Tropanol (amino alcohol), 233, 336
Tropine (amino alcohol), 336
Tryptamine, 255, 340
Tryptamine alkaloids, 299–300
Tumors, 321, 349, 350
Turkey "X" disease, 15

Tympany, primary ruminal, 365
Tyramine, 260, 330
Tyrosine, 328, 342, 343, 344

Umbelliferone (coumarin deriva-
    tive), 320
Umbellulone (essential oil), 184,
    349
Urushiol (phenolic resin), 66–67,
    350

Valepotriate (monoterpene), 360
Valerian, oil of, 360
Verline (alkaloid), 360
Vermifuge, 97, 349
Vertine (alkaloid), 188, 338
Vicine (glycoside), 172
Vinblastine (indole alkaloid), 340

Vincristine (indole alkaloid), 340
Visnagin (fuochromone), 70
Vitiligo, 321

Warfarin, 320
Waxes, 348
Wistarin (glycoside), 173
Wormwood oil, 349, 355

Xanthone, 17, 326

Yohimbine (indole alkaloid), 356
Yohimbinine (indole alkaloid), 356
Yuccagenin (hemolytic saponin),
    266

Zoom, 358
Zygadenine (steroid alkaloid), 291

Designer:	Rick Chafian
Compositor:	G&S Typesetters, Inc.
Text:	10/12 Times Roman
Display:	Helvetica
Printer:	Consolidated Printers
Binder:	Mt. States Bindery

STATEWIDE

Cacti of California, *Dawson*
California Amphibians and Reptiles, *Stebbins*
California Butterflies, *Garth/Tilden*
California Insects, *Powell/Hogue*
California Landscape, *Hill*
California Mammals, *Jameson/Peeters*
Early Uses of California Plants, *Balls*
Edible and Useful Plants of California, *Clarke*
Ferns and Fern Allies of California, *Grillos*
Freshwater Fishes of California, *McGinnis*
Geologic History of Middle California, *Howard*
Geology of the Sierra Nevada, *Hill*
Grasses in California, *Crampton*
Introduced Trees of Central California, *Metcalf*
Introduction to California Plant Life, *Ornduff*
Marine Food and Game Fishes of California, *Fitch/Lavenberg*
Marine Mammals of California, *Orr*
Mushrooms of Western North America, *Orr/Orr*
Native Shrubs of the Sierra Nevada, *Thomas/Parnell*
Native Trees of the Sierra Nevada, *Peterson*
Natural History of Vacant Lots, *Vessel/Wong*
Poisonous Plants of California, *Fuller/McClintock*
Seashore Plants of California, *Dawson/Foster*
Sierra Wildflowers: Mt. Lassen to Kern Canyon, *Niehaus*
Teaching Science in an Outdoor Environment, *Gross/Railton*
Water Birds of California, *Cogswell*

SAN FRANCISCO AND NORTHERN CALIFORNIA

Introduction to the Natural History of the S.F. Bay Region, *Smith*
Introduction to Seashore Life of the S.F. Bay Region and Coast
    of Northern California, *Hedgpeth*
Mammals of the S.F. Bay Region, *Berry/Berry*
Native Shrubs of the S.F. Bay Region, *Ferris*
Native Trees of the S.F. Bay Region, *Metcalf*
Rocks and Minerals of the S.F. Bay Region, *Bowen*
Spring Wildflowers of the S.F. Bay Region, *Sharsmith*
Weather of the S.F. Bay Region, *Gilliam*

SOUTHERN CALIFORNIA

Climate of Southern California, *Bailey*
Fossil Vertebrates of Southern California, *Downs*
Introduction to the Natural History of Southern California,
    Jaeger/Smith
Native Shrubs of Southern California, *Raven*
Native Trees of Southern California, *Peterson*
Seashore Life of Southern California, *Hinton*